MEASURE, INTEGRATION AND FUNCTION SPACES

MEASURE, INTEGRATION AND FUNCTION SPACES

Charles Swartz
Department of Mathematical Sciences
New Mexico State University
USA

World Scientific
Singapore • New Jersey • London • Hong Kong

Published by

World Scientific Publishing Co. Pte. Ltd.
P O Box 128, Farrer Road, Singapore 9128
USA office: Suite 1B, 1060 Main Street, River Edge, NJ 07661
UK office: 73 Lynton Mead, Totteridge, London N20 8DH

MEASURE, INTEGRATION AND FUNCTION SPACES

Copyright © 1994 by World Scientific Publishing Co. Pte. Ltd.

All rights reserved. This book, or parts thereof, may not be reproduced in any form or by any means, electronic or mechanical, including photocopying, recording or any information storage and retrieval system now known or to be invented, without written permission from the Publisher.

For photocopying of material in this volume, please pay a copying fee through the Copyright Clearance Center, Inc., 27 Congress Street, Salem, MA 01970, USA.

ISBN 981-02-1610-6

Printed in Singapore by JBW Printers & Binders Pte. Ltd.

To The Memory

of

John DePree

PREFACE

These notes are based on courses in measure and integration theory that the author has taught for a number of years in the mathematics department of New Mexico State University. Although the presentation may be somewhat different, the notes contain the basic information on measures and the Lebesgue integral contained in such standard introductory texts as Royden, Hewitt and Stromberg or Aliprantis and Burkinshaw; however, we try to emphasize the role played by countable additivity by including a substantial amount of material on finitely additive set functions which is not contained in most introductory texts. The material on finitely additive set functions can easily be skipped if the reader wishes to concentrate on studying the basic properties of the Lebesgue integral.

We motivate the introduction to measure and integration theory by discussing Lebesgue's original description of the Lebesgue integral. Following this historical introduction we present an abstract set-up which is sufficient to discuss measure theory and following a discussion of outer measures we give a general method which can be used to construct measures including Lebesgue and Lebesgue-Stieltjes measures. The important properties of Lebesgue measure are then discussed in detail. We then define and develop the basic properties of the class of measurable functions. The Lebesgue integral is defined, and its major properties are derived. Several special topics such as convergence in measure, the Riesz-Fischer Theorem and the relationship between the Riemann and Lebesgue integrals are discussed. We then define the product of measures and use an interesting characterization of the Lebesgue integral due to J. Mikusinski to prove Fubini's Theorem on the equality of double and iterated integrals. The section of the notes devoted to measure and integration closes with a discussion of the Hahn and Jordan decompositions and the Radon-Nikodym Theorem. A chapter on the relationship between differentiation and integration follows.

The second part of the notes is devoted to the study of some of the important spaces of functions encountered in analysis. In order to have a general framework in which to discuss function spaces, we introduce normed spaces and study enough of their important properties to facilitate our study. In particular, we discuss what Dunford and Schwartz refer to as the three basic principles of functional analysis: the Uniform Boundedness Principle, the Hahn-Banach Theorem and the Open Mapping/Closed Graph Theorems. As noted by F. Riesz many of the most important function spaces carry a natural order structure so we follow the example of Aliprantis and Burkinshaw and also study the basic properties of ordered normed spaces. After the discussion of abstract spaces, we proceed to study the L^p spaces, spaces of finitely and countable additive set functions, the space of continuous functions on a compact space and abstract Hilbert space along with the Fourier transform. For the convenience of the reader not familiar with them we discuss the topics of functions of bounded variation, the Baire Category Theorem, the Arzela-Ascoli Theorem and the Stone-Weierstrass Theorem in appendices

It is assumed that the reader of these notes has had an introductory course in real analysis at the level of the texts by Rudin, *Principles of Real Analysis*, Bartle, *The Elements of Real Analysis*, Apostol, *Mathematical Analysis* or DePree and Swartz, *Introduction to Real Analysis*. The term "topological space" is used several times in the text; however, for students with no background in topology, "topological space" can be replaced by "metric space" and in the measure/integration section "topological space" can even be replaced by $\mathbf{R}^n$ except for parts of §3.5 and §3.6.

The notes contain more material than I have ever been able to cover in a one year course. In order to develop the properties of the integral quickly, I cover Chapter 1, 2.1, 2.3 (2.3.1 being optional at this point), 2.4, 2.5 (2.6 being optional), 3.1-3.9 (section 3.10 and 3.11 being optional), return to 2.2.1, 2.2.2 to develop the material necessary to cover the Radon-Nikodym Theorem in 3.12 and then do 3.13. The other sections cover more specialized topics, many dealing with properties of finitely additive set functions, which can either be covered or left to the student to read on his/her own. Chapter 4 covers basic topics in differentiation and integration including the Fundamental Theorem of Calculus; the material in §2.6 on Lebesgue-Stieltjes measures is used in this chapter. In Chapter 5, sections 5.3 on the Uniform Boundedness Principle and 5.5 on the Open Mapping/Closed Graph Theorems can be skipped if desired along with sections 5.6.2 and 5.6.3. Chapter 6 contains a discussion of classical function spaces. The results on ordered spaces in §5.7 are not used in sections 6.1, 6.2 and 6.3 so these sections on L^p spaces and their duals can be covered after §5.6.1. The material in 6.4 and 6.5 relies on some of the results in §5.7. The material on Hilbert space is also independent of §5.7.

I would like to thank the many graduate students at New Mexico State University who took the graduate course in measure and integration while these notes were evolving. I particularly would like to thank Dan Gagliardi, Jillian Lee, Diane Martinez and Debra Zarret for reading through this final version of the notes and correcting many of my errors. Special thanks go to Valerie Reed for doing her usual exemplary job of translating my hand-written notes into manuscript form.

TABLE OF CONTENTS

Preface	vii

1.	Introduction	1
	1.1 Preliminaries	1
	1.2 Extended Real Numbers ($\mathbf{R}^*$) and $\mathbf{R}^n$	4
	1.3 Lebesgue's Definition of the Integral	9
2.	Measure Theory	15
	2.1 Semi-rings and Algebras of Sets	15
	2.2 Additive Set Functions	20
	2.2.1 Jordan Decomposition	28
	2.2.2 Hahn Decomposition	33
	2.2.3 Drewnowski's Lemma	35
	2.3 Outer Measures	36
	2.3.1 Metric Outer Measures	39
	2.4 Extensions of Premeasures	41
	2.4.1 Hewitt-Yosida Decomposition	49
	2.5 Lebesgue Measure	50
	2.6 Lebesgue-Stieltjes Measure	57
	2.6.1 Hewitt-Yosida Decomposition for Lebesgue-Stieltjes Measures	60
	2.7 Regular Measures	62
	2.8 The Nikodym Convergence and Boundedness Theorems	66
3.	Integration	71
	3.1 Measurable Functions	71
	3.1.1 Approximation of Measurable Functions	78
	3.2 The Lebesgue Integral	81
	3.3 The Riemann and Lebesgue Integrals	96
	3.4 Integrals Depending on a Parameter	100
	3.5 Convergence in Mean	103
	3.6 Convergence in Measure	107
	3.7 Comparison of Modes of Convergence	112
	3.8 Mikusinski's Characterization of the Lebesgue Measure	115
	3.9 Product Measures and Fubini's Theorem	118
	3.10 A Geometric Interpretation of the Integral	126
	3.11 Convolution Product	127
	3.12 The Radon-Nikodym Theorem	132
	3.12.1 The Radon-Nikodym Theorem for Finitely Additive Set Functions	137
	3.13 Lebesgue Decomposition	141
	3.14 The Vitali-Hahn-Saks Theorem	144

4. Differentiation and Integration 147
   4.1 Differentiating Indefinite Integrals 147
   4.2 Differentiation of Monotone Functions 152
   4.3 Integrating Derivatives 158
   4.4 Absolutely Continuous Functions 160

5. Introduction to Functional Analysis 165
   5.1 Normed Linear Spaces (NLS) 165
   5.2 Linear Mappings between Normed Linear Spaces 174
   5.3 The Uniform Boundedness Principle 178
   5.4 Quotient Spaces 181
   5.5 The Closed Graph/Open Mapping Theorems 183
   5.6 The Hahn-Banach Theorem 186
      5.6.1 Applications of the Hahn-Banach Theorem in NLS 189
      5.6.2 Extension of Bounded, Finitely Additive Set Functions 193
      5.6.3 A Translation Invariant, Finitely Additive Set Function 194
   5.7 Ordered Linear Spaces 197

6. Function Spaces 207
   6.1 L^p-spaces, $1 \le p < \infty$ 207
   6.2 The Space $L^\infty(\mu)$ 218
   6.3 The Space of Finitely Additive Set Functions 223
   6.4 The Space of Countably Additive Set Functions 226
   6.5 The Space of Continuous Functions 228
   6.6 Hilbert Space 235
      6.6.1 Fourier Transform 249

Appendix 255
   A1. Functions of Bounded Variation 255
   A2. The Baire Category Theorem 258
   A3. The Arzela-Ascoli Theorem 260
   A4. The Stone-Weierstrass Theorem 262

References 267

Index 272

MEASURE, INTEGRATION AND FUNCTION SPACES

Chapter 1

Introduction

1.1 Preliminaries

In this introductory section we set down some of the basic notations which will be used in the sequel. It will be assumed throughout that the reader is familiar with basic introductory real analysis as set forth in such texts as [Ap], [Ba], [DeS] or [R1].

Set Theory

We denote the positive integers, integers, rationals, reals and complex numbers by $\mathbf{N}, \mathbf{Z}, \mathbf{Q}, \mathbf{R}$ and $\mathbf{C}$, respectively. We denote the set of all non-negative real numbers by $\mathbf{R}_+ = \{t \in \mathbf{R} : t \geq 0\}$. The n-dimensional Euclidean space of all ordered n-tuples of real [complex] numbers will be denoted by $\mathbf{R}^n$ [$\mathbf{C}^n$].

We use standard set theoretic notation: Let S be a non-empty set. If $\mathcal{E} = \{E_a : a \in I\}$ is a family of subsets of S, we denote the *union* and *intersection* of the family $\mathcal{E}$ by

$$\cup_{a \in I} E_a = \cup \mathcal{E} = \{x \in S : x \in E_a \text{ for some } a \in I\}$$

and

$$\cap_{a \in I} E_a = \cap \mathcal{E} = \{x \in S : x \in E_a \text{ for all } a \in I\}.$$

If $I = \{1, \ldots, n\}$, we write

$$\cup_{i \in I} E_i = E_1 \cup \cdots \cup E_n = \cup_{i=1}^n E_i$$

and

$$\cap_{i \in I} E_i = E_1 \cap \cdots \cap E_n = \cap_{i=1}^n E_i.$$

Similarly, for $I = \mathbf{N}$, we write $\cup_{i \in \mathbf{N}} E_i = \cup_{i=1}^\infty E_i$ and $\cap_{i \in \mathbf{N}} E_i = \cap_{i=1}^\infty E_i$. The family $\mathcal{E}$ is said to be *pairwise disjoint* if $E_a \cap E_b = \emptyset$ whenever $a, b \in I$ and $a \neq b$. If $A, B \subset S$, the *complement* of B in A is $A \backslash B = \{x \in A : x \in A, x \notin B\}$, and we set $A^c = S \backslash A$. A *permutation* of a set S is a 1-1, onto function $f : S \to S$. The *power set* of S is denoted by $\mathcal{P}(S) = 2^S$. If $A \subset S$, the *characteristic function* of A is the real-valued function C_A defined by $C_A(t) = 1$ if $t \in A$ and $C_A(t) = 0$ for $t \notin A$.

A *sequence* in S is a function $f : \mathbf{N} \to S$; we write $f(k) = f_k$ and denote the sequence by $\{f_k\}_{k=1}^{\infty}$ or, simply, $\{f_k\}$. If $\{E_k\}$ is a sequence of subsets of S, we say that $\{E_k\}$ is *increasing [decreasing]* if $E_k \subset E_{k+1}$ $[E_k \supset E_{k+1}]$ for each k; we write $E_k \uparrow$ or $E_k \uparrow \cup_{k=1}^{\infty} E_k$ $[E_k \downarrow$ or $E_k \downarrow \cap_{k=1}^{\infty} E_k]$. If $\{E_k\}$ is a sequence of subsets of S, the *limit superior [limit inferior]* of $\{E_k\}$ is defined by

$$\overline{\lim} E_k = \cap_{j=1}^{\infty} \cup_{k=j}^{\infty} E_k \quad [\underline{\lim} E_k = \cup_{j=1}^{\infty} \cap_{k=j}^{\infty} E_k];$$

since the sequence $\{\cup_{k=j}^{\infty} E_k\}_{j=1}^{\infty}$ $[\cap_{k=j}^{\infty} E_k]$ is decreasing [increasing],

$$\cup_{k=j}^{\infty} E_k \downarrow \overline{\lim} E_k \quad [\cap_{k=j}^{\infty} E_k \uparrow \underline{\lim} E_k].$$

If $\overline{\lim} E_k = \underline{\lim} E_k$, we say that $\{E_k\}$ *converges* and set

$$\lim E_k = \overline{\lim} E_k = \underline{\lim} E_k.$$

Note any increasing [decreasing] sequence $\{E_k\}$ converges to $\cup_{k=1}^{\infty} E_k$ $[\cap_{k=1}^{\infty} E_k]$. If $\mathcal{C}$ is any family of subsets of S, $\mathcal{C}_\sigma$ $[\mathcal{C}_\delta]$ denotes the family of all subsets of S which are countable unions [countable intersections] of members of $\mathcal{C}$.

A non-empty subset $E \subset \mathbf{R}$ is *bounded above [below]* if there exists $b \in \mathbf{R}$ such that $t \leq b$ $[b \leq t]$ for every $t \in E$; b is called an *upper [lower] bound* for E. We say that E is *bounded* if it is bounded from above and below. If E is bounded above [below], an upper [lower] bound, b, for E is called a *least upper bound [greatest lower bound]* or *supremum [infimum]* if b is an upper [lower] bound for E and if a is any other upper [lower] bound for E, then $b \leq a$ $[a \leq b]$; we write $b = \sup E$ $[\inf E]$. The real numbers possess the important *order completeness* property: every non-empty subset of $\mathbf{R}$ which is bounded above [below] has a supremum [infimum]; see [A], [Ba], [DeS] or [R1].

A real-valued sequence $\{t_k\}$ is *increasing [decreasing]* if $t_k \leq t_{k+1}$ $[t_k \geq t_{k+1}]$ for all k; we write $t_k \uparrow$ $[t_k \downarrow]$. If $t_k \uparrow$ or $t_k \downarrow$, we say that $\{t_k\}$ is *monotone*. From the order completeness property of the real numbers, it follows that if $\{t_k\}$ is an increasing [decreasing] sequence in $\mathbf{R}$ which is bounded above [below], then $\{t_k\}$ converges to $\sup\{t_k : k \in \mathbf{N}\}$ $[\inf\{t_k : k \in \mathbf{N}\}]$.

If $\{t_k\}$ is a bounded sequence in $\mathbf{R}$, the *limit superior [limit inferior]* of $\{t_k\}$ is defined by

$$\overline{\lim} t_k = \inf_{j \geq 1} \sup_{k \geq j} t_k \quad [\underline{\lim} t_k = \sup_{j \geq 1} \inf_{k \geq j} t_k]. \tag{1.1}$$

A bounded real sequence, $\{t_k\}$, converges if and only if $\overline{\lim} t_k = \underline{\lim} t_k$, and in this case $\lim t_k = \overline{\lim} t_k = \underline{\lim} t_k$ [DeS].

Let $a, b \in \mathbf{R}$ with $a < b$. We use standard notation for the *intervals* generated by

$$a, b : [a, b] = \{t \in \mathbf{R} : a \leq t \leq b\},$$

$$(a, b) = \{t \in \mathbf{R} : a < t < b\},$$

1.1. PRELIMINARIES

$$[a, b) = \{t \in \mathbf{R} : a \leq t < b\},$$
$$(a, b] = \{t \in \mathbf{R} : a < t \leq b\},$$
$$[a, \infty) = \{t \in \mathbf{R} : t \geq a\},$$
$$(a, \infty) = \{t \in \mathbf{R} : t > a\},$$
$$(-\infty, a) = \{t \in \mathbf{R} : t < a\}$$

and

$$(-\infty, a] = \{t \in \mathbf{R} : t \leq a\}.$$

Intervals of the form (a, b), $(-\infty, a)$, (a, ∞) and $(-\infty, \infty) = \mathbf{R}$ $[[a, b], (-\infty, a], [a, \infty), (-\infty, \infty)]$ are all called *open intervals [closed intervals]*.

If I is an interval in $\mathbf{R}$, a real-valued function $f : I \to \mathbf{R}$ is *increasing [decreasing]* if $t, s \in I$, $t < s$, implies $f(t) \leq f(s)$ $[f(t) \geq f(s)]$; we write $\underline{f \uparrow}$ $[\underline{f \downarrow}]$ on I. f is said to be *monotone* if f is either increasing or decreasing. Recall that any monotone function has both right and left hand limits at every point of I [DeS].

If $f_k : S \to \mathbf{R}$, the sequence $\{f_k\}$ *converges pointwise* to the function $f : S \to \mathbf{R}$ if $\lim f_k(t) = f(t)$ for every $t \in S$; we write $f_k \to f$ pointwise on S. If for every $\epsilon > 0$ there exists N such that $|f_k(t) - f(t)| < \epsilon$ for every $k \geq N$, $t \in S$, then $\{f_k\}$ *converges uniformly* to f on S, and we write $f_k \to f$ uniformly on S. Basic properties of pointwise and uniformly convergent sequences of functions are given in [DeS].

If $f, g : S \to \mathbf{R}$, the max and min of f are defined by $\underline{f \vee g}(t) = \max\{f(t), g(t)\}$, $\underline{f \wedge g}(t) = \min\{f(t), g(t)\}$. We have the important formulas for $f \vee g$ and $f \wedge g$:

$$f \vee g = f + g + |f - g|/2, \quad f \wedge g = f + g - |f - g|/2. \tag{1.2}$$

We write $\underline{f^+} = f \vee 0$ and $\underline{f^-} = (-f) \vee 0$ so $f = f^+ - f^-$, and if $|f|(t) = |f(t)|$ for $t \in S$, $|f| = f^+ + f^-$.

Exercise 1. Prove (1.2).

Exercise 2. Show $A = \overline{\lim} E_k$ if and only if $C_A = \overline{\lim} C_{E_k}$; $A = \underline{\lim} E_k$ if and only if $C_A = \underline{\lim} C_{E_k}$; $A = \lim A_k$ if and only if $C_A = \lim C_{A_k}$, where the limit statements concerning functions are pointwise.

1.2 The Extended Real Numbers

Particularly in measure and integration theory, it is convenient to extend the real numbers by adjoining two additional elements, denoted by ∞ and $-\infty$ [distinct and not belonging to $\mathbf{R}$] with the order properties $-\infty < t < \infty$ for $t \in \mathbf{R}$. We denote $\mathbf{R}$ plus the elements ∞, $-\infty$ by $\mathbf{R}^*$. If $E \subset \mathbf{R}$ is non-empty and not bounded from above [below], we define $\sup E = \infty$ [$\inf E = -\infty$]. With this convention, every non-empty subset of $\mathbf{R}$ has a supremum [infimum] in $\mathbf{R}^*$. Also, with this convention, every sequence in $\mathbf{R}^*$ has a $\overline{\lim}$ and a $\underline{\lim}$ in $\mathbf{R}^*$ [§1, (1)].

A sequence $\{t_k\}$ in $\mathbf{R}^*$ has limit ∞ [$-\infty$] or converges to ∞ [$-\infty$] if for every $r \in \mathbf{R}$ there exists N such that $k \geq N$ implies $t_k \geq r$ [$t_k \leq r$]. We write $\lim t_k = \infty$ or $t_k \to \infty$ [$\lim t_k = -\infty$ or $t_k \to -\infty$].

We adopt the following conventions for algebraic operations on $\mathbf{R}^*$.

$\infty + t = t + \infty$ unless $t = -\infty$;

$(-\infty) + t = t + (-\infty)$ unless $t = \infty$;

if $a > 0$, then $a \cdot \infty = \infty \cdot a = \infty$ and $a \cdot (-\infty) = (-\infty) \cdot a = -\infty$;

if $a < 0$, then $a \cdot \infty = \infty \cdot a = -\infty$ and $a \cdot (-\infty) = (-\infty) \cdot a = \infty$.

The motivation for these definitions is the limit theorems from basic analysis. As usual we do not define $\infty + (-\infty)$ or $(-\infty) + \infty$.

We also adopt the convention that $0 \cdot \infty = 0$; this definition may appear to be somewhat unorthodox, but it will be apparent when we study integration theory that it is very useful.

With these definitions, we have

Theorem 1 *If $\{t_k\}$ is an increasing [decreasing] sequence in $\mathbf{R}^*$, then $\{t_k\}$ converges to $\sup\{t_k : k \in \mathbf{N}\}$ [$\inf\{t_k : k \in \mathbf{N}\}$]. A sequence $\{t_k\}$ in $\mathbf{R}^*$ converges in $\mathbf{R}^*$ if and only if $\overline{\lim} t_k = \underline{\lim} t_k$ and in this case $\lim t_k = \overline{\lim} t_k = \underline{\lim} t_k$. [Here, $\overline{\lim} t_k = \inf_{j \geq 1} \sup_{k \geq j} t_k$ and $\underline{\lim} t_k = \sup_{j \geq 1} \inf_{k \geq j} t_k$.]*

The proof is left to Exercise 1; see also Exercise 10.

A similar result is valid for increasing or decreasing real-valued functions defined on intervals in $\mathbf{R}$. We leave it to the reader to formulate the statements.

For intervals in $\mathbf{R}^*$, we adopt the following notation: $[-\infty, \infty] = \mathbf{R}^*$; $(-\infty, \infty) = \mathbf{R}$; if $a \in \mathbf{R}$,

$$[a, \infty] = \{t \in \mathbf{R}^* : a \leq t \leq \infty\},$$
$$[a, \infty) = \{t \in \mathbf{R} : a \leq t < \infty\},$$
$$(a, \infty) = \{t \in \mathbf{R} : a < t < \infty\},$$

1.2. THE EXTENDED REAL NUMBERS

$$(a, \infty] = \{t \in \mathbf{R}^* : a < t \leq \infty\};$$

and similarly for $[-\infty, a]$, $[-\infty, a)$, $(-\infty, a]$, $(-\infty, a)$.

We have the following important structure theorem for open subsets of $\mathbf{R}$.

Theorem 2 *Let $G \subset \mathbf{R}$ be open. Then G is a countable union of pairwise disjoint open intervals in $\mathbf{R}$.*

Proof: For each $x \in G$ we associate an open interval containing x as follows: let $\mathcal{I}$ be the family of all open intervals which are contained in G and contain x and set $I_x = \cup \mathcal{I}$. If $b = \sup\{\beta : (\alpha, \beta) \in \mathcal{I}\}$ and $a = \inf\{\alpha : (\alpha, \beta) \in \mathcal{I}\}$, then $I_x = (a, b)$ so I_x is an open interval.

The family $\{I_x : x \in G\}$ is pairwise disjoint. For suppose that $z \in I_x \cap I_y$. Then $I_x \cup I_y$ is an open interval so $I_x \cup I_y \subset I_x$ by construction. Hence $I_x = I_x \cup I_y$ and, similarly, $I_y = I_x \cup I_y$ so $\{I_x : x \in G\}$ is pairwise disjoint.

The family $\{I_x : x \in G\}$ is clearly countable since each I_x contains a rational. Since $G = \cup\{I_x : x \in G\}$ the result follows.

For later use we establish an important result for series in $\mathbf{R}^*$.

Definition 3 *Let $\{t_k\} \subset \mathbf{R}^*$ be such that $s_n = \sum\limits_{k=1}^{n} t_k$ is defined for every n; s_n is called the nth partial sum of the sequence $\{t_k\}$. The (formal) series $\sum\limits_{k=1}^{\infty} t_k$ converges in $\mathbf{R}^*$ if and only if the sequence $\{s_n\}$ converges in $\mathbf{R}^*$. We write $s = \lim\limits_{n} s_n = \sum\limits_{k=1}^{\infty} t_k$ and call s the sum of the series.*

The series $\sum\limits_{k=1}^{\infty} t_k$ is called *unconditionally convergent* or *rearrangement convergent* if for every permutation $\sigma : \mathbf{N} \to \mathbf{N}$, the series $\sum\limits_{k=1}^{\infty} t_{\sigma(k)}$ converges. The series $\sum\limits_{k=1}^{\infty} t_{\sigma(k)}$ is called a rearrangement of $\sum\limits_{k=1}^{\infty} t_k$. It is established in elementary analysis that a real-valued series is unconditionally convergent in $\mathbf{R}$ if and only if it is absolutely convergent [i.e., $\sum\limits_{k=1}^{\infty} |t_k|$ converges in $\mathbf{R}$], and in this case all rearrangements converge to the same limit [see, for example, [DeS] 4.21].

For series with non-negative terms, we have

Proposition 4 *If $t_k \in \mathbf{R}^*$, $t_k \geq 0$, then $\sum\limits_{k=1}^{\infty} t_k$ is rearrangement convergent and every rearrangement converges to $\sum\limits_{k=1}^{\infty} t_k$.*

Proof: $\sum\limits_{k=1}^{\infty} t_k$ converges in $\mathbf{R}^*$ since the sequence of partial sums $\{s_n\}$ is increasing. Set $a = \sum\limits_{k=1}^{\infty} t_k$. Let $\sigma : \mathbf{N} \to \mathbf{N}$ be a permutation of $\mathbf{N}$. Set $b = \sum\limits_{k=1}^{\infty} t_{\sigma(k)}$. It suffices to show $b \leq a$ or $\sum_{k=1}^{n} t_{\sigma(k)} \leq a$ for every n. For $n \in \mathbf{N}$, set $m = \max\{\sigma(1), \ldots, \sigma(n)\}$ and note $\sum_{k=1}^{n} t_{\sigma(k)} \leq \sum_{k=1}^{m} t_k \leq a$.

We next establish a result for double series of non-negative terms which is very important in measure theory. Let $t : \mathbf{N} \times \mathbf{N} \to \mathbf{R}^*$; we call t a *double sequence* and denote the value $t(i,j) = t_{ij}$. The series $\sum_{i=1}^{\infty}(\sum_{j=1}^{\infty} t_{ij}) = \sum_{i=1}^{\infty} \sum_{j=1}^{\infty} t_{ij} \ [\sum_{j=1}^{\infty}(\sum_{i=1}^{\infty} t_{ij}) = \sum_{j=1}^{\infty} \sum_{i=1}^{\infty} t_{ij}]$ is called an *iterated series*.

Theorem 5 *Let $t_{ij} \in \mathbf{R}^*$, $t_{ij} \geq 0$. Then*

(i) $\sum_{i=1}^{\infty} \sum_{j=1}^{\infty} t_{ij} = \sum_{j=1}^{\infty} \sum_{i=1}^{\infty} t_{ij}$ *and*

(ii) *if $\sigma : \mathbf{N} \to \mathbf{N} \times \mathbf{N}$ is $1-1$, onto [so $\{t_{\sigma(i)}\}$ is an enumeration of $\{t_{ij}\}$], then*
$\sum_{i=1}^{\infty} t_{\sigma(i)} = \sum_{i=1}^{\infty} \sum_{j=1}^{\infty} t_{ij} \ [= \sum_{j=1}^{\infty} \sum_{i=1}^{\infty} t_{ij}]$.

Proof: (i): Since $t_{ij} \geq 0$, the series $\sum_{j=1}^{\infty} t_{ij}$ always converges to a non-negative element of $\mathbf{R}^*$ so the iterated series $\sum_{i=1}^{\infty} \sum_{j=1}^{\infty} t_{ij}$ always converges. Set $a = \sum_{i=1}^{\infty} \sum_{j=1}^{\infty} t_{ij}$ and $b = \sum_{j=1}^{\infty} \sum_{i=1}^{\infty} t_{ij}$. For each m, n,

$$\sum_{i=1}^{m}\sum_{j=1}^{n} t_{ij} \leq \sum_{i=1}^{m}\sum_{j=1}^{\infty} t_{ij} \leq a.$$

Hence $b \leq a$ and by symmetry $a \leq b$.

(ii): Set $c = \sum_{k=1}^{\infty} t_{\sigma(i)}$ and for each i set $\sigma(i) = (m_i, n_i)$. Then

$$\sum_{i=1}^{k} t_{\sigma(i)} = \sum_{i=1}^{k} t_{m_i n_i} \leq \sum_{i=1}^{k}\sum_{j=1}^{\infty} t_{m_i n_j} \leq a$$

so $c \leq a$.

Fix m, n. There exists N such that if $1 \leq i \leq m$, $1 \leq j \leq n$, then there exists $1 \leq r \leq N$ such that $\sigma(r) = (i, j)$. Then

$$\sum_{i=1}^{m}\sum_{j=1}^{n} t_{ij} \leq \sum_{r=1}^{N} t_{\sigma(r)} \leq c.$$

Hence, $a \leq c$.

For series whose terms are not positive, see Exercises 2 and 3.

Exercise 1. Prove Theorem 1.

1.2. THE EXTENDED REAL NUMBERS

Exercise 2. Let $t_{ij} = 1$ if $i - j = 1$, $t_{ij} = -1$ if $i - j = -1$ and $t_{ij} = 0$ otherwise. Compute

$$\sum_{i=1}^{\infty} \sum_{j=1}^{\infty} t_{ij}$$

and

$$\sum_{j=1}^{\infty} \sum_{i=1}^{\infty} t_{ij}.$$

Find an enumeration $\{t_i\}$ of $\{t_{ij}\}$ such that $\sum_{i=1}^{\infty} t_i$ diverges.

Exercise 3. An iterated series $\sum_{i=1}^{\infty} \sum_{j=1}^{\infty} t_{ij}$ *converges absolutely* if $\sum_{i=1}^{\infty} \sum_{j=1}^{\infty} |t_{ij}| < \infty$. Show that if $\sum_{i=1}^{\infty} \sum_{j=1}^{\infty} t_{ij}$ converges absolutely, it converges; moreover, if $\{t_i\}$ is any enumeration of $\{t_{ij}\}$, show $\sum_{i=1}^{\infty} t_i$ converges to $\sum_{i=1}^{\infty} \sum_{j=1}^{\infty} t_{ij} \ [= \sum_{j=1}^{\infty} \sum_{i=1}^{\infty} t_{ij}]$.

Exercise 4. A real-valued series $\sum_{k=1}^{\infty} t_k$ is *subseries convergent* in **R** if for every subsequence $\{t_{n_k}\}$, the series $\sum_{k=1}^{\infty} t_{n_k}$ converges in **R**. Show $\sum_{k=1}^{\infty} t_k$ is subseries convergent in **R** if and only if $\sum_{k=1}^{\infty} |t_k| < \infty$.

Exercise 5. If $\sum_{k=1}^{\infty} t_k$ converges absolutely in **R**, show $\sum_{k=1}^{\infty} t_k^2$ converges in **R**. Can absolute convergence be replaced by convergence? Does the convergence of $\sum_{k=1}^{\infty} t_k^2$ (in **R**) imply the convergence of $\sum_{k=1}^{\infty} t_k$?

Exercise 6. If $\sum_{k=1}^{\infty} t_k^2$ and $\sum_{k=1}^{\infty} s_k^2$ converge in **R**, show $\sum_{k=1}^{\infty} t_k s_k$ converges absolutely in **R**.

Exercise 7. If $t_k \to 0$, show $\{t_k\}$ has a subsequence $\{t_{n_k}\}$ such that the series $\sum_{k=1}^{\infty} t_{n_k}$ converges absolutely in **R**.

Exercise 8. Let σ be a permutation of $\mathbf{N} \times \mathbf{N}$ and let $0 \leq t_{ij} \leq \infty$. Show $\sum_{i=1}^{\infty} \sum_{j=1}^{\infty} t_{ij} = \sum_{i=1}^{\infty} \sum_{j=1}^{\infty} t_{\sigma(i,j)}$.

Exercise 9. Show that a sequence $\{t_j\} \subset \mathbf{R}$ is bounded if and only if the series $\sum_{j=1}^{\infty} t_j s_j$ converges for every absolutely convergent series $\sum_{j=1}^{\infty} s_j$.

Exercise 10. Show $\overline{\lim}(a_k + b_k) \leq \overline{\lim} a_k + \overline{\lim} b_k$, $\underline{\lim}(a_k + b_k) \geq \underline{\lim} a_k + \underline{\lim} b_k$ and $\overline{\lim}(-a_k) = -\underline{\lim} a_k$. If $\lim a_k = a$, then $\overline{\lim}(a_k + b_k) = a + \overline{\lim} b_k$, $\underline{\lim}(a_k + b_k) = a + \underline{\lim} b_k$.

1.3 Lebesgue's Definition of the Integral

Even shortly after Riemann introduced the integral which now carries his name the mathematicians of the era realized that the integral had serious shortcomings. In particular, the integral was only defined for functions which were bounded and defined on bounded intervals; in order to integrate unbounded functions or functions defined on unbounded intervals required a special definition and led to an integral often referred to as the Cauchy-Riemann integral. It was also recognized that the Riemann integral has poor convergence properties; for example, a function which is the pointwise limit of a uniformly bounded sequence of integrable functions need not be Riemann integrable (Exer. 1). In his 1902 Ph.D. thesis H. Lebesgue ([L1]) introduced an integral which now bears his name and which overcame most of the shortcomings of the Riemann integral. Since its inception the Lebesgue integral has continued to evolve and in its present form it is recognized as the most useful and powerful theory of integration which is available. In order to explain and motivate our presentation of the Lebesgue integral we now give a brief sketch of Lebesgue's introduction to the Lebesgue integral.

Lebesgue began with what he referred to as the "problem of integration" which can be paraphrased as follows: Assign to each bounded function f defined on a bounded interval $I = [a, b]$ a number, called its integral and denoted by $\int_a^b f(t)dt = \int_a^b f = \int_I f$, satisfying the properties

(a) $\int_a^b f(t)dt = \int_{a+h}^{b+h} f(t-h)dt$

(b) $\int_a^b f + \int_b^c f + \int_c^a f = 0$

(c) $\int_a^b (f+g) = \int_a^b f + \int_a^b g$

(d) $f \geq 0$, $b \geq a$ implies $\int_a^b f \geq 0$

(e) $\int_0^1 1 = 1$

(f) if $f_k \uparrow f$, then $\int_a^b f_k \to \int_a^b f$.

Lebesgue made the assumption that such an integral existed and then proceeded to deduce what additional properties it would have to possess.

For example, from (c) it follows immediately by taking $f = g = 0$ that $\int_a^b 0 = 0$, and then it follows that $\int_a^b (-f) = -\int_a^b f$. From this property, (c) and (d) imply that when $f \geq g$ on $[a, b]$, then $\int_a^b f \geq \int_a^b g$, and whence $|\int_a^b f| \leq \int_a^b |f|$, since $\int_a^b f \leq \int_a^b |f|$ and $\int_a^b (-f) = -\int_a^b f \leq \int_a^b |f|$.

From (c) it follows that if n is a positive integer, then $n \int_a^b f = \int_a^b (nf)$, and from the observation above this also holds if n is a negative integer. Also, we have

$$\int_a^b f = \int_a^b n(\frac{1}{n}f) = n \int_a^b \frac{1}{n} f$$

so
$$\frac{1}{n}\int_a^b f = \int_a^b \frac{1}{n} f$$

for any non-zero integer. It follows that $\int_a^b qf = q\int_a^b f$ for any rational q. If $r \in \mathbf{R}$, let q be a rational and choose a rational p, $0 < p < 1$, such that $|r - q|(M + p)$ is rational, where $M = \sup\{|f(x)| : a \leq x \leq b\}$. Then

$$\left|\int_a^b rf - q\int_a^b f\right| \leq \int_a^b |r - q||f| \leq |r - q|(M + p)\int_a^b 1$$

and by letting q approach r, we obtain $\int_a^b rf = r\int_a^b f$. That is, if axioms (c) and (d) hold, then the integral (if it exists) must be homogeneous.

It can also be shown using (a)-(e) that $\int_a^b 1 = b - a$ (Exer. 2). Note that in deriving these properties of the integral condition (f) has not been employed.

It follows from these properties that any integral satisfying axioms (a)-(e) must agree with the Riemann integral for Riemann integrable functions. For consider any interval $I = [a, b]$ and subdivide I into subintervals $\{I_i : i = 1, ..., n\}$. If f is a bounded function on I, let $m_i = \inf\{f(x) : x \in I_i\}$, $M_i = \sup\{f(x) : x \in I_i\}$ and let $\ell(I)$ be the length of I. Then

$$\sum_{i=1}^n m_i \ell(I_i) \leq \sum_{i=1}^n \int_{I_i} f = \int_I f \leq \sum_{i=1}^n M_i \ell(I_i)$$

so
$$\underline{\int_a^b} f \leq \int_a^b f \leq \overline{\int_a^b} f \, ,$$

where $\underline{\int_a^b} f$ and $\overline{\int_a^b} f$ denote the lower and upper Riemann integrals of f, respectively. Recall, however, that the Riemann integral does not satisfy property (f) (Exer. 1).

Now Lebesgue made a crucial observation using condition (f). Suppose $f : [a, b] \to \mathbf{R}$ is bounded with $\ell \leq f(x) < L$ for $a \leq x \leq b$. Consider a partition $\pi : \ell = \ell_0 < \ell_1 < ... < \ell_n = L$ of the range of f and let

$$E_i = \{x : \ell_{i-1} \leq f(x) < \ell_i\}, i = 1, ..., n. \tag{1.1}$$

If we set $\varphi = \sum_{i=1}^n \ell_{i-1} C_{E_i}$, $\psi = \sum_{i=1}^n \ell_i C_{E_i}$, where C_E denotes the characteristic function of E, then we have $\varphi \leq f \leq \psi$ so

$$\int_a^b \varphi = \sum_{i=1}^n \ell_{i-1} \int_a^b C_{E_i} \leq \int_a^b f \leq \int_a^b \psi = \sum_{i=1}^n \ell_i \int_a^b C_{E_i}. \tag{1.2}$$

The inequality (2) leads to the important observation that in order to define the integral of the bounded function f it is only necessary to define the integral for certain characteristic functions. For suppose that $\int_a^b C_E$ has been defined for all characteristic functions arising from (1). The functions φ and ψ depend on the partition π and as

$$\max\{\ell_i - \ell_{i-1} : i = 1, ..., n\} \to 0, \tag{1.3}$$

1.3. LEBESGUE'S DEFINITION OF THE INTEGRAL

the functions φ and ψ converge to f uniformly. If a sequence of integrable functions $\{f_n\}$ converge uniformly to an integrable function f, then $\int_a^b f_n \to \int_a^b f$ since

$$\left|\int_a^b f_n - \int_a^b f\right| \le \int_a^b |f_n - f| \le (b-a)\sup\{|f_n(x) - f(x)| : a \le x \le b\}.$$

Thus, it follows that the integrals of φ and ψ must converge to $\int_a^b f$ if (3) holds. Note that in order to define the integral of the function f it is only necessary to define the integral of characteristic functions of sets of the form (1) which are generated by the function f. The integral of f can then be defined as the common limit of the integrals of the functions φ and ψ as (3) holds. It is often remarked that the key difference between the Riemann integral and Lebesgue integral is that in the Riemann integral the domain $[a, b]$ is partitioned while in the Lebesgue integral the range $[\ell, L]$ is partitioned. Note, however, that the sets $\{E_i : i = 1, ..., n\}$ from a "partition" of $[a, b]$ in the sense that the sets are pairwise disjoint and their union is $[a, b]$ so another view of the Lebesgue integral is that the domain $[a, b]$ is "partitioned" by using sets which are more general than subintervals.

Thus, to define the integral, Lebesgue was led to what might be called the "problem of measure":

Assign to each bounded subset E of $\mathbf{R}$ a non-negative number, called the measure of E and denoted by $m(E)$, satisfying:

(i) any two congruent subsets have the same measure [two subsets of $\mathbf{R}$ are congruent if one can be obtained from the other by reflection and translation]

(ii) if E is either a finite or countably infinite union of pairwise disjoint sets $\{E_i\}$, then
$m(E) = \sum m(E_i)$ [this condition is called *countable additivity*]

(iii) $m([0, 1]) = 1$.

Since $m([a, b]) = b - a$ for any interval (Exer. 3), we are seeking an extension of the length function to the class of all bounded subsets of $\mathbf{R}$.

Using the Axiom of Choice, Vitali showed that the problem of measure has no solution in $\mathbf{R}$.

Example 1 For $x, y \in [0, 1]$, say that $x \sim y$ if and only if $x - y$ is rational. Then $\sim$ is an equivalence relation on $[0, 1]$. Let P be a subset of $[0, 1]$ which consists of 1 point from each equivalence class of $\sim$; the Axiom of Choice is being used here.

We first claim that if r, s are distinct rationals, then $(P+r) \cap (P+s) = \emptyset$. For if $x \in (P+r) \cap (P+s)$, then $x = p + r = q + s$ where $p, q \in P$. Then $p - q = s - r \ne 0$ so $p \sim q$ with $p \ne q$. This violates the construction of P.

Next we claim that $[0, 1] \subseteq \cup \{P + r : r \in \mathbf{Q}_0\}$, where $\mathbf{Q}_0 = \mathbf{Q} \cap [-1, 1]$. For suppose that $x \in [0, 1]$. Then x is in some equivalence class of $\sim$ so $x \sim p$ for some $p \in P$. Then $x - p$ is a rational r in $[-1, 1]$ so $x \in P + r$.

By (i), $m(P+r) = m(P)$ for any r and by (ii), (iii) and Exer. 3,

$$m([0,1]) = 1 \leq m(\cup\{P+r : r \in \mathbf{Q}_0\}) = \sum_{r \in \mathbf{Q}_0} m(P) \leq m([-1,2]) = 3.$$

Whether $m(P) = 0$ or $m(P) > 0$, this equation is clearly impossible.

Remark 2 Note that the full force of (iii) was not used to reach the conclusion above; it is only necessary to assume that $m([0,1]) > 0$.

Since the problem of measure as posed above has no solution, there are two obvious ways to weaken the statement of the problem so that it is possible that a solution might exist. First, we might consider relaxing condition (ii) to require only finite additivity of the measure m. That is, replace (ii) by the weaker condition:

(ii)' if E is a finite union of pairwise disjoint sets $E_1, ..., E_n$, then $m(E) = \sum_{i=1}^{n} m(E_i)$.

This version of the problem of measure is sometimes referred to as the "easy" problem of measure while the original version of the problem of measure is referred to as the "difficult" problem of measure ([Na] III.7; this terminology is not often used but is convenient in this discussion). Banach showed that the easy problem of measure in **R** has a solution ([B1]); see 5.6.3.

Both problems of measure have obvious generalizations from **R** to $\mathbf{R}^n$ where the unit interval $I = [0,1]$ in (iii) is replaced by the unit square $I \times ... \times I$, and in (i) two sets are said to be congruent if one can be obtained from the other by reflections, translations and rotations. Example 1 can be used to show that the difficult problem of measure has no solution in $\mathbf{R}^n$ (see Exer. 3.9.4). Banach also showed that the "easy" problem of measure has a solution in $\mathbf{R}^2$ ([B1]). However, even the "easy" problem of measure has no solution in $\mathbf{R}^n$ for $n \geq 3$. Indeed, in $\mathbf{R}^n$ with $n \geq 3$ we have the remarkable Banach-Tarski Paradox:

Theorem 3 *If U and V are bounded subsets of $\mathbf{R}^n$, $n \geq 3$, with non-empty interiors, then there exist $k \in \mathbf{N}$ and partitions $\{E_1, \cdots, E_k\}$ and $\{F_1, \cdots, F_k\}$ of U and V, respectively, such that E_j is congruent to F_j for $j = 1, \cdots, k$.*

That is, one can take a golf ball, cut it into a finite number of pieces and reassemble the pieces into a basketball! Theorem 3 clearly precludes the existence of a finitely additive measure on $\mathbf{R}^n$, $n \geq 3$, which solves the "easy" problem of measure.

For discussions of the Banach-Tarski Paradox see [St1] or [Fr].

The other obvious weakening of the "difficult" problem of measure is to retain condition (ii) but to seek a measure which is defined on some proper subfamily of the family of all bounded subsets of **R**. This is the approach that we adopt. As we will see when we study the Lebesgue integral the countable additivity in condition (ii) leads to very powerful convergence properties for the Lebesgue integral.

1.3. LEBESGUE'S DEFINITION OF THE INTEGRAL

We now consider how we might construct a measure m satisfying conditions (i)-(iii) on some family of subsets of $\mathbf{R}$. Obviously, what we are seeking is an extension of the length function, $\ell((a, b)) = b - a$, from the class of all bounded intervals which satisfies conditions (i) and (ii) on some appropriate subfamily of subsets of $\mathbf{R}$. To see how we might construct such an extension assume that a measure m exists satisfying condition (ii). If G is an open set which is contained in some bounded interval, then by (ii) $m(G)$ must be the sum of the lengths of the open subintervals which make up G (1.2.2). That is, if G is the union of open intervals $\{I_i\}$, then $m(G) = \sum_i \ell(I_i)$; this shows how the length function can be extended to bounded open sets. If E is a bounded subset and G is a bounded open set containing E, then $m(G) \geq m(E)$ (Exer. 3) so

$$\inf\{m(G) : G \text{ open and bounded}, G \supseteq E\} = m^*(E) \geq m(E); \quad (1.4)$$

$m^*(E)$ is called the *outer measure* of E and gives an extension of m from the bounded open sets to the family of all bounded subsets of $\mathbf{R}$. As Example 1 points out, m^* cannot satisfy condition (ii) for all bounded sets so we seek a "nice" subfamily of bounded sets on which m^* will satisfy (ii). In order to isolate an appropriate subfamily, Lebesgue also defined the inner measure of a bounded set. Suppose that E is bounded and contained in the interval $I = [a, b]$. The *inner measure* of E, $m_*(E)$, is defined by computing the outer measure of the complement of E in I and setting

$$m_*(E) = m(I) - m^*(I \setminus E). \quad (1.5)$$

A set E is called (Lebesgue) measurable if $m^*(E) = m_*(E)$ and the (Lebesgue) measure of E, $m(E)$, is defined to be this common value. As we will see in §2.5 Lebesgue measure on the family of Lebesgue measurable sets satisfies conditions (i)-(iii) and so furnishes a solution to the weakened "difficult" problem of measure.

It is desirable to have the measure extended to subsets of $\mathbf{R}$ which are not bounded. The measure of an open set, even if unbounded, can be defined as above and the formula (4) defining the outer measure is still meaningful. However, if I is an unbounded interval and $E \subseteq I$ is such that $m^*(I \setminus E) = \infty$, then the formula (5) defining the inner measure of E is no longer meaningful. There are means of defining the inner measure of an arbitrary subset of $\mathbf{R}$ ([Roy], [Wi]), but, fortunately, there is a characterization of measurability due to Caratheodory which involves only the outer measure of a set and this characterization can be used to define the measurability of a set even if the set is unbounded. Namely, we say that a subset $E \subseteq \mathbf{R}$ is (Lebesgue) measurable if and only if

$$m^*(A) = m^*(A \cap E) + m^*(A \setminus E) \text{ for any } A \subseteq \mathbf{R} \text{ ([Ca])}. \quad (1.6)$$

Thus, a set is measurable if no matter how it is divided, the measure of the set is the sum of the measures of the pieces into which it is divided. As will be seen in §2.3, this definition can be used to define measurability in a very abstract setting.

As observed by Lebesgue, in order to define the integral of a function f, it is only necessary to define the integral of the characteristic function for sets of the form (1); i.e., it is only necessary to define the measure of such sets. [Functions with the property that inverse images of the form (1) are measurable are called measurable functions and are studied in §3.1.] In part 2 of these notes, we begin by developing the basic properties of measures. We then give an abstract treatment of outer measures and measurable sets by using the Caratheodory characterization of measurability given in (6). We show how outer measures can be constructed in general, and then construct Lebesgue and Lebesgue-Stieltjes measures and derive their most important properties. The Lebesgue integral is then defined and studied in part 3. In part 4 we study the relationship between differentiation and integration, and Parts 5 and 6 of the notes are devoted to the study of function spaces, many of which arise through integration processes.

For a beautiful exposition on integration theory by H. Lebesgue see [L2], especially part II. A brief biography of Lebesgue is given by K.O. May in [L2]. Historical developments of integration theory, including the Lebesgue integral, are given in [Ha] and [Pe].

Exercise 1. Let $\{r_i\}$ be an enumeration of the rationals in $[0, 1] = I$. Define f_n on I by $f_n(t) = 1$ if $t = r_1, ..., r_n$ and $f_n(t) = 0$ otherwise. Show each f_n is Riemann integrable but converges pointwise to a function which is not Riemann integrable.

Exercise 2. Use (a)-(e) to show that $\int_a^b 1 = b - a$.

Exercise 3. Show that if (i), (ii)' and (iii) hold, then $m(\{a\}) = 0$ for any a and $m(E) \geq m(F)$ when $E \supset F$. Show that if (i)-(iii) hold, $m(E) = 0$ for any countable set and $m([a, b]) = b - a$.

Chapter 2
Measure Theory

2.1 Semi-rings and Algebras of Sets

In our description of Lebesgue measure in §1.3 , we observed that the length function in $\mathbf{R}$ was to be extended to Lebesgue measure defined on some subfamily of subsets of $\mathbf{R}$. In this section, we describe and develop some of the basic properties of the types of subfamilies on which measures and other set functions are defined.

Let S be a non-void set and $\mathcal{S} \subseteq \mathcal{P}(S)$, the power set of S.

Definition 1 $\mathcal{S}$ *is a* semi-ring *if and only if*

(i) $\emptyset \in \mathcal{S}$

(ii) $A, B \in \mathcal{S}$ *implies* $A \cap B \in \mathcal{S}$

(iii) $A, B \in \mathcal{S}$ *implies* $A \backslash B = \cup_{j=1}^n S_j$, *where* $S_j \in \mathcal{S}$ *and* $\{S_j : j = 1, \ldots, n\}$ *are pairwise disjoint.*

  If in addition to (i)-(iii), $\mathcal{S}$ satisfies:

(iv) $S \in \mathcal{S}$,

  then $\mathcal{S}$ is called a semi-algebra.

Definition 2 $A \subseteq S$ *is called a* σ-set *with respect to $\mathcal{S}$ if $A = \cup_{j=1}^\infty S_j$, $S_j \in \mathcal{S}$, $\{S_j\}$ pairwise disjoint.*

Proposition 3 *Let $\mathcal{S}$ be a semi-ring.*

(i) *If $A \in \mathcal{S}$ and $A_1, \ldots, A_n \in \mathcal{S}$, then $A \backslash \cup_{j=1}^n A_j$ is a finite, pairwise, disjoint union of elements of $\mathcal{S}$ (and, hence, a σ-set with respect to $\mathcal{S}$).*

(ii) $\{A_i : i \in \mathbf{N}\} \subseteq \mathcal{S}$ *implies* $\cup_{i=1}^\infty A_i$ *is a σ-set with respect to $\mathcal{S}$.*

(iii) *Countable unions and finite intersections of σ-sets are σ-sets.*

Proof: (i): We use induction on n. For $n = 1$, this is Definition 1(iii). Assume (i) holds for n. Let $A_1, ..., A_{n+1} \in \mathcal{S}$. By the induction hypothesis

$$B = A \setminus \cup_{j=1}^n A_j = \cup_{i=1}^k B_i,$$

where $B_i \in \mathcal{S}$, $\{B_i\}$ pairwise disjoint. Then

$$A \setminus \cup_{j=1}^{n+1} A_j = B \setminus A_{n+1} = \cup_{i=1}^k (B_i \setminus A_{n+1})$$

and by Definition 1(iii) each $B_i \setminus A_{n+1}$ is a finite, pairwise disjoint union of elements of $\mathcal{S}$ so $A \setminus \cup_{j=1}^{n+1} A_j$ is likewise.

(ii): Set $A = \cup_{i=1}^\infty A_i$. Define $B_1 = A_1$ and $B_{k+1} = A_{k+1} \setminus \cup_{j=1}^k A_j$ for $k \geq 1$. Then $\{B_k\}$ are pairwise disjoint and $A = \cup_{k=1}^\infty B_k$. [This is a standard construction in measure theory.] Each B_k is a σ-set by (i) so A is a σ-set.

(iii) follows from (ii) and Definition 1(ii).

Definition 4 $\mathcal{S}$ is a *ring* (of subsets of S) if $\mathcal{S}$ is a semi-ring and

(iii)' $A, B \in \mathcal{S}$ implies $A \setminus B \in \mathcal{S}$.

If $\mathcal{S}$ is a ring with $S \in \mathcal{S}$, then $\mathcal{S}$ is called an *algebra*. Note an algebra is closed under complementation.

Proposition 5 *Let $\mathcal{A}$ be an algebra. Then $\mathcal{A}$ is closed under finite unions and intersections.*

Proof: By DeMorgan's Laws, $A \cup B = (A^c \cap B^c)^c$.

Note that $\mathcal{A}$ is an algebra if and only if $\emptyset \in \mathcal{A}$ and $\mathcal{A}$ is closed under complementation and finite unions (intersections).

We now give some examples.

Example 6 $\mathcal{A} = \{\emptyset, S\}$ is an algebra.

Example 7 $\mathcal{A} = \mathcal{P}(S)$, the power set of S, is an algebra.

Example 8 Write $[a, a) = \emptyset$ for $a \in \mathbf{R}$. Let $\mathcal{S}$ consist of all intervals in $\mathbf{R}$ of the form $[a, b)$ with $a \leq b$. Then $\mathcal{S}$ is a semi-ring and is not a semi-algebra.

Example 9 Let $a < b$. Let $\mathcal{S}$ consist of all subintervals of the form $[c, d)$ with $a \leq c \leq d \leq b$. Then $\mathcal{S}$ is a semi-algebra.

2.1. SEMI-RINGS AND ALGEBRAS OF SETS

Example 10 Let $\mathcal{S}_n$ be the collection of all subsets $A \subseteq \mathbf{R}^n$ of the form

$$A = [a_1, b_1) \times \ldots \times [a_n, b_n)$$

with $-\infty < a_i \leq b_i < \infty$. Then $\mathcal{S}_n$ is a semi-ring of subsets of $\mathbf{R}^n$. [For $n = 1$, this is Example 8. Conditions (i) and (ii) of Definition 1 are clear. We establish (iii) by induction on n. For this note

$$A \times B \backslash C \times D = \{(A \backslash C) \times B\} \cup \{(A \cap C) \times (B \backslash D)\}$$

and apply this to

$$\{[a_1, b_1) \times \ldots \times [a_n, b_n)\} \times [a_{n+1}, b_{n+1}) \backslash \{[c_1, d_1) \times \ldots \times [c_n, d_n)\} \times [c_{n+1}, d_{n+1})$$

and use the induction hypothesis.]

Since the intersection of algebras (rings) is an algebra (ring), given any family $\mathcal{S}$ of subsets of S there is always a smallest algebra (ring) containing $\mathcal{S}$, called the algebra (ring) generated by $\mathcal{S}$. For semi-algebras (semi-rings), we have

Proposition 11 *Let $\mathcal{S}$ be a semi-algebra (semi-ring). The algebra (ring) $\mathcal{A}$ generated by $\mathcal{S}$ consists of all the pairwise disjoint finite unions of elements of $\mathcal{S}$.*

The proof is left to Exercise 1.

Definition 12 *An algebra $\sum$ of subsets of S is a σ-algebra if*

$$\{A_j : j \in \mathbf{N}\} \subset \sum \text{ implies } \cup_{j=1}^{\infty} A_j \in \sum.$$

From DeMorgan's Laws, we have

Proposition 13 *A σ-algebra is closed under countable intersections.*

Note that $\sum$ is a σ-algebra if and only if $\emptyset \in \sum$ and $\mathcal{A}$ is closed under complementation and countable unions (intersections).

Since the intersection of σ-algebra is a σ-algebra, any family of subsets $\mathcal{S}$ of S has a smallest σ-algebra containing $\mathcal{S}$, called the σ-algebra generated by $\mathcal{S}$.

We next establish a useful criterion, called the Monotone Class Lemma, for establishing when an algebra is a σ-algebra. This result is not used in the sequel and may be skipped.

Definition 14 *Let $\mathcal{M} \subseteq \mathcal{P}(S)$. $\mathcal{M}$ is a* monotone class *if $\{A_i\} \subseteq \mathcal{M}$ and if $A_i \uparrow$ (or $A_i \downarrow$), then $\cup A_i \in \mathcal{M} (\cap A_i \in \mathcal{M})$.*

Again the intersection of monotone classes is a monotone class so any family of subsets $\mathcal{S}$ of S has a smallest monotone class containing it, called the monotone class generated by $\mathcal{S}$.

Lemma 15 *(Monotone Class Lemma). Let $\mathcal{A}$ be an algebra. Then the σ-algebra, $\sum$, generated by $\mathcal{A}$ coincides with the monotone class, $\mathcal{M}$, generated by $\mathcal{A}$.*

Proof: By Exercise 6, $\sum \supset \mathcal{M}$. Hence, by Exercise 7, it suffices to show that $\mathcal{M}$ is an algebra.

If $E \in \mathcal{M}$, define $\mathcal{M}_E$ to be the collection of all $F \in \mathcal{M}$ such that $E \backslash F$, $E \cap F$ and $F \backslash E$ belong to $\mathcal{M}$. Clearly, $\emptyset \in \mathcal{M}_E$ and $\mathcal{M}_E$ is a monotone class containing $\mathcal{A}$. Moreover, $F \in \mathcal{M}_E$ if and only if $E \in \mathcal{M}_F$.

If $E \in \mathcal{A}$, then $\mathcal{A} \subseteq \mathcal{M}_E$ for $E \in \mathcal{A}$. Hence, if $E \in \mathcal{A}$ and $F \in \mathcal{M}$, then $E \in \mathcal{M}_F$ so that $\mathcal{A} \subseteq \mathcal{M}_F$ for $F \in \mathcal{M}$. The minimality of $\mathcal{M}$ implies $\mathcal{M}_F = \mathcal{M}$ for $F \in \mathcal{M}$. Thus, $\mathcal{M}$ is closed under intersections and relative complements; but $S \in \mathcal{M}$ so $\mathcal{M}$ is an algebra.

Corollary 16 *If a monotone class $\mathcal{M}$ contains an algebra $\mathcal{A}$, then $\mathcal{M}$ contains the σ-algebra generated by $\mathcal{A}$.*

Exercise 1. Prove Proposition 11.

Exercise 2. Let S be uncountable. Let $\sum$ be all subsets, A, of S with either A or A^c at most countable. Show $\sum$ is a σ-algebra and is the σ-algebra generated by the singleton subsets of S.

Exercise 3. Let $\mathcal{A}$ consist of all subsets, A, of S with either A or A^c finite. Show $\mathcal{A}$ is an algebra, and if S is infinite, $\mathcal{A}$ is not a σ-algebra.

Exercise 4. Let $\sum$ be a σ-algebra of subsets of S and $E \in \sum$. Show

$$\sum\nolimits_E = \{F \cap E : F \in \sum\}$$

is a σ-algebra of subsets of E.

Exercise 5. Show the families in Examples 8, 9 and 10 are not algebras. Show the σ-algebras generated by $\mathcal{S}$ in Examples 8, 9 and 10 contain all of the open and closed sets.

2.1. SEMI-RINGS AND ALGEBRAS OF SETS

Exercise 6. If $S \subseteq \mathcal{P}(S)$, $\mathcal{M}$ is the monotone class generated by S and $\sum$ is the σ-algebra generated by S, show $\sum \supset \mathcal{M}$.

Exercise 7. If $\mathcal{M}$ is a monotone class and an algebra, show $\mathcal{M}$ is a σ-algebra.

Exercise 8. Let $\mathcal{R}$ consist of all subsets A of $[a, b]$ which are such that C_A is Riemann integrable. Show $\mathcal{R}$ is an algebra which is not a σ-algebra.

2.2 Additive Set Functions

In this section we study the additivity properties of set functions associated with Lebesgue measure. [A *set function* is a function defined on some family of subsets of a set.] That is, if $S \subseteq \mathcal{P}(S)$ and $\emptyset \in S$ and $\mu : S \to \mathbf{R}^*$, we are concerned with properties of the form $\mu(A \cup B) = \mu(A) + \mu(B)$ when $A, B, A \cup B \in S$ and $A \cap B = \emptyset$. To avoid arithmetic problems (*i.e.*, $\infty + (-\infty)$), we always agree that any set function with values in the extended real numbers, $\mathbf{R}^*$, can take on only one of the values ∞ or $-\infty$.

Let $S \subseteq \mathcal{P}(S)$ satisfy $\emptyset \in S$ and let $\mu : S \to \mathbf{R}^*$.

Definition 1 μ *is finitely additive if*

(i) $\mu(\emptyset) = 0$

(ii) $\{A_i : i = 1, ..., n\} \subseteq S$ *pairwise disjoint and* $\bigcup_{i=1}^{n} A_i \in S$ *implies* $\mu(\bigcup_{i=1}^{n} A_i) = \sum_{i=1}^{n} \mu(A_i)$.

μ *is countably additive if* (i) *holds and*

(ii)' $\{A_i : i \in \mathbf{N}\} \subseteq S$ *pairwise disjoint and* $\bigcup_{i=1}^{\infty} A_i \in S$ *implies* $\mu(\bigcup_{i=1}^{\infty} A_i) = \sum_{i=1}^{\infty} \mu(A_i)$.

Note that the series in (ii)' must be absolutely convergent since the left hand side is independent of the ordering of the $\{A_i\}$.

A non-negative countably additive set function defined on a σ-algebra Σ (semi-ring) of subsets of S is called a *measure* (*premeasure*) on Σ; a countably additive set function defined on a σ-algebra Σ is called a *signed measure* on Σ. An ordered triple (S, Σ, μ), where μ is a measure on the σ-algebra Σ, is called a *measure space*.

We first make some elementary observations concerning finitely additive set functions.

Proposition 2 *Let S be a semi-ring and $\mu : S \to [0, \infty]$.*

(i) *If μ is finitely additive and $A, B \in S$ with $A \subseteq B$, then $\mu(A) \leq \mu(B)$ [set functions with this property are called monotone].*

(ii) *If μ is countably additive, then μ is finitely additive.*

Proof: (i): $B \setminus A = \bigcup_{i=1}^{n} A_i$, where $\{A_i : i = 1, ..., n\} \in S$ pairwise disjoint. Therefore, $\mu(B) = \mu(A) + \sum_{i=1}^{n} \mu(A_i) \geq \mu(A)$.

(ii) is clear since $\emptyset \in S$ and $\mu(\emptyset) = 0$.

2.2. ADDITIVE SET FUNCTIONS

Proposition 3 *Let $\mathcal{A}$ be an algebra, $\mu : \mathcal{A} \to \mathbf{R}^*$ finitely additive and $A, B \in \mathcal{A}$.*

(i) *If $A \supseteq B$ and $\mu(B)$ infinite, then $\mu(A)$ is infinite.*

(ii) *If $A \supseteq B$ and $\mu(A)$ is finite, then $\mu(B)$ is finite.*

(iii) *If $A \supseteq B$ and $\mu(B)$ is finite, then $\mu(A \backslash B) = \mu(A) - \mu(B)$.*

Proof: By finite additivity, we have $\mu(B) + \mu(A \backslash B) = \mu(A)$ so (i), (ii) and (iii) are immediate.

We next give some characterizations of countable additivity.

Proposition 4 *Let $\mathcal{S}$ be a semi-ring and $\mu : \mathcal{S} \to [0, \infty]$. Then μ is countably additive if and only if*

(i) $\mu(\emptyset) = 0$,

(ii) $A \in \mathcal{S}$ and $\{A_i : i = 1, \ldots, n\} \subseteq \mathcal{S}$ pairwise disjoint with $\bigcup_{i=1}^{n} A_i \subseteq A$ implies $\sum_{i=1}^{n} \mu(A_i) \leq \mu(A)$ and

(iii) $A \in \mathcal{S}$ and $\{A_i : i \in \mathbf{N}\} \subseteq \mathcal{S}$ with $A \subseteq \bigcup_{i=1}^{\infty} A_i$ implies $\mu(A) \leq \sum_{i=1}^{\infty} \mu(A_i)$. [$\mu$ is said to be countably subadditive].

Proof: $\Leftarrow$: (ii) and (iii) clearly imply that μ is countably additive.
$\Rightarrow$: (i) is clear.

(ii): There exist $\{B_i : i = 1, \ldots, m\} \subseteq \mathcal{S}$ pairwise disjoint such that $A \backslash \bigcup_{i=1}^{n} A_i = \bigcup_{i=1}^{m} B_i$ by Proposition 2.1.3. Then

$$\mu(A) = \mu((\bigcup_{i=1}^{n} A_i) \cup (\bigcup_{i=1}^{m} B_i)) = \sum_{n=1}^{m} \mu(A_i) + \sum_{i=1}^{m} \mu(B_i) \geq \sum_{i=1}^{n} \mu(A_i).$$

(iii): We disjointify the $\{A_i\}$ by setting $B_1 = A_1$ and $B_{k+1} = A_{k+1} \backslash \bigcup_{i=1}^{k} A_i$ for $k \geq 1$. Then $\{B_i\}$ are pairwise disjoint and $\bigcup_{i=1}^{\infty} B_i = \bigcup_{i=1}^{\infty} A_i$, $B_i \subseteq A_i$. By Proposition 2.1.3 each $B_i = \bigcup_{j=1}^{k_i} C_{ij}$ where $\{C_{ij} : j = 1, \ldots, k_i\}$ are pairwise disjoint. By (ii), $\sum_{j=1}^{k_i} \mu(C_{ij}) \leq \mu(A_i)$. Now

$$A = \bigcup_{i=1}^{\infty} B_i \cap A = \bigcup_{i=1}^{\infty} \bigcup_{j=1}^{k_i} C_{ij} \cap A$$

is a pairwise disjoint union so countable additivity and (ii) imply

$$\mu(A) = \sum_{i=1}^{\infty}\sum_{j=1}^{k_i}\mu(C_{ij}\cap A) \le \sum_{i=1}^{\infty}\sum_{j=1}^{k_i}\mu(C_{ij}) \le \sum_{i=1}^{\infty}\mu(A_i).$$

Proposition 5 *Let $\sum$ be a σ-algebra and $\mu : \sum \to \mathbf{R}^*$ be finitely additive.*

(i) *μ is countably additive if and only if for every increasing sequence $\{E_j\} \subseteq \sum$, $\mu(\bigcup_{j=1}^{\infty} E_j) = \lim \mu(E_j)$.*

(ii) *If μ is countably additive and $\{E_j\} \subseteq \sum$ is a decreasing sequence with $|\mu(E_j)| < \infty$ for some j, then $\mu(\bigcap_{j=1}^{\infty} E_j) = \lim \mu(E_j)$.*

(iii) *If μ is finite ($\mathbf{R}$) valued, then μ is countably additive if and only if for every decreasing sequence $\{E_j\} \subseteq \sum$, $\mu(\bigcap_{j=1}^{\infty} E_j) = \lim \mu(E_j)$.*

Proof: (i): $\Rightarrow$: Put $E_0 = \emptyset$. Since $E_j \uparrow$, $E = \bigcup_{j=1}^{\infty} E_j = \bigcup_{j=1}^{\infty}(E_j \backslash E_{j-1})$ and since μ is countably additive,

$$\mu(E) = \sum_{j=1}^{\infty}\mu(E_j\backslash E_{j-1}) = \lim \sum_{j=1}^{n}\mu(E_j\backslash E_{j-1}) = \lim_n \mu(E_n).$$

$\Leftarrow$: Let $\{A_j\} \subseteq \sum$ be pairwise disjoint and set $E_k = \bigcup_{j=1}^{k} A_j$. Then $E_k \uparrow \bigcup_{j=1}^{\infty} A_j$ so

$$\lim_k \mu(E_k) = \lim_k \sum_{j=1}^{k}\mu(A_j) = \mu(\bigcup_{j=1}^{\infty} A_j)$$

by hypothesis.

(ii): We may as well assume $|\mu(E_1)| < \infty$. Since $E_1 \supseteq \bigcap_{k=1}^{\infty} E_k = E$ and $E_1 \supseteq E_k$, $\mu(E)$ and $\mu(E_k)$ are finite by Proposition 3. By (i) and Proposition 3,

$$\begin{aligned}\mu(\bigcup_{k=1}^{\infty}(E_1\backslash E_k)) &= \mu(E_1 \backslash \bigcap_{k=1}^{\infty} E_k) = \mu(E_1) - \mu(\bigcap_{k=1}^{\infty} E_k)\\ &= \lim \mu(E_1\backslash E_k) = \lim(\mu(E_1) - \mu(E_k)) = \mu(E_1) - \lim \mu(E_k).\end{aligned}$$

Hence, $\mu(E) = \lim \mu(E_k)$.

(iii): $\Rightarrow$ follows from (ii).

$\Leftarrow$: Let $\{A_j\}$ be a pairwise disjoint sequence from $\sum$ and set $A = \bigcup_{j=1}^{\infty} A_j$. Then by Proposition 3,

$$\mu(A) - \sum_{j=1}^{k}\mu(A_j) = \mu(\bigcup_{j=k+1}^{\infty} A_j) \to 0$$

2.2. ADDITIVE SET FUNCTIONS

since $\bigcup_{j=k+1}^{\infty} A_j \downarrow \emptyset$.

For a sequence of subsets, $\{A_j\}$, from S, recall $\underline{\lim} A_j = \bigcup_{k=1}^{\infty} \bigcap_{j=k}^{\infty} A_j$ and $\overline{\lim} A_j = \bigcap_{k=1}^{\infty} \bigcup_{j=k}^{\infty} A_j$. We say the sequence $\{A_j\}$ converges if $\underline{\lim} A_j = \overline{\lim} A_j$ and define the limit of $\{A_j\}$, denoted by $\lim A_j$, to be this common value. For example, any increasing (decreasing) sequence $\{A_j\}$ converges to $\bigcup_{j=1}^{\infty} A_j (\bigcap_{j=1}^{\infty} A_j)$. (See §1.1.) From Proposition 5, we have

Corollary 6 *Let $\sum$ be a σ-algebra and $\mu : \sum \to [0, \infty]$ be countably additive and let $\{E_j\} \subseteq \sum$.*

(i) $\mu(\underline{\lim} E_j) \leq \underline{\lim} \mu(E_j)$

(ii) *If* $\mu(\bigcup_{j=1}^{\infty} E_j) < \infty$, *then* $\mu(\overline{\lim} E_j) \geq \overline{\lim} \mu(E_j)$.

(iii) *If $\{E_j\}$ converges and* $\mu(\bigcup_{j=1}^{\infty} E_j) < \infty$, *then* $\mu(\lim E_j) = \lim \mu(E_j)$.

Proof: (i): Set $A_k = \bigcap_{j=k}^{\infty} E_j$. So $\{A_k\}$ is increasing and $A_k \supseteq E_k$. By Propositions 2 and 5(i),

$$\lim \mu(A_k) = \mu(\bigcup_{k=1}^{\infty} A_k) = \mu(\underline{\lim} E_k) \geq \underline{\lim} \mu(E_k).$$

(ii) follows from (i) by taking complements and using Proposition 3(iii).

(iii) follows from (i) and (ii).

The finiteness condition in Proposition 5(ii) or Corollary 6(ii) cannot be dropped (Exercise 3).

We now give some examples of measures and premeasures.

Example 7 (Counting Measure). If S is a non-void set, the counting measure on S is the set function μ defined on the power set of S by $\mu(A)$ is the number of points in A if A is finite and $\mu(A) = \infty$ if A is infinite. μ is a measure on $\mathcal{P}(S)$.

Example 8 (Point Mass or Dirac Measure). Fix $x \in S$. Define $\delta_x : \mathcal{P}(S) \to \mathbf{R}$ by $\delta_x(A) = 1$ if $x \in A$ and $\delta_x(A) = 0$ if $x \notin A$. δ_x is called the point mass or Dirac measure at x; δ_x is a measure on $\mathcal{P}(S)$.

We next give the construction of the Lebesgue-Stieltjes premeasure.

Example 9 Let $f : \mathbf{R} \to \mathbf{R}$ be increasing and $\mathcal{S}$ the semi-ring of Example 2.1.8. Define $\mu_f : \mathcal{S} \to \mathbf{R}$ by $\mu_f([a, b)) = f(b) - f(a)$. Then μ_f is finitely additive on $\mathcal{S}$ [Exer. 10], and we consider what conditions on f are equivalent to the countable additivity of μ_f.

First suppose that μ_f is countably additive. Let $a \in \mathbf{R}$ and let $\{a_k\}$ be an arbitrary sequence with $a_k \uparrow a$. Then $[a_1,a) = \bigcup_{i=1}^{\infty} [a_i, a_{i+1})$ so

$$\mu_f[a_1, a) = f(a) - f(a_1) = \sum_{i=1}^{\infty}(-f(a_i) + f(a_{i+1})) = \lim f(a_{i+1}) - f(a_1)$$

so $\lim f(a_i) = f(a)$. Hence, if μ_f is countably additive, f must be left continuous (i.e., $\lim_{t \to a^-} f(t) = f(a)$).

We show conversely that if f is left continuous, then μ_f is countably additive on $\mathcal{S}$. Let $[a, b) = \bigcup_{i=1}^{\infty} [a_i, b_i)$ with the union pairwise disjoint.

First, we claim that

$$\sum_{i=1}^{\infty} \mu_f[a_i, b_i) \leq f(b) - f(a) = \mu_f[a, b):$$

For each n, $[a, b) \supseteq \bigcup_{i=1}^{n} [a_i, b_i)$ and we may assume, by relabeling if necessary, that $a \leq a_1 < b_1 \leq a_2 < b_2 < \ldots \leq a_n < b_n \leq b$. Since f is increasing,

$$\sum_{i=1}^{n}(f(b_i) - f(a_i)) \leq f(b_n) - f(a_1) \leq f(b) - f(a).$$

Since n is arbitrary, $\sum_{i=1}^{\infty}(f(b_i) - f(a_i)) \leq f(b) - f(a)$.

Next, we claim that $\mu_f[a, b) \leq \sum_{i=1}^{\infty} \mu_f[a_i, b_i)$: Let $\epsilon > 0$. For each i there exists $\eta_i > 0$ such that $0 \leq f(a_i) - f(a_i - \eta_i) < \epsilon/2^i$ and there exists $\delta > 0$ such that $0 \leq f(b) - f(b - \delta) < \epsilon$. Then $\{(a_i - \eta_i, b_i) : i \in \mathbf{N}\}$ cover $[a, b - \delta]$ so a finite number cover. By discarding and relabeling, we may assume

$$a_1 - \eta_1 < a < a_2 - \eta_2 < b_1 < \ldots < a_n - \eta_n < b - \delta < b_n.$$

Then

$$\begin{aligned} f(b-\delta) - f(a) &\leq \sum_{i=1}^{n}(f(b_i) - f(a_i - \eta_i)) \leq \sum_{i=1}^{\infty}(f(b_i) - f(a_i - \eta_i)) \\ &\leq \sum_{i=1}^{\infty}(f(b_i) - f(a_i)) + \sum_{i=1}^{\infty} \epsilon/2^i \end{aligned}$$

so

$$f(b) - f(a) + f(b - \delta) - f(b) \leq \sum_{i=1}^{\infty}(f(b_i) - f(a_i)) + \epsilon$$

and

$$f(b) - f(a) \leq \sum_{i=1}^{\infty}(f(b_i) - f(a_i)) + 2\epsilon.$$

Thus, $f(b) - f(a) \leq \sum_{i=1}^{\infty}(f(b_i) - f(a_i))$ as desired.

The premeasure μ_f is called the *Lebesgue-Stieltjes* measure induced by f. If f is the identity function, then μ_f is called *Lebesgue measure* and is denoted by m; note $m([a, b)) = b - a$ so the Lebesgue measure of a half-closed interval is just its length.

2.2. ADDITIVE SET FUNCTIONS

Example 10 (Lebesgue measure on $\mathbf{R}^n$). Let $\mathcal{S}(=\mathcal{S}_n)$ be the semi-ring of Example 2.1.10. Define $m(=m_n)$ on $\mathcal{S}$ by $m(I) = \prod_{i=1}^{n}(b_i - a_i)$, where

$$I = [a_1, b_1) \times ... \times [a_n, b_n).$$

In $\mathbf{R}^2$, $m(I)$ is the area of I and in $\mathbf{R}^3$, $m(I)$ is the volume of I. m is called *Lebesgue measure* on $\mathbf{R}^n$.

m_n is countably additive on $\mathcal{S}_n$. A geometric proof of this fact along the lines of the proof in Example 9 is possible, but is surprisingly difficult (see [Si] §6.3, [Z1] 2.5.3). We give a proof of this fact in Appendix II of §3.2 by using the Monotone Convergence Theorem for the Lebesgue integral.

We give an example of a countably additive set function which takes on both positive and negative values.

Example 11 Let $\{a_k\} \subset \mathbf{R}$. Set $p_k = a_k$ if $a_k \geq 0$, $p_k = 0$ if $a_k < 0$ and $q_k = a_k$ if $a_k \leq 0$ and $q_k = 0$ if $a_k > 0$. Consider the following cases:

(I) $\sum p_k$ and $\sum q_k$ both converge,

(II) one of the series $\sum p_k$, $\sum q_k$ converges while the other diverges.

Now define $\mu : \mathcal{P}(\mathbf{N}) \to \mathbf{R}^*$ by $\mu(A) = \sum_{k \in A} a_k$. In Case I, $\sum a_k$ is subseries convergent in $\mathbf{R}$ and μ is a real-valued, bounded, countably additive set function. In case II, every subseries of $\sum a_k$ converges in $\mathbf{R}^*$ and μ is a countably additive $\mathbf{R}^*$-valued set function.

Examples of finitely additive set functions which are not countably additive are given in the exercises.

Exercise 1. Let $\mathcal{S}$ be a semi-ring and $\mu : \mathcal{S} \to [0, \infty]$ be finitely additive. Show that if μ is countably subadditive, then μ is countably additive.

Exercise 2. Let $\mathcal{S}$ be a semi-ring (semi-algebra) and $\mu : \mathcal{S} \to \mathbf{R}$ finitely additive. Show μ has a unique finitely additive extension to the ring (algebra) generated by $\mathcal{S}$ (see Proposition 2.1.11).

Exercise 3. Show the finiteness condition in Proposition 5(ii) and Corollary 6(ii) cannot be dropped.

Exercise 4. Let S be uncountable and $\sum$ the σ-algebra of sets which are at most countable or have complements which are at most countable (Exer. 2.1.2). Define μ on $\sum$ by $\mu(E) = 0$ if E is at most countable and $\mu(E) = 1$ if E^c is at most countable. Show μ is countably additive.

Exercise 5. Let $\{a_k\}$ be non-negative. Define $\mu : \mathcal{P}(\mathbf{N}) \to [0, \infty]$ by $\mu(E) = \sum_{k \in E} a_k$. Show μ is countably additive.

Exercise 6. Let $S \neq \emptyset$ and $f : S \to [0, \infty]$. Define $\mu : \mathcal{P}(S) \to [0, \infty]$ by $\mu(E) = \sum_{x \in E} f(x)$ if E is at most countable and $\mu(E) = \infty$ otherwise. Show μ is countably additive.

Exercise 7. Let $\mathcal{A}$ be all subsets A of S which are either finite or have finite complements (see Exer. 2.1.3). Define μ on $\mathcal{A}$ by $\mu(A)$ is the number of points in A when A is finite and $\mu(A)$ is the negative of the number of points in A^c when A^c is finite. Show μ is finitely additive and if S is infinite, μ is not countably additive.

Exercise 8. Let $\mathcal{A}$ be the semi-algebra of all subintervals $[a, b)$, $a \leq b$, of $[0, 1)$ (see Example 2.1.9). Fix $t \in (0, 1)$ and define $\lambda_t : \mathcal{A} \to \{0, 1\}$ by

$$\lambda_t(A) = \begin{cases} 1 & \text{if } [t-\delta, t) \subseteq A \text{ for some } \delta > 0 \\ 0 & \text{otherwise}. \end{cases}$$

Show λ_t is finitely additive but not countably additive.

Exercise 9. Let $\mathcal{A}$ be the algebra in Exercise 7. Let $\mu(A) = 0$ if A is finite and $\mu(A) = 1$ if A^c is finite. If S is countable, show μ is not countably additive, while if S is uncountable, μ is countably additive.

Exercise 10. Let $f : \mathbf{R} \to \mathbf{R}$ and $\mathcal{S}$ be the semi-ring of Example 2.1.8. Define $\tau_f : \mathcal{S} \to \mathbf{R}$ by $\tau_f[a, b) = f(b) - f(a)$. Show τ_f is finitely additive on $\mathcal{S}$.

Exercise 11. Repeat Exercise 10 for the semi-algebra of Example 2.1.9.

Exercise 12. Let $S = \mathbf{N}$ and $\mathcal{A}$ the algebra of Exercise 7. Let $\{a_k\} \subset \mathbf{R}$ and set $\mu(A) = \sum_{k \in A} a_k$ when A is finite and $\mu(A) = -\sum_{k \in A^c} a_k$ when A is infinite. Show μ is finitely additive.

Exercise 13. Suppose Σ is a σ-algebra and $\mu_i : \Sigma \to [0, \infty]$ is a measure for each i and $\mu_i(E) \uparrow \mu(E)$ for each $E \in \Sigma$. Show μ is a measure.

Exercise 14 (Borel-Cantelli). Let μ be a measure on the σ-algebra Σ. Let $\{E_k\} \subset \Sigma$ be such that $\sum \mu(E_k) < \infty$. Show $\overline{\lim} E_k$ has μ-measure 0.

2.2. ADDITIVE SET FUNCTIONS

Exercise 15. Let $\{t_n\} \subset \mathbf{R}$ and set $\mu = \sum_{m=1}^{\infty} \delta_{t_n}$ [Example 8 and Exercise 13]. Show μ assigns finite measure to bounded intervals if and only if $\lim |t_n| = \infty$. When is μ finite?

Exercise 16. Show $\mu : \mathcal{P}(S) \to [0, \infty]$ defined by $\mu(\emptyset) = 0$ and $\mu(A) = \infty$ otherwise is a measure.

2.2.1 Jordan Decomposition

In this section we show that any signed measure is the difference of two (positive) measures. This result is called the Jordan Decomposition Theorem. We first develop a similar decomposition for certain finitely additive set functions.

The material in this section is not used until section 3.12 and may be skipped at this time.

Let $\mathcal{A}$ be an algebra of subsets of S and $\mu : \mathcal{A} \to \mathbf{R}^*$ finitely additive. For $E \subseteq S$, set $\mu^+(E) = \sup\{\mu(A) : A \in \mathcal{A}, A \subseteq E\} \geq 0$ and $\mu^-(E) = -\inf\{\mu(A) : A \in \mathcal{A}, A \subseteq E\} \geq 0$. μ^+ is called the *positive part (upper variation)* of μ, and μ^- is called the *negative part (lower variation)* of μ.

Proposition 1 (i) $\mu^+ \geq \mu$, $-\mu \leq \mu^-$ on $\mathcal{A}$,

(ii) $\mu^+(\mu^-)$ is increasing and $\mu^- = (-\mu)^+$,

(iii) If μ is finitely additive, then μ^+ and μ^- are finitely additive.

(iv) If $\mathcal{A}$ is a σ-algebra and μ is countably additive, then μ^+ and μ^- are countably additive.

Proof: (i) and (ii) are clear.

(iii): Let $E_1, E_2 \in \mathcal{A}$ be pairwise disjoint and set $E = E_1 \cup E_2$. If $A \subseteq E$, $A \in \mathcal{A}$, then
$$\mu(A) = \mu(A \cap E_1) + \mu(A \cap E_2) \leq \mu^+(E_1) + \mu^+(E_2)$$
so
$$\mu^+(E) \leq \mu^+(E_1) + \mu^+(E_2).$$

If either $\mu^+(E_1)$ or $\mu^+(E_2) = \infty$, then for every $r > 0$ there exists $A_i \subseteq E_i$, $A_i \in \mathcal{A}$ such that $\mu(A_i) > r$ for $i = 1$ or 2; $A_i \subseteq E$ implies $\mu^+(E) \geq r$ so $\mu^+(E) = \infty$ and $\mu^+(E) = \mu^+(E_1) + \mu^+(E_2)$. If both $\mu^+(E_1), \mu^+(E_2) < \infty$, let $\epsilon > 0$ and pick $A_i \in \mathcal{A}$, $A_i \subseteq E_i$, such that $\mu(A_i) \geq \mu^+(E_i) - \epsilon/2$. Then $A_1 \cup A_2 \subseteq E$ and $A_1 \cup A_2 \in \mathcal{A}$ so
$$\mu^+(E) \geq \mu(A_1 \cup A_2) = \mu(A_1) + \mu(A_2) \geq \mu^+(E_1) + \mu^+(E_2) - \epsilon.$$

Hence, $\mu^+(E) \geq \mu^+(E_1) + \mu^+(E_2)$.

(iv): Let $\{E_i\} \subseteq \mathcal{A}$ be pairwise disjoint and $E = \bigcup_{i=1}^{\infty} E_i$. From (ii) and (iii), for each n
$$\mu^+(E) \geq \mu^+(\bigcup_{i=1}^{n} E_i) = \sum_{i=1}^{n} \mu^+(E_i)$$
so
$$\mu^+(E) \geq \sum_{i=1}^{\infty} \mu^+(E_i).$$

2.2. ADDITIVE SET FUNCTIONS

Let $A \in \mathcal{A}$ and $A \subset E$. Then

$$\mu(A) = \sum_{i=1}^{\infty} \mu(A \cap E_i) \le \sum_{i=1}^{\infty} \mu^+(E_i)$$

so

$$\mu^+(E) \le \sum_{i=1}^{\infty} \mu^+(E_i).$$

The statements about μ^- in (iii) and (iv) follow from (ii).

Lemma 2 *Let $E \in \mathcal{A}$. If one of the numbers $\mu^+(E)$, $\mu^-(E)$ is finite, then $\mu(E) = \mu^+(E) - \mu^-(E)$.*

Proof: If $\mu(E) = \infty$, then $\mu^+(E) = \infty$ so $\mu^-(E) < \infty$ and $\mu(E) = \mu^+(E) - \mu^-(E)$. Similarly, if $\mu(E) = -\infty$, then $\mu^-(E) = \infty$ so $\mu(E) = \mu^+(E) - \mu^-(E)$.

Assume $\mu(E)$ is finite. Then $\mu(A)$ is finite for every $A \subset E$, $A \in \mathcal{A}$ (2.2.3). By 2.2.3,

$$\begin{aligned}\mu^+(E) - \mu(E) &= \sup\{\mu(A) - \mu(E) : A \subset E, A \in \mathcal{A}\} \\ &= \sup\{-\mu(E \setminus A) : A \subset E, A \in \mathcal{A}\} \\ &= \mu^-(E)\end{aligned}$$

and the result follows.

Theorem 3 (Jordan Decomposition) *If $\mu : \mathcal{A} \to \mathbf{R}$ is bounded, then $\mu = \mu^+ - \mu^-$.*

Proof: Immediate from Lemma 2.

The boundedness condition in Theorem 3 is important; see Exercise 7 or 8.

We next show that Theorem 3 is applicable to any finite signed measure on a σ-algebra.

Lemma 4 *Let $\mathcal{A}$ be a σ-algebra and μ countably additive. If $E \in \mathcal{A}$ is such that $\mu^+(E) = \infty (\mu^-(E) = \infty)$, then $\mu(E) = \infty$ ($\mu(E) = -\infty$).*

Proof: By Proposition 2.2.3, we need only find a subset $A \subseteq E$, $A \in \mathcal{A}$, satisfying $\mu(A) = \infty$. Suppose $\mu^+(E) = \infty$. Set $E_0 = E$. Then there exists $A_1 \in \mathcal{A}$, $A_1 \subseteq E_0$, such that $\mu(A_1) > 1$. Either $\mu^+(A_1)$ or $\mu^+(E_0 \setminus A_1)$ equals ∞ so pick one and label it E_1.

For each positive integer n suppose $A_n \in \mathcal{A}$, $A_n \subseteq E_{n-1}$, satisfies $\mu(A_n) > n$ and either $\mu^+(A_n)$ or $\mu^+(E_{n-1} \setminus A_n) = \infty$. Pick one which satisfies this last condition and label it E_n. There are two cases:

Case I: $E_n = E_{n-1} \setminus A_n$ for infinitely many values of n and

Case II: there exists N such that $E_n = A_n$ for $n \ge N$.

In Case II, $A_N, A_{N+1}, \ldots$ is a decreasing sequence from $\mathcal{A}$. If $\mu(A_N) = \infty$, we are through; if not, by Proposition 2.2.5, $\mu(\bigcap_{n=N}^{\infty} A_n) = \lim_n \mu(A_n) \ge \lim_n n = \infty$.

In Case I, there is a subsequence $\{A_{n_k}\}$ with

$$\mu(\bigcup_{k=1}^{\infty} A_{n_k}) = \sum_{k=1}^{\infty} \mu(A_{n_k}) \geq \sum_{k=1}^{\infty} n_k = \infty.$$

Immediately from Lemma 4, we have an important property of signed measures defined on σ-algebras.

Theorem 5 *If $\mathcal{A}$ is a σ-algebra and $\mu : \mathcal{A} \to \mathbf{R}$ is countably additive, then μ is bounded.*

The countable additivity assumption in Theorem 5 is important; we give an example of a real-valued, finitely additive set function on a σ-algebra which is unbounded in Example 3.2.21. The σ-algebra assumption is also important; see Exercises 5 and 8. The conclusion of Theorem 5 can also be improved; see [Hah] 3.3.1, p. 17.

Theorem 6 (Jordan Decomposition) *Let $\mathcal{A}$ be a σ-algebra and μ a signed measure. Then $\mu = \mu^+ - \mu^-$ and μ^+, μ^- are measures.*

Proof: If $E \in \mathcal{A}$, then by Lemma 4 one of $\mu^+(E)$, $\mu^-(E)$ must be finite so $\mu(E) = \mu^+(E) - \mu^-(E)$ (Lemma 2). Proposition 1 gives the last statement.

We address the uniqueness of the Jordan Decomposition for signed measures in §2.2.2.

The *total variation* or *variation* of μ is defined to be $|\mu| = \mu^+ + \mu^-$. Note $|\mu|$ is an additive set function on $\mathcal{A}$ by Proposition 1, and if μ is a signed measure on a σ-algebra, then $|\mu|$ is a measure by Proposition 1. We have the following properties of the variation.

Proposition 7 (i) $|\mu(A)| \leq |\mu|(A)$ for $A \in \mathcal{A}$.

(ii) If $A, A_i \in \mathcal{A}$ and $A \supseteq \bigcup_{i=1}^{\infty} A_i$ with $\{A_i\}$ pairwise disjoint, then

$$\sum_{i=1}^{\infty} |\mu(A_i)| \leq |\mu|(A).$$

(iii) $|\mu|(A) = \sup\{|\mu(B)| + |\mu(A\setminus B)| : B \subseteq A, B \in \mathcal{A}\}$.

(iv) $|\mu|(A) = \sup\{\sum_{i=1}^{n} |\mu(A_i)| : \{A_i\}$ pairwise disjoint from $\mathcal{A}$ with $\bigcup_{i=1}^{n} A_i = A\}$.

(v) $\sup\{|\mu(B)| : B \subseteq A, B \in \mathcal{A}\} \leq |\mu|(A) \leq 2 \sup\{|\mu(B)| : B \subseteq A, B \in \mathcal{A}\}$.

2.2. ADDITIVE SET FUNCTIONS

Proof: (i): $\mu(A) \leq \mu^+(A)$ and $-\mu(A) \leq \mu^-(A)$.

(ii): Fix n. Set $\sigma^+ = \{i : 1 \leq i \leq n, \mu(A_i) \geq 0\}$, $\sigma^- = \{i : 1 \leq i \leq n, \mu(A_i) < 0\}$. Then

$$\sum_{i\in\sigma^+} \mu(A_i) = \sum_{i\in\sigma^+} |\mu(A_i)| \leq \sum_{i\in\sigma^+} |\mu|(A_i)$$

and

$$-\sum_{i\in\sigma^-} \mu(A_i) = \sum_{i\in\sigma^-} |\mu(A_i)| \leq \sum_{i\in\sigma^-} |\mu|(A_i)$$

so

$$\sum_{i=1}^{n} |\mu(A_i)| = \sum_{i\in\sigma^+} \mu(A_i) - \sum_{i\in\sigma^-} \mu(A_i) \leq \sum_{i=1}^{n} |\mu|(A_i) = |\mu|(\bigcup_{i=1}^{n} A_i) \leq |\mu|(A).$$

Hence, $\sum_{i=1}^{\infty} |\mu(A_i)| \leq |\mu|(A)$.

(iii): Set s equal to the term on the right hand side of (iii). Then $|\mu|(A) \geq s$ by (ii). There exist $A_k \in \mathcal{A}$, $A_k \subseteq A$ such that $\mu(A_k) \to \mu^+(A)$ so $|\mu(A_k)| \to \mu^+(A)$. If $\mu^+(A) = \infty$, then $s = \infty$ so $\mu^+(A) = |\mu|(A) = s$. If $\mu^+(A) < \infty$, then $\mu(A) = \mu(A_k) + \mu(A \setminus A_k)$ so $\mu(A \setminus A_k) \to -\mu^-(A)$ by Lemma 2. Hence, $|\mu(A \setminus A_k)| \to \mu^-(A)$ and

$$|\mu(A_k)| + |\mu(A \setminus A_k)| \to \mu^+(A) + \mu^-(A) = |\mu|(A)$$

so $s = |\mu|(A)$.

(iv) follows from (ii) and (iii).

(v) follows from (i) and (iii).

From (v), we have the important observation.

Corollary 8 *A finitely additive set function $\mu : \mathcal{A} \to \mathbf{R}$ is bounded if and only if μ has finite variation.*

Remark 9 *The formula in (iv) can be used to define the total variation of a set function directly without recourse to the Jordan Decomposition. It can also be used to define the variation of complex-valued set functions.*

The reader who has covered the chapters on integration before this chapter should do Exercise 3.2.26 at this point.

Exercise 1. Suppose $\mu = \mu_1 + \mu_2$ with μ_1, μ_2 finitely additive. Show $\mu^+ \leq \mu_1^+ + \mu_2^+$, $\mu^- \leq \mu_1^- + \mu_2^-$, $|\mu| \leq |\mu_1| + |\mu_2|$.

Exercise 2. Suppose $\mu = \mu_1 - \mu_2$ with μ_1, μ_2 positive and finitely additive. Show $\mu_1 \geq \mu^+$, $\mu_2 \geq \mu^-$.

Exercise 3. Show that if $\mu : \Sigma \to \mathbf{R}$ is countably additive and Σ is a σ-algebra, then $|\mu|$ is a finite measure.

Exercise 4. If μ is countably additive and $\mathcal{A}$ is a σ-algebra, show either μ^+ or μ^- is a finite measure.

Exercise 5. Let $\mathcal{A}$ be the algebra which consists of all subsets of $\mathbf{N}$ which are either finite or have finite complements. Define μ on $\mathcal{A}$ by $\mu(A) = \sum_{n \in A} (-1)^n/n$ if A is finite and $\mu(A) = - \sum_{n \in A^c} (-1)^n/n$ if A^c is finite and $\mu(\emptyset) = \mu(\mathbf{N}) = 0$. Show μ is finitely additive but not bounded; compare with Theorem 5.

Exercise 6. Show $|\mu|(E) = \sup\{|\sum_{i=1}^n a_i \mu(A_i)| : \{A_1, ..., A_n\}$ is a partition of E, $|a_i| \leq 1\}$.

Exercise 7. Let $f(t) = 1/t$ for $t \neq 0$ and $f(0) = 0$. Let τ_f be the set function of Exer. 2.2.10 and assume that τ_f has been extended to the algebra $\mathcal{A}$ generated by $\mathcal{S}$ (Exer. 2.2.2). Show the conclusion of Lemma 2 and Theorem 3 fails for τ_f.

Exercise 8. Let $\mathcal{A}$ be the algebra of subsets of $\mathbf{N}$ which are either finite or have finite complements. Define μ on $\mathcal{A}$ by $\mu(A)$ is the number of points in A when A is finite and $\mu(A)$ is the negative of the number of points in A^c when A^c is finite. Show μ is finitely additive and find μ^+, μ^-. Is $\mu = \mu^+ - \mu^-$?

2.2.2 Hahn Decomposition

In this section we use the Jordan Decomposition derived in §2.2.1 to develop another decomposition theorem for signed measures called the Hahn Decomposition. This material is not used until section §3.12 and may be skipped at this time.

Let $\sum$ be a σ-algebra of subsets of S and let $\mu : \sum \to \mathbf{R}^*$ be a signed measure.

Definition 1 *A subset $P \in \sum$ is μ-positive (μ-negative) if $E \in \sum$, $E \subseteq P$ implies $\mu(E) \geq 0 (\mu(E) \leq 0)$. A set which is both μ-positive and μ-negative is called μ-null.*

Theorem 2 (Hahn Decomposition) *There exist a μ-positive set $P \in \sum$ and a μ-negative set $N \in \sum$ such that $S = P \cup N$ and $P \cap N = \emptyset$.*

Proof: We may assume that $+\infty$ is the infinite value not assumed by μ so $\mu^+(S) < \infty$ [2.2.1.4]. For each n choose $E_n \in \sum$ such that $\infty > \mu(E_n) > \mu^+(S) - 1/2^n$. For $A \in \sum$, $A \subseteq E_n^c$, we have $\mu(A \cup E_n) = \mu(A) + \mu(E_n) \leq \mu^+(S)$ so $\mu(A) \leq \mu^+(S) - \mu(E_n) < 1/2^n$ and $\mu^+(E_n^c) \leq 1/2^n$. By the Jordan Decomposition,

$$\mu^-(E_n) = \mu^+(E_n) - \mu(E_n) \leq \mu^+(S) - \mu(E_n) < 1/2^n.$$

Set $P = \overline{\lim} E_n$, $N = P^c = \underline{\lim} E_n^c$. By Corollary 2.2.6

$$0 \leq \mu^+(N) \leq \underline{\lim} \mu^+(E_n^c) \leq \lim 1/2^n = 0$$

so N is μ-negative. Now μ^- is a measure so for each k

$$0 \leq \mu^-(P) \leq \mu^-(\bigcup_{n=k}^{\infty} E_n) \leq \sum_{n=k}^{\infty} \mu^-(E_n) \leq \sum_{n=k}^{\infty} 1/2^n = 1/2^{n-1}$$

so $\mu^-(P) = 0$. Hence, P is μ-positive.

A pair of sets (P, N) satisfying the conclusion of Theorem 2 is called a *Hahn Decomposition* for μ. Such decompositions are obviously not unique since if $Z \in \sum$ is μ-null, then $(P \cup Z, N \setminus Z)$ is also a Hahn Decomposition. However, such decompositions are unique up to μ-null sets in the following sense.

Proposition 3 *Let (P_i, N_i), $i = 1, 2$, be Hahn Decompositions for μ. Then for every $E \in \sum$, $\mu(E \cap P_1) = \mu(E \cap P_2)$, $\mu(E \cap N_1) = \mu(E \cap N_2)$ and $\mu(P_1 \triangle P_2) = \mu(N_1 \triangle N_2) = 0$. (Here $A \triangle B = (A \setminus B) \cup (B \setminus A)$, the symmetric difference.)*

Proof: $E \cap (P_1 \setminus P_2) \subseteq E \cap P_1$ and $E \cap (P_1 \setminus P_2) \subseteq E \cap N_2$ imply $\mu(E \cap (P_1 \setminus P_2)) \geq 0$ and $\mu(E \cap (P_1 \setminus P_2)) \leq 0$ so $\mu(E \cap (P_1 \setminus P_2)) = 0$. By symmetry $\mu(E \cap (P_2 \setminus P_1)) = 0$. Thus, $\mu(P_1 \triangle P_2) = 0$ and

$$\mu(E \cap P_1) = \mu(E \cap (P_1 \cup P_2)) - \mu(E \cap (P_2 \setminus P_1)) = \mu(E \cap (P_1 \cup P_2))$$
$$= \mu(E \cap (P_1 \cup P_2)) - \mu(E \cap (P_1 \setminus P_2)) = \mu(E \cap P_2).$$

The other equalities are similar.

We can use the Hahn Decomposition to obtain formulas for the positive and negative variations of μ.

Proposition 4 *Let (P, N) be a Hahn Decomposition for μ. Then $\mu^+(E) = \mu(P \cap E)$ and $\mu^-(E) = -\mu(E \cap N)$ for every $E \in \Sigma$.*

Proof: Certainly $\mu^+(E) \geq \mu(E \cap P)$. If $A \in \Sigma$, $A \subseteq E$, then

$$\mu(A) = \mu(A \cap P) + \mu(A \cap N) \leq \mu(A \cap P) \leq \mu(E \cap P)$$

so $\mu^+(E) \leq \mu(E \cap P)$.

The other equality is similar.

Remark 5 The Hahn Decomposition can be established independent of the Jordan Decomposition and then the formulas in Proposition 4 can be used to establish the Jordan Decomposition. See, for example, [Roy].

Definition 6 *Two measures μ and ν on Σ are mutually singular or singular, written $\mu \perp \nu$, if there exist $A, B \in \Sigma$ such that $A \cap B = \emptyset$, $S = A \cup B$, $\mu(A) = \nu(B) = 0$.*

Proposition 7 (Uniqueness of Jordan Decomposition). *$\mu^+ \perp \mu^-$ and if $\mu = \lambda_1 - \lambda_2$, where λ_1 and λ_2 are mutually singular measures on Σ, then $\lambda_1 = \mu^+$, $\lambda_2 = \mu^-$.*

Proof: That $\mu^+ \perp \mu^-$ follows from Proposition 4 since $\mu^+(N) = \mu^-(P) = 0$.

Let $A, B \in \Sigma$, $S = A \cup B$, $A \cap B = \emptyset$ and $\lambda_1(B) = \lambda_2(A) = 0$. Then we claim (A, B) is a Hahn Decomposition for μ. First, A is μ-positive since if $E \subseteq A$, $E \in \Sigma$, then

$$\mu(E) = \lambda_1(E) - \lambda_2(E) = \lambda_1(E) \geq 0,$$

and, similarly, B is μ-negative.

Let $E \in \Sigma$. By Propositions 3 and 4,

$$\mu(E \cap P) = \mu(E \cap A) = \mu^+(E) = \lambda_1(E \cap A) - \lambda_2(E \cap A) = \lambda_1(E \cap A)$$
$$= \lambda_1(E) - \lambda_1(E \cap B) = \lambda_1(E)$$

so $\mu^+ = \lambda_1$.

Similarly, $\mu^- = \lambda_2$.

Remark 8 In general, there is not a Hahn Decomposition for finitely additive set functions defined on algebras; see Exercise 2. There are known necessary and sufficient conditions for a bounded, finitely additive set function defined on an algebra to have a Hahn Decomposition; see [Cob]. For a simple proof of the Hahn Decomposition which does not use the Jordan Decomposition, see [Do].

Exercise 1. Show $E \in \Sigma$ is μ-null if and only if $|\mu|(E) = 0$.

Exercise 2. Give an example of a finitely additive set function on an algebra with no Hahn Decomposition. [Hint: Use Exercise 2.2.12 with $a_k = (-1)^k/2^k$.]

2.2.3 Drewnowski's Lemma

In this section we establish a remarkable result of Drewnowski which asserts that a bounded, finitely additive set function defined on a σ-algebra is not "too far" from being countably additive.

Let Σ be a σ-algebra of subsets of S and $\mu : \Sigma \to \mathbf{R}$ bounded and finitely additive. If $\{E_j\} \subseteq \Sigma$ is pairwise disjoint, then it follows from Proposition 2.2.1.7 that the series $\sum \mu(E_j)$ is absolutely convergent. Thus, if μ is to fail to be countably additive, there must be a pairwise disjoint sequence $\{E_j\}$ in Σ such that the series $\sum \mu(E_j)$ fails to converge to the "correct value", $\mu(\bigcup_{j=1}^{\infty} E_j)$.

Lemma 1 (Drewnowski) [Dr]. *If $\{E_j\}$ is a pairwise disjoint sequence from Σ, then there exists a subsequence $\{E_{n_j}\}$ such that μ is countably additive on the σ-algebra generated by $\{E_{n_j}\}$.*

Proof: Partition $\mathbf{N}$ into a pairwise disjoint sequence of infinite sets $\{K_j^1\}_{j=1}^{\infty}$. By the observation above, $|\mu|(\bigcup_{j \in K_i^1} E_j) \to 0$ as $i \to \infty$. So $\exists i$ such that $|\mu|(\bigcup_{j \in K_i^1} E_j) < 1/2$. Let $N_1 = K_i^1$ and $n_1 = \inf N_1$. Now partition $N_1 \setminus \{n_1\}$ into a pairwise disjoint sequence of infinite sets $\{K_j^2\}_{j=1}^{\infty}$. As before $\exists i$ such that $|\mu|(\bigcup_{j \in K_i^2} E_j) < 1/2^2$. Let $N_2 = K_i^2$ and $n_2 = \inf N_2$. Note $n_2 > n_1$ and $N_2 \subseteq N_1$. Continuing produces a subsequence $n_j \uparrow \infty$ and a sequence of infinite subsets of $\mathbf{N}$, $\{N_j\}$, such that $N_{j+1} \subseteq N_j$ and $|\mu|(\bigcup_{i \in N_j} E_i) < 1/2^j$. Let Σ_0 be the σ-algebra generated by $\{E_{n_j}\}$.

We claim that μ is countably additive on Σ_0. If $\{H_k\} \subseteq \Sigma_0$ and $H_k \downarrow \emptyset$, then $|\mu(H_k)| \leq |\mu|(H_k) \downarrow 0$ so μ is countably additive by Proposition 2.2.5 [given j, $\exists H_i$ such that $\min H_i > n_j$ so $|\mu|(H_i) \leq |\mu|(\bigcup_{k \geq j} E_{n_k}) < 1/2^j$].

Corollary 2 *Let $\mu_i : \Sigma \to \mathbf{R}$ be bounded and finitely additive for $i \in \mathbf{N}$. If $\{E_j\}$ is a pairwise disjoint sequence from Σ, then there exists a subsequence $\{E_{n_j}\}$ such that each μ_i is countably additive on the σ-algebra generated by $\{E_{n_j}\}$.*

Proof: Set $\mu(E) = \sum_{i=1}^{\infty} \frac{1}{2^i} \frac{|\mu_i|(E)}{1+|\mu_i|(S)}$ for $E \in \Sigma$. Then μ is bounded and finitely additive so by Lemma 1 there is a subsequence $\{E_{n_j}\}$ such that μ is countably additive on the σ-algebra, Σ_0, generated by $\{E_{n_j}\}$. If $\{H_j\} \subseteq \Sigma_0$ and $H_j \downarrow \emptyset$, then

$$\lim \mu(H_j) = 0$$

so

$$\lim_j \mu_i(H_j) = 0$$

for each i, and μ_i is countably additive on Σ_0 by Proposition 2.2.5.

2.3 Outer Measures

Recall in Lebesgue's construction of the integral described in §1.3, Lebesgue constructed a set function on $\mathcal{P}(\mathbf{R})$ called Lebesgue outer measure which gave an extension of the length function in $\mathbf{R}$. In this section we give an abstract treatment of outer measures due to Caratheodory and then apply the abstract theory to Lebesgue measure in $\mathbf{R}$ and $\mathbf{R}^n$ in section 2.5.

Let $S \neq \emptyset$.

Definition 1 $\mu^* : \mathcal{P}(S) \to [0, \infty]$ *is an* outer measure *(on S) if*

(i) $\mu^*(\emptyset) = 0$

(ii) $A \subseteq B$ *implies* $\mu^*(A) \leq \mu^*(B)$ *[μ^* is monotone]*

(iii) $A_j \subset S$, $j \in \mathbf{N}$, *implies* $\mu^*(\bigcup_{j=1}^{\infty} A_j) \leq \sum_{j=1}^{\infty} \mu^*(A_j)$ *[μ^* is countably subadditive].*

Following Caratheodory, we define measurability with respect to μ^* (see (6) of §1.3).

Definition 2 $E \subseteq S$ *is μ^*-measurable if and only if*

(c)
$$\mu^*(A) = \mu^*(A \cap E) + \mu^*(A \backslash E) \text{ for any subset } A \subseteq S.$$

The set A in (c) is called a *test set* for the μ^*-measurability condition. Note in order to test measurability in (c) it is only necessary to use test sets with $\mu^*(A) < \infty$ since (c) always holds if $\mu^*(A) = \infty$ by (iii).

The family of all μ^*-measurable sets will be denoted by $\mathcal{M}(\mu^*)$. We proceed to study the properties of $\mathcal{M}(\mu^*)$ and μ^* restricted to this family.

Definition 3 *A set* $E \subseteq S$ *is μ^*-null if $\mu^*(E) = 0$.*

The same term was used for signed measures in §2.2.2, but this should cause no difficulties.

From (ii) and (iii), we have

Proposition 4 *A countable union of μ^*-null sets is μ^*-null and subsets of μ^*-null sets are μ^*-null.*

Proposition 5 *Every μ^*-null set is μ^*-measurable.*

Proof: Let $\mu^*(E) = 0$ and $A \subseteq S$. Then

$$\mu^*(A) \leq \mu^*(A \cap E) + \mu^*(A \backslash E) = \mu^*(A \backslash E) \leq \mu^*(A).$$

2.3. OUTER MEASURES

Lemma 6 *Let $E_1, ..., E_n$ be pairwise disjoint and μ^*-measurable. Then for any $A \subset S$, $\mu^*(A \cap \bigcup_{i=1}^{n} E_i) = \sum_{i=1}^{n} \mu^*(A \cap E_i)$.*

Proof: We proceed by induction on n. The result is obvious for $n = 1$ so assume that it holds for n. By the μ^*-measurability of E_{n+1} and the induction hypothesis,

$$\begin{aligned}
\mu^*(A \cap \bigcup_{i=1}^{n+1} E_i) &= \mu^*([A \cap \bigcup_{i=1}^{n+1} E_i] \cap E_{n+1}) + \mu^*([A \cap \bigcup_{i=1}^{n+1} E_i] \setminus E_{n+1}) \\
&= \mu^*(A \cap E_{n+1}) + \mu^*(A \cap \bigcup_{i=1}^{n} E_i) \\
&= \mu^*(A \cap E_{n+1}) + \sum_{i=1}^{n} \mu^*(A \cap E_i).
\end{aligned}$$

Theorem 7 $\mathcal{M}(\mu^*)$ *is a σ-algebra.*

Proof: If $E \in \mathcal{M}(\mu^*)$, then $E^c \in \mathcal{M}(\mu^*)$ by the symmetry in (c) and clearly $\emptyset \in \mathcal{M}(\mu^*)$.

We next claim that if $E_1, E_2 \in \mathcal{M}(\mu^*)$, then $E = E_1 \cup E_2 \in \mathcal{M}(\mu^*)$. Let $A \subseteq S$. Note $E = E_1 \cup (E_2 \cap E_1^c)$ so

$$\begin{aligned}
\mu^*(A) &\leq \mu^*(A \cap E) + \mu^*(A \cap E^c) \\
&\leq \mu^*(A \cap E_1) + \mu^*([A \cap E_1^c] \cap E_2) + \mu^*([A \cap E_1^c] \cap E_2^c) \\
&= \mu^*(A \cap E_1) + \mu^*(A \setminus E_1) = \mu^*(A).
\end{aligned}$$

Thus, $\mathcal{M}(\mu^*)$ is an algebra.

Let $\{E_i : i \in \mathbb{N}\} \subseteq \mathcal{M}(\mu^*)$ and set $E = \bigcup_{i=1}^{\infty} E_i$. We disjointify the $\{E_i\}$ by setting $F_1 = E_1$, $F_{k+1} = E_{k+1} \setminus \bigcup_{j=1}^{k} E_j$ for $k \geq 1$. Then $E = \bigcup_{k=1}^{\infty} F_k$.

For any $A \subseteq S$, we have by the part above and Lemma 6,

$$\begin{aligned}
\mu^*(A) &= \mu^*(A \cap \bigcup_{j=1}^{n} F_j) + \mu^*(A \setminus \bigcup_{j=1}^{n} F_j) \\
&\geq \mu^*(A \cap \bigcup_{j=1}^{n} F_j) + \mu^*(A \setminus E) \\
&= \sum_{j=1}^{n} \mu^*(A \cap F_j) + \mu^*(A \setminus E)
\end{aligned}$$

so

$$\mu^*(A) \geq \sum_{j=1}^{\infty} \mu^*(A \cap F_j) + \mu^*(A \setminus E) \geq \mu^*(A \cap E) + \mu^*(A \setminus E) \geq \mu^*(A).$$

Hence, E is μ^*-measurable.

Theorem 8 μ^* *restricted to $\mathcal{M}(\mu^*)$ is a measure.*

Proof: Let $\{E_j\} \subseteq \mathcal{M}(\mu^*)$ be pairwise disjoint and set $E = \bigcup_{j=1}^{\infty} E_j$. By countable subadditivity, $\mu^*(E) \leq \sum_{j=1}^{\infty} \mu^*(E_j)$. By Lemma 6, for each n

$$\sum_{j=1}^{n} \mu^*(E_j) = \mu^*(\bigcup_{j=1}^{n} E_j) \leq \mu^*(E)$$

so

$$\sum_{j=1}^{\infty} \mu^*(E_j) \leq \mu^*(E)$$

and μ^* is countably additive on $\mathcal{M}(\mu^*)$.

We denote μ^* restricted to $\mathcal{M}(\mu^*)$ by μ and call μ the measure generated by the outer measure μ^*.

Examples of outer measures are given in the exercises.

We show how outer measures can be constructed from premeasures in section 2.4.

Exercise 1. For subsets $A \subseteq S$ let $\mu^*(A)$ be the number of points in A if A is finite and $\mu^*(A) = \infty$ if A is infinite. Show μ^* is an outer measure and describe $\mathcal{M}(\mu^*)$.

Exercise 2. For $A \subseteq S$ let $\mu^*(A) = 1$ if $A \neq \emptyset$ and $\mu^*(\emptyset) = 0$. Show μ^* is an outer measure and describe $\mathcal{M}(\mu^*)$.

Exercise 3. Fix $x \in S$. Define $\mu^*(A) = 1$ if $x \in A$ and $\mu^*(A) = 0$ if $x \notin A$. Show μ^* is an outer measure and describe $\mathcal{M}(\mu^*)$.

Exercise 4. Let S be uncountable. Set $\mu^*(A) = 1$ if A is uncountable and $\mu^*(A) = 0$ otherwise. Show μ^* is an outer measure and describe $\mathcal{M}(\mu^*)$.

Exercise 5. Let S be a metric space. Set $\mu^*(A) = 1$ if A is second category and $\mu^*(A) = 0$ if A is first category. Show μ^* is an outer measure and describe $\mathcal{M}(\mu^*)$.

Exercise 6. Show that if an outer measure is finitely additive, it is countably additive.

Exercise 7. Show that a subset A is μ^*-measurable if and only if for every $\epsilon > 0$ there exists a μ^*-measurable set E such that $E \subseteq A$ and $\mu^*(A\backslash E) < \epsilon$.

Exercise 8. If E is μ^*measurable, then for every $A \subset S$ show

$$\mu^*(E \cup A) + \mu^*(E \cap A) = \mu^*(E) + \mu^*(A).$$

Exercise 9. Let $\{E_j\}$ be a pairwise disjoint sequence of μ^*-measurable sets. If $A \subset S$ show $\mu^*(A \cap \bigcup_{j=1}^{\infty} E_j) = \sum_{j=1}^{\infty} \mu^*(A \cap E_j)$.

2.3.1 Metric Outer Measures

Often, as in the case of Lebesgue measure, an outer measure is defined on a metric space (or, more generally, a topological space), and it is natural to ask if there are conditions on the outer measure which guarantee that the open sets are measurable. We now describe such a condition, which is often easy to check, due to Caratheodory.

Let (S, d) be a metric space. Recall that for $A, B \subseteq S$, the distance from A to B, $d(A, B)$ is defined by $d(A, B) = \inf\{d(a, b) : a \in A, b \in B\}$.

Let μ^* be an outer measure on S. μ^* is said to be a *metric outer measure* if

$$\mu^*(A \cup B) = \mu^*(A) + \mu^*(B)$$

when $d(A, B) > 0$.

Lemma 1 (Caratheodory) *Let μ^* be a metric outer measure on S, let $G \subseteq S$ be open and let $A_0 \subseteq G$. For each $n \geq 1$ let $A_n = \{x \in A_0 : d(x, G^c) \geq 1/n\}$. Then $\lim \mu^*(A_n) = \mu^*(A_0)$.*

Proof: Since $A_n \uparrow$ and $A_n \subseteq A_0$, we need to show that $\lim \mu^*(A_n) \geq \mu^*(A_0)$. Since G is open and $A_0 \supset \bigcup_{n=1}^{\infty} A_n$, each point of A_0 is an interior point of G and so must belong to A_n for large n. That is, $A_0 \subseteq \bigcup_{n=1}^{\infty} A_n$ so $A_0 = \bigcup_{n=1}^{\infty} A_n$.

Set $D_n = A_{n+1} \setminus A_n$. Then for each n,

$$A_0 = A_{2n} \cup \left(\bigcup_{k=2n}^{\infty} D_k\right) = A_{2n} \cup \left(\bigcup_{k=n}^{\infty} D_{2k}\right) \cup \left(\bigcup_{k=n}^{\infty} D_{2k+1}\right)$$

so

$$\mu^*(A_0) \leq \mu^*(A_{2n}) + \sum_{k=n}^{\infty} \mu^*(D_{2k}) + \sum_{k=n}^{\infty} \mu^*(D_{2k+1}).$$

If both of the last two series converge, then letting $n \to \infty$ implies

$$\mu^*(A_0) \leq \lim \mu^*(A_{2n}) = \lim \mu^*(A_n).$$

Otherwise at least one of the two series diverges so for definiteness assume that it is the first. Since $d(D_{2k}, D_{2k+2}) \geq \frac{1}{2k+1} - \frac{1}{2k+2} > 0$ and μ^* is a metric outer measure,

$$\mu^*(A_{2n}) \geq \mu^*\left(\bigcup_{k=1}^{n-1} D_{2k}\right) = \sum_{k=1}^{n-1} \mu^*(D_{2k}) \to \infty$$

so $\lim \mu^*(A_n) = \infty \geq \mu^*(A_0)$.

If S is any metric space (topological space), the σ-algebra generated by the open subsets of S is called the family of *Borel sets* of S and is denoted by $\mathcal{B}(S)$.

Theorem 2 μ^* *is a metric outer measure if and only if every Borel set is μ^*-measurable.*

Proof: $\Rightarrow$: It suffices to show that any closed set F is μ^*-measurable. Let $A \subseteq S$ be any test set. Then $A \backslash F$ is contained in the open set F^c so by Lemma 1 there is a sequence $\{A_k\}$ of subsets of $A \backslash F$ such that $d(A_k, F) \geq 1/k$ and $\lim \mu^*(A_k) = \mu^*(A \backslash F)$. Since μ^* is a metric outer measure,

$$\mu^*(A) \geq \mu^*((A \cap F) \cup A_k) = \mu^*(A \cap F) + \mu^*(A_k) \to \mu^*(A \cap F) + \mu^*(A \backslash F)$$

and F is μ^*-measurable.

$\Leftarrow$: Let $d(A, B) > 0$. Pick G open such that $G \supset A$ and $G \cap B = \emptyset$. Now G is μ^*-measurable so

$$\begin{aligned} \mu^*(A \cup B) &= \mu^*((A \cup B) \cap G) + \mu^*((A \cup B) \backslash G) \\ &= \mu^*(A) + \mu^*(B). \end{aligned}$$

Exercise 1. Let $E = \{1/n : n \in \mathbf{N}\}$. Define μ^* on $\mathcal{P}(\mathbf{R})$ be setting $\mu^*(A)$ equal to the number of points in $A \cap E$. Show μ^* is a metric outer measure on $\mathbf{R}$.

Exercise 2. For $S = \mathbf{R}$ which of the outer measures in Exercises 2.3.1-2.3.4 are metric outer measures?

2.4 Extensions of Premeasures

In this section we show that any premeasure on a semi-ring generates an outer measure which then gives an extension of the premeasure to a measure on the σ-algebra of subsets which are measurable with respect to the outer measure. In particular, this construction can be used to generate Lebesgue- Stieltjes measures on $\mathbf{R}$ and Lebesgue measure on $\mathbf{R}^n$. Some parts of the construction are valid for finitely additive set functions so we begin with this case.

Let $\mathcal{S}$ be a semi-ring of subsets of S and $\mu : \mathcal{S} \to [0, \infty]$ be finitely additive.

Definition 1 $\mu^*(A) = \inf\{\sum_{i=1}^{\infty} \mu(A_i) : A \subseteq \bigcup_{i=1}^{\infty} A_i, A_i \in \mathcal{S}\}$, $A \subseteq S$. [Here, we use the convention $\inf \emptyset = \infty$.]

Theorem 2 μ^* is an outer measure on S.

Proof: Clearly $\mu^* \geq 0$, $\mu^*(\emptyset) = 0$ and μ^* is monotone. We need to check countable subadditivity. Let $A_i \subseteq S$ and set $A = \bigcup_{i=1}^{\infty} A_i$. If $\sum_{i=1}^{\infty} \mu^*(A_i) = \infty$, clearly

$$\mu^*(A) \leq \sum_{i=1}^{\infty} \mu^*(A_i)$$

so assume $\sum_{i=1}^{\infty} \mu^*(A_i) < \infty$. Let $\epsilon > 0$. For each i choose $\{B_{ij} : j \in \mathbf{N}\} \subseteq \mathcal{S}$ such that $A_i \subseteq \bigcup_{j=1}^{\infty} B_{ij}$ and $\mu^*(A_i) + \epsilon/2^i \geq \sum_{j=1}^{\infty} \mu(B_{ij})$. Then $A \subset \bigcup_{i=1}^{\infty} \bigcup_{j=1}^{\infty} B_{ij}$ so

$$\mu^*(A) \leq \sum_{i=1}^{\infty} \sum_{j=1}^{\infty} \mu^*(B_{ij}) \leq \sum_{i=1}^{\infty} (\mu^*(A_i) + \epsilon/2^i) = \sum_{i=1}^{\infty} \mu^*(A_i) + \epsilon.$$

Hence, $\mu^*(A) \leq \sum_{i=1}^{\infty} \mu^*(A_i)$.

Proposition 3 If $A \in \mathcal{S}$, then $\mu^*(A) \leq \mu(A)$ and if μ is a premeasure on $\mathcal{S}$, then $\mu^*(A) = \mu(A)$.

Proof: The first statement is clear. Suppose μ is a premeasure and $\{A_i\} \subseteq \mathcal{S}$ and $A \subseteq \bigcup_{i=1}^{\infty} A_i$. By countable subadditivity (Proposition 2.2.4), $\mu(A) \leq \sum_{i=1}^{\infty} \mu(A_i)$ so $\mu(A) \leq \mu^*(A)$.

Proposition 4 If $\mathcal{S}$ is an algebra, then μ^* is finitely additive on $\mathcal{S}$.

Proof: Let A, $B \in \mathcal{S}$ with $A \cap B = \emptyset$ and set $C = A \cup B$. By Theorem 2 we may assume that $\mu^*(C) < \infty$. Let $\epsilon > 0$. There exists $\{A_i\} \subseteq \mathcal{S}$ such that $\bigcup_{i=1}^{\infty} A_i \supseteq C$ and $\sum_{i=1}^{\infty} \mu(A_i) < \mu^*(C) + \epsilon$. Then $\{A_i \cap A\}[\{A_i \cap B\}]$ covers A [B] so since $\mathcal{S}$ is an algebra

$$\begin{aligned}\mu^*(A) + \mu^*(B) &\leq \sum_{i=1}^{\infty}(\mu(A_i \cap A) + \mu(A_i \cap B)) \\ &= \sum_{i=1}^{\infty} \mu(A_i \cap C) \leq \sum_{i=1}^{\infty} \mu(A_i) < \mu^*(C) + \epsilon\end{aligned}$$

Hence, $\mu^*(A) + \mu^*(B) \leq \mu^*(C)$ and Theorem 2 gives the reverse inequality.

For the outer measure μ^* we have simpler tests for measurability given by the following result.

Proposition 5 *For $E \subseteq S$, the following are equivalent:*

(i) *E is μ^*-measurable.*

(ii) *$\mu^*(A) = \mu^*(E \cap A) + \mu^*(A \backslash E)$ for $A \in \mathcal{S}$ with $\mu(A) < \infty$.*

(iii) *$\mu^*(A) \geq \mu^*(E \cap A) + \mu^*(A \backslash E)$ for $A \in \mathcal{S}$ with $\mu(A) < \infty$.*

(iv) *$\mu^*(A) \geq \mu^*(E \cap A) + \mu^*(A \backslash E)$ for $A \subseteq S$.*

Proof: Clearly (i) implies (ii) implies (iii).

(iii) $\Rightarrow$ (iv): Let $A \subseteq S$. If $\mu^*(A) = \infty$, trivial so assume $\mu^*(A) < \infty$. Let $\epsilon > 0$. Choose $\{A_i\} \subseteq \mathcal{S}$ such that $A \subseteq \bigcup_{i=1}^{\infty} A_i$ and $\sum_{i=1}^{\infty} \mu(A_i) < \mu^*(A) + \epsilon$. Since $\mu^*(A_i) < \infty$, by (iii) $\mu^*(A_i) \geq \mu^*(A_i \cap E) + \mu^*(A_i \backslash E)$. Hence,

$$\begin{aligned}\mu^*(A \cap E) + \mu^*(A\backslash E) &\leq \mu^*(\bigcup_{i=1}^{\infty} A_i \cap E) + \mu^*((\bigcup_{i=1}^{\infty} A_i)\backslash E) \\ &\leq \sum_{i=1}^{\infty}\mu^*(A_i \cap E) + \sum_{i=1}^{\infty} \mu^*(A_i\backslash E) = \sum_{i=1}^{\infty}\{\mu^*(A_i \cap E) + \mu^*(A_i\backslash E)\} \\ &\leq \sum_{i=1}^{\infty} \mu^*(A_i) < \mu^*(A) + \epsilon.\end{aligned}$$

Hence, $\mu^*(A \cap E) + \mu^*(A\backslash E) \leq \mu^*(A)$.

$(iv) \Rightarrow (i)$ follows by subadditivity.

The importance of conditions (ii) and (iii) is that we need only test the Caratheodory measurability condition for sets from the semi-ring $\mathcal{S}$.

We use Proposition 5 to show that the elements of $\mathcal{S}$ are μ^*-measurable when μ^* is finitely additive.

Theorem 6 *If μ^* is finitely additive on $\mathcal{S}$, then every element of $\mathcal{S}$ is μ^*-measurable.*

2.4. EXTENSIONS OF PREMEASURES

Proof: Let $E, A \in \mathcal{S}$. Then there exists $\{B_i : 1 \leq i \leq n\} \subseteq \mathcal{S}$ pairwise disjoint such that $A \setminus E = \bigcup_{i=1}^{n} B_i$. Since $A = (A \cap E) \cup \left(\bigcup_{i=1}^{n} B_i \right)$,

$$\mu^*(A \cap E) + \mu^*(A \setminus E) \leq \mu^*(A \cap E) + \sum_{i=1}^{n} \mu^*(B_i) = \mu^*(A)$$

and E is μ^*-measurable by Proposition 5 (ii). ∎

Remark 7 Note that μ^* is finitely additive on $\mathcal{S}$ if either μ is a premeasure (Proposition 3) or $\mathcal{S}$ is an algebra (Proposition 4) so Theorem 6 is applicable in either case. In particular, from Proposition 3 we have

Theorem 8 *Let μ be a premeasure on $\mathcal{S}$. Then every element of $\mathcal{S}$ is μ^*-measurable and μ^* restricted to the σ-algebra $\mathcal{M}(\mu^*)$ of μ^*-measurable sets is a countably additive extension of μ.*

If m_n is the Lebesgue premeasure on $\mathbf{R}^n$ (Example 2.2.10), then the countably additive extension to the σ-algebra of m_n^*-measurable sets is called *Lebesgue measure* on $\mathbf{R}^n$; we will study Lebesgue measure in §2.5. If $f: \mathbf{R} \to \mathbf{R}$ is increasing and left continuous and μ_f is the Lebesgue-Stieltjes premeasure of Example 2.2.9, then the countably additive extension of μ_f to the class of μ_f^*-measurable sets is called the *Lebesgue-Stieltjes measure* induced by f.

Remark 9 It is also the case that a bounded, finitely additive set function defined on an algebra has a bounded, finitely additive extension to the σ-algebra generated by the algebra. See section 5.6.2; see also [Bi], p. 185.

Henceforth, we assume that μ is a premeasure on $\mathcal{S}$ and denote the countably additive extension, μ^*, of μ to the class of μ^*-measurable sets, $\mathcal{M}(\mu^*)$, by μ.

Recall in section 1.3 we pointed out that Lebesgue defined the inner measure of a bounded subset E of $\mathbf{R}$ to be $m_*(E) = \ell(I) - m^*(I \setminus E)$, where I is a bounded interval containing E. Lebesgue then called a set E measurable if and only if $m^*(E) = m_*(E)$. For finite premeasures, we show that Lebesgue's definition is equivalent to Caratheodory's.

Theorem 10 *If $\mu(S) < \infty$, then $E \subset S$ is μ^*-measurable if and only if $\mu^*(E) + \mu^*(E^c) = \mu(S)$.*

Proof: $\Rightarrow$: Clear by taking S to be the test set in the definition of measurability.
$\Leftarrow$: Let $A \in \mathcal{S}$ be a test set. Then A is μ^*-measurable so $\mu^*(E) = \mu^*(E \cap A) + \mu^*(E \setminus A)$ and $\mu^*(E^c) = \mu^*(A \setminus E) + \mu^*(E^c \cap A^c)$. Adding these two equalities gives

$$\begin{aligned} \mu(S) &= \{\mu^*(E \cap A) + \mu^*(E^c \cap A)\} + \{\mu^*(E \cap A^c) + \mu^*(E^c \cap A^c)\} \\ &\geq \mu^*(A) + \mu^*(A^c) \geq \mu(S). \end{aligned}$$

Hence,

$$\mu(S) = \mu^*(A) + \mu^*(A^c) = \mu^*(E \cap A) + \mu^*(E^c \cap A) + \mu^*(E \cap A^c) + \mu^*(E^c \cap A^c).$$

But

$$\mu^*(A^c) \leq \mu^*(E \cap A^c) + \mu^*(E^c \cap A^c)$$

and $\mu(S) < \infty$ implies $\mu^*(A) \geq \mu^*(E \cap A) + \mu^*(A^c \cap E)$ so E is μ^*-measurable by Proposition 5.

We now address the uniqueness of the countably additive extension of the premeasure μ.

Lemma 11 *For $E \subset S$, $\mu^*(E) = \inf\{\sum_{i=1}^{\infty} \mu(A_i) : A_i \in \mathcal{S}, \{A_i\}$ pairwise disjoint, $E \subseteq \bigcup_{i=1}^{\infty} A_i\}$.*

Proof: Denote the right hand side of the equation above by $\mu'(E)$. Clearly $\mu^* \leq \mu'$. Suppose $E \subseteq \bigcup_{i=1}^{\infty} A_i$ with $A_i \in \mathcal{S}$. Disjointify the $\{A_i\}$ by setting $B_1 = A_1$ and $B_{j+1} = A_{j+1} \setminus \bigcup_{i=1}^{j} A_i$. By Proposition 2.1.3 each $B_i = \bigcup_{j=1}^{k_i} D_{ij}$, where $\{D_{ij} : 1 \leq j \leq k_i\}$ are pairwise disjoint from $\mathcal{S}$. Then $\mu(A_i) \geq \mu^*(B_i) = \sum_{j=1}^{k_i} \mu(D_{ij})$ and $E \subset \bigcup_{i=1}^{\infty} \bigcup_{j=1}^{k_i} D_{ij}$ is a disjoint union with

$$\sum_{i=1}^{\infty} \mu(A_i) \geq \sum_{i=1}^{\infty} \sum_{j=1}^{k_i} \mu(D_{ij}) \geq \mu'(E)$$

so $\mu^*(E) \geq \mu'(E)$.

A premeasure ν on a semi-ring $\mathcal{S}$ (measure ν on a σ-algebra Σ) is *finite* if $\nu(S) < \infty$ and *σ-finite* if $S = \bigcup_{j=1}^{\infty} A_j$, $A_j \in \mathcal{S}(A_j \in \Sigma)$ with $\nu(A_j) < \infty$. By the proof of Proposition 2.1.3 if ν is σ-finite, we may assume that the $\{A_j\}$ above are pairwise disjoint. Lebesgue-Stieltjes measure on $\mathbf{R}$ and Lebesgue measure on $\mathbf{R}^n$ are examples of σ-finite measures. Counting measure on a set S is σ-finite if and only if S is at most countable.

Theorem 12 *Let μ be a σ-finite premeasure. Let $\mathcal{T}$ be a semi-ring such that*

$$\mathcal{S} \subseteq \mathcal{T} \subseteq \mathcal{M}(\mu^*)$$

and ν a premeasure on $\mathcal{T}$. If $\nu = \mu$ on $\mathcal{S}$, then $\nu = \mu^*$ on $\mathcal{T}$.

2.4. EXTENSIONS OF PREMEASURES

Proof: Let ν^* be the outer measure induced by ν. If $E \subset S$ and $E \subseteq \bigcup_{i=1}^{\infty} A_i$, $A_i \in \mathcal{S}$, then $\nu^*(E) \leq \sum_{i=1}^{\infty} \nu(A_i) = \sum_{i=1}^{\infty} \mu(A_i)$ so $\nu^* \leq \mu^*$.

Let $A \in \mathcal{T}$ and $\mu^*(A) < \infty$. We claim that $\mu^*(A) \leq \nu^*(A)$ [so $\mu^*(A) = \nu^*(A)$]. Let $\epsilon > 0$. Choose $\{A_i\} \subset \mathcal{S}$ pairwise disjoint such that $A \subset \bigcup_{i=1}^{\infty} A_i$ and $\mu^*(A) + \epsilon \geq \sum_{i=1}^{\infty} \mu(A_i)$ [Lemma 11]. Set $B = \bigcup_{i=1}^{\infty} A_i$ so $\mu^*(A) \leq \mu^*(B) = \sum_{i=1}^{\infty} \mu(A_i) \leq \mu^*(A) + \epsilon$. Then

$$\nu^*(B \setminus A) \leq \mu^*(B \setminus A) = \mu^*(B) - \mu^*(A) \leq \epsilon.$$

Since ν^* is countably additive on the σ-algebra generated by $\mathcal{S}$,

$$\mu^*(A) \leq \mu^*(B) = \nu^*(B) = \nu^*(A) + \nu^*(B \setminus A) \leq \nu^*(A) + \epsilon,$$

and $\mu^*(A) \leq \nu^*(A)$.

Suppose $S = \bigcup_{j=1}^{\infty} E_j$, $E_j \in \mathcal{S}$, $\{E_j\}$ pairwise disjoint with $\mu(E_j) < \infty$. If $A \in \mathcal{T}$, then by the equality above

$$\mu^*(A) = \sum_{j=1}^{\infty} \mu^*(A \cap E_j) = \sum_{j=1}^{\infty} \nu(A \cap E_j) = \nu(A).$$

In particular, if μ is a σ-finite premeasure, then μ has a unique countably additive extension to the σ-algebra generated by $\mathcal{S}$. The σ-finiteness assumption in Theorem 12 is important.

Example 13 Let $S = \mathbf{R}$ and let $\mathcal{S}$ be the semi-ring of Example 2.1.8. Define μ on $\mathcal{S}$ by $\mu(\emptyset) = 0$, $\mu[a, b) = \infty$ if $a < b$. Then $\mu^*(A) = \infty$ if $A \neq \emptyset$ and $\mathcal{M}(\mu^*) = \mathcal{P}(\mathbf{R})$. Note counting measure ν on $\mathcal{S}$ agrees with m on $\mathcal{S}$. In fact, any $t\nu$, $t > 0$, agrees with μ on $\mathcal{S}$ so μ has an infinite number of countably additive extensions to $\mathcal{M}(\mu^*)$.

Another uniqueness result is given by

Theorem 14 Let $(S, \mathcal{S}_1, \mu_1)$, $(S, \mathcal{S}_2, \mu_2)$ be premeasure spaces. Then μ_1 and μ_2 generate the same outer measure if and only if $\mu_2^* = \mu_1$ on $\mathcal{S}_1$ and $\mu_1^* = \mu_2$ on $\mathcal{S}_2$.

Proof: $\Rightarrow$: $\mu_1^* = \mu_1$ on $\mathcal{S}_1$ and $\mu_2^* = \mu_2$ on $\mathcal{S}_2$ (Proposition 3) so if $\mu_1^* = \mu_2^*$, the result follows.

$\Leftarrow$: Let $A \subseteq S$. It suffices to show $\mu_2^*(A) \leq \mu_1^*(A)$ so we may assume $\mu_1^*(A) < \infty$. Then given $\epsilon > 0$ there exist $\{B_i\} \subseteq \mathcal{S}_1$ such that $A \subseteq \bigcup_{i=1}^{\infty} B_i$ and

$$\infty > \mu_1^*(A) > \sum_{i=1}^{\infty} \mu_1(B_i) - \epsilon = \sum_{i=1}^{\infty} \mu_2^*(B_i) - \epsilon$$

so $\mu_2^*(B_i) < \infty$ for each i. Therefore, by our standard $\epsilon/2^k$ construction there exist $\{A_{ik} : k \in \mathbf{N}\} \subseteq \mathcal{S}_2$ such that $A \subseteq \bigcup_{i=1}^{\infty} \bigcup_{k=1}^{\infty} A_{ik}$ and $\mu_1^*(A) > \sum_{i=1}^{\infty} \sum_{k=1}^{\infty} \mu_2(A_{ik}) - 2\epsilon$. Hence, $\mu_1^*(A) > \mu_2^*(A) - 2\epsilon$ and $\mu_1^*(A) \geq \mu_2^*(A)$.

Remark 15 The equality $\mu_2^* = \mu_1$ on S_1 is not enough to guarantee that $\mu_1^* = \mu_2^*$. Let S_1 be all $[a, b)$ with integer a, b and set $\mu_1[a, b) = b - a$. Let S_2 be all $[a, b)$ with rational a, b and set $\mu_2[a, b) = b - a$. Then $S_1 \subseteq S_2$ and $\mu_1 = \mu_2$ on S_1. However, for $I = [0, 1/2)$, $\mu_2^*(I) = 1/2$ and $\mu_1^*(I) = 1$.

Remark 16 As noted in Remark 9 a bounded, finitely additive set function defined on an algebra has a bounded, finitely additive extension to the generating σ-algebra, but this extension may fail to be unique ([Hah]).

Approximation:

We now consider how measurable sets can be approximated by elements of the semi-ring S.

If $\mathcal{H}$ is any family of subsets of S, then $\mathcal{H}_\sigma(\mathcal{H}_\delta)$ denotes the family of all finite or countable unions (intersections) of members of $\mathcal{H}$. We write $\mathcal{H}_{\sigma\delta} = (\mathcal{H}_\sigma)_\delta$, etc.

Theorem 17 *Let $E \subset S$.*

(i) *For every $\epsilon > 0$ there exists $A \in S_\sigma$ such that $E \subset A$ and $\mu^*(E) + \epsilon \geq \mu(A)$.*

(ii) *There exists $A \in S_{\sigma\delta}$ such that $E \subset A$ and $\mu^*(E) = \mu(A)$.*

Proof: (i): There exist $\{A_i\} \subset S$ such that $E \subset \bigcup_{i=1}^{\infty} A_i$ and $\mu^*(E) + \epsilon \geq \sum_{i=1}^{\infty} \mu(A_i)$. Put $A = \bigcup_{i=1}^{\infty} A_i$.

(ii): By (i) for each i there exists $A_i \in S_\sigma$ such that $E \subset A_i$ and $\mu^*(E) + 1/i \geq \mu(A_i)$. Put $A = \bigcap_{i=1}^{\infty} A_i$ so $A \in S_{\sigma\delta}$ and $E \subset A$. Then

$$\mu^*(E) \leq \mu(A) \leq \mu(A_i) < \mu^*(E) + 1/i$$

for each i so $\mu^*(E) = \mu(A)$.

Theorem 18 *Assume μ is a σ-finite premeasure. A subset $E \subset S$ is μ^*-measurable if and only if $E = A \backslash B$, where $A \in S_{\sigma\delta}$ and $\mu^*(B) = 0$.*

Proof: $\Leftarrow$: $S_{\sigma\delta} \subset \mathcal{M}(\mu^*)$ and $B \in \mathcal{M}(\mu^*)$.

$\Rightarrow$: Let $S = \bigcup_{i=1}^{\infty} E_i$, $E_i \in S$, and $\mu(E_i) < \infty$. Then $E = \bigcup_{i=1}^{\infty} E \cap E_i$ with $\mu(E_i \cap E) < \infty$. Set $B_i = E \cap E_i$. By Theorem 17(i) for each i, j there exists $A_{ij} \in S_\sigma$ such that $B_i \subset A_{ij}$ and $\mu(B_i) + 1/2^i j > \mu(A_{ij})$. Set $A_j = \bigcup_{i=1}^{\infty} A_{ij}$. Then $E \subset A_j \in S_\sigma$ and $A_j \backslash E \subset \bigcup_{i=1}^{\infty} (A_{ij} \backslash E)$ implies

$$\mu(A_j \backslash E) \leq \sum_{i=1}^{\infty} \mu(A_{ij} \backslash E) \leq \sum_{i=1}^{\infty} 1/2^i j = 1/j.$$

2.4. EXTENSIONS OF PREMEASURES

Set $A = \bigcap_{j=1}^{\infty} A_j$ so $A \in \mathcal{S}_{\sigma\delta}$ and $E \subset A$, $A \backslash E \subseteq A_j \backslash E$ for each j. Hence, $\mu(A \backslash E) \leq \mu(A_j \backslash E) \leq 1/j$ for each j and $\mu(A \backslash E) = 0$. If $B = A \backslash E$, then $E = A \backslash B$.

Completeness of measures:

Definition 19 *Let ν be a measure on the σ-algebra Σ. ν is said to be a* complete *measure (or complete) or (S, Σ, ν) is a complete measure space if all subsets of ν measure 0 are in Σ.*

From Proposition 2.3.5, $(S, \mathcal{M}(\mu^*), \mu)$ is always a complete measure space.

Theorem 20 *Let $\mathcal{S}$ be a σ-algebra. Then there exists a complete measure space $(S, \overline{\mathcal{S}}, \bar{\mu})$ such that*

(i) $\mathcal{S} \subset \overline{\mathcal{S}}$,

(ii) $\bar{\mu} = \mu$ *on* $\mathcal{S}$,

(iii) $\overline{A} \in \overline{\mathcal{S}}$ *if and only if* $\overline{A} = A \cup Z$ *where* $A \in \mathcal{S}$, $Z \subset N$ *for some* $N \in \mathcal{S}$ *with* $\mu(N) = 0$.

$(S, \overline{\mathcal{S}}, \bar{\mu})$(or $\bar{\mu}$) is called the completion of $(S, \mathcal{S}, \mu)$ (or μ).
Let $\mathcal{Z} = \{Z \subset S : Z \subset N \text{ for some } N \in \mathcal{S} \text{ with } \mu(N) = 0\}$ and

$$\overline{\mathcal{S}} = \{E \cup Z : E \in \mathcal{S}, \ Z \in \mathcal{Z}\}.$$

Define $\bar{\mu}$ on $\overline{\mathcal{S}}$ by $\bar{\mu}(E \cup Z) = \mu(E)$. It is straightforward to check that $\overline{\mathcal{S}}$ is a σ-algebra, $\bar{\mu}$ is well-defined on $\overline{\mathcal{S}}$ and gives a countably additive extension of μ. We leave the details to the reader.

Theorem 21 *Let μ be a σ-finite premeasure. Let μ_1 be the unique countably additive extension of μ to Σ, the σ-algebra generated by $\mathcal{S}$ (Theorem 12). Then the completion of μ_1 is identical to μ^* restricted to $\mathcal{M}(\mu^*)$. That is, $(S, \overline{\Sigma}, \bar{\mu}_1) = (S, \mathcal{M}(\mu^*), \mu^*)$.*

Example 22 *The σ-finiteness condition in Theorem 21 is important. Let $S = \mathbf{R}$ and Σ the σ-algebra of subsets of S which are either countable or have countable complements. Let μ be counting measure restricted to Σ. Then (S, Σ, μ) is complete but $\mathcal{M}(\mu^*) = \mathcal{P}(S)$.*

Exercise 1. Define $\nu : \mathcal{P}(\mathbf{N}) \to [0, \infty]$ by $\nu(E) = \sum_{i \in E} 1/2^i$ if E is finite and $\nu(E) = \infty$ if E is infinite. Show ν is finitely additive but not countably additive. Compute ν^*, $\mathcal{M}(\nu^*)$. Show $\nu^* \neq \nu$. Compare with Proposition 3.

Exercise 2. Let $S = [0, 1)$ and $\mathcal{A}$ the algebra generated by the subintervals $[a, b)$. Give an example of a non-negative finitely additive set function μ on $\mathcal{A}$ which has two distinct finitely additive extensions to the σ-algebra, $\mathcal{B}(S)$, generated by $\mathcal{A}$. [Hint: Example 13.]

Exercise 3. Show $E \subset S$ is μ^*-measurable if and only if $E \cap A$ is μ^*-measurable for every μ^*-measurable A with $\mu^*(A) < \infty$.

Exercise 4. Give an example of a non-complete measure.

Exercise 5. An outer measure μ^* on S is called regular if for every $A \subset S$ there is a μ^*-measurable $E \supset A$ such that $\mu(E) = \mu^*(A)$ [see Theorem 17]. Show that if μ^* is regular and $A_j \subset S$, then $\mu^*(\underline{\lim} A_j) \leq \underline{\lim} \mu^*(A_j)$.

2.4. EXTENSIONS OF PREMEASURES

2.4.1 Hewitt-Yosida Decomposition

The construction of section 2.4 can be used to obtain a decomposition of a non-negative finitely additive set function into a countably additive part and another part which is called purely finitely additive. We give the details. The material in this section is only used in §2.6.1 and can be skipped at this time.

Let $\mathcal{A}$ be an algebra of subsets of S.

Definition 1 *A non-negative, finitely additive set function* $\nu : \mathcal{A} \to [0, \infty)$ *is* purely finitely additive *if and only if whenever* $\mu : \mathcal{A} \to [0, \infty)$ *is countably additive and* $0 < \mu \leq \nu$ *on* $\mathcal{A}$ *, then* $\mu = 0$. *A bounded, finitely additive* $\nu : \mathcal{A} \to \mathbf{R}$ *is purely finitely additive if and only if* ν^+ *and* ν^- *are purely finitely additive.*

Theorem 2 (Hewitt-Yosida) *If* $\nu : \mathcal{A} \to [0, \infty)$ *is finitely additive, then there exist a countably additive* $\nu_c : \mathcal{A} \to [0, \infty)$ *and a purely finitely additive*

$$\nu_f : \mathcal{A} \to [0, \infty)$$

such that $\nu = \nu_c + \nu_f$. *Moreover, the decomposition is unique.*

Proof: Let ν^* be the outer measure on S induced by ν and let ν_c be ν^* restricted to $\mathcal{A}$. By Remark 2.4.7 and Proposition 2.4.3, ν_c is countably additive and $\nu_c \leq \nu$. Set $\nu_f = \nu - \nu_c$ so ν_f is non-negative and finitely additive. Suppose $\mu : \mathcal{A} \to \mathbf{R}$ is countably additive with $0 < \mu \leq \nu_f$. For any $A \in \mathcal{A}$,

$$0 \leq \mu(A) \leq \nu_f(A) = \nu(A) - \nu_c(A)$$

so $\nu_c(A) + \mu(A) \leq \nu(A)$. By Proposition 2.4.3, $\nu_c(A) + \mu(A) \leq \nu_c(A)$ so $\mu = 0$ and ν_f is purely finitely additive.

Suppose $\nu = \nu_c + \nu_f = \mu_c + \mu_f$ with μ_c non-negative, countably additive and μ_f non-negative, purely finitely additive. Then for any $A \in \mathcal{A}$,

$$\nu_c(A) + \nu_f^*(A) = \mu_c(A) + \mu_f^*(A)$$

and $\nu_f^* \leq \nu_f$, $\mu_f^* \leq \mu_f$. Since ν_f^* and μ_f^* are countably additive on $\mathcal{A}$ and dominated by ν_f and μ_f, respectively, $\nu_c = \mu_c$ and, consequently, $\nu_f = \mu_f$.

We give examples of purely finitely set funtions and Hewitt-Yosida decompositions for Lebesgue-Stieltjes "measures" in §2.6.1.

Exercise 1. Let $\nu : \mathcal{A} \to [0, \infty)$ be finitely additive. Show ν is purely finitely additive if and only if for every $\epsilon > 0$ there exist pairwise disjoint $\{A_i\} \subset \mathcal{A}$ such that $S = \bigcup_{i=1}^{\infty} A_i$ and $\sum_{i=1}^{\infty} \nu(A \cap A_i) < \epsilon$ for every $A \in \mathcal{A}$.

2.5 Lebesgue Measure

In this section we develop some of the basic properties of the most important measure, Lebesgue measure. Let S be the semi-ring in $\mathbf{R}^n$ consisting of the half-closed n-dimensional rectangles of the form $I = [a_1, b_1) \times \cdots \times [a_n, b_n)$ [Example 2.1.10] and let $m(= m_n)$ be the n-dimensional volume of such rectangles, $m(I) = (b_1-a_1)\ldots(b_n-a_n)$ [Example 2.2.10]. Since m is a σ-finite premeasure on S, m has a unique countably additive extension to a complete measure, called *Lebesgue measure* on $\mathbf{R}^n$ and still denoted by $m(= m_n)$, defined on a σ-algebra, $\mathcal{M}(= \mathcal{M}_n)$, containing S. The elements of $\mathcal{M}$ are called the *Lebesgue measurable* subsets of $\mathbf{R}^n$. We show below that the σ-algebra $\mathcal{M}$ contains the Borel sets, $\mathcal{B}(\mathbf{R}^n)$, of $\mathbf{R}^n$. For this it suffices to show that any open set belongs to $\mathcal{M}$. This follows from the result below which gives the structure of open sets in $\mathbf{R}^n$ analogous to the characterization of open sets in $\mathbf{R}$ given in 1.2.2.

Call a half-closed interval in S of the form $[a_1, a_1+\delta) \times \cdots \times [a_n, a_n+\delta)$, $\delta > 0$, a brick with vertex $(a_1, \cdots, a_n)$ and side length δ.

Lemma 1 *Every open set G in $\mathbf{R}^n$ is a countable pairwise disjoint union of bricks.*

Proof: Let $\mathcal{B}^k$ be the family of all bricks whose vertices are integral multiples of 2^{-k} with side length 2^{-k}. Note that $x \in \mathbf{R}^n$ lies in exactly one member of $\mathcal{B}^k$, and if $A \in \mathcal{B}^j$, $B \in \mathcal{B}^k$ where $j < k$, then either $A \subset B$ or $A \cap B = \emptyset$. If $x \in G$, then x belongs to an open sphere contained in G so x belongs to a brick contained in G belonging to some $\mathcal{B}^k$. That is, G is the union of all bricks belonging to $\bigcup_k \mathcal{B}^k$ which are contained in G. From this collection of bricks choose all of those belonging to $\mathcal{B}^1$ and remove those in $\bigcup_{k \geq 2} \mathcal{B}^k$ which lie in any of the bricks chosen from $\mathcal{B}^1$. From the remaining bricks choose those belonging to $\mathcal{B}^2$. Continuing this construction produces a countable pairwise disjoint family of bricks whose union is G.

Remark 2 Note the bricks in Lemma 1 all have side length 2^{-k} for some $k \in \mathbf{N}$.

From Lemma 1 it follows that $\mathcal{B}(\mathbf{R}^n) \subseteq \mathcal{M}_n \subseteq \mathcal{P}(\mathbf{R}^n)$. We show later (Examples 9 and 1.3.1) that each of these containments is proper.

We next consider what is called the regularity of Lebesgue measure. Each half-closed interval $I = [a_1, b_1) \times \cdots \times [a_n, b_n)$ may be approximated by an open interval $J = (\alpha_1, \beta_1) \times \cdots \times (\alpha_n, \beta_n)$ containing I whose volume is arbitrarily close to the volume of I so the Lebesgue outer measure of a subset A of $\mathbf{R}^n$ can also be computed by

$$m^*(A) = \inf\{\sum_{j=1}^{\infty} m(I_j) : \text{ each } I_j \text{ an open interval with } \bigcup_{j=1}^{\infty} I_j \supset A\}.$$

We use this observation below.

If S is any topological space, a subset E of S is called a $\underline{\mathcal{G}_\delta \text{ set}}$ ($\underline{\mathcal{F}_\sigma \text{ set}}$) if E is the countable intersection (union) of open (closed) subsets of S.

2.5. LEBESGUE MEASURE

Theorem 3 *Let $A \subset \mathbf{R}^n$.*

(i) *For every $\epsilon > 0$ there exists open $G \supset A$ such that $m(G) \leq m^*(A) + \epsilon$.*

(ii) *There exists a $\mathcal{G}_\delta$ $H \supset A$ such that $m(H) = m^*(A)$.*

Proof: (i): By the observation there exist open intervals $\{I_j\}$ such that $A \subset \bigcup_{j=1}^\infty I_j$ and
$$\sum_{j=1}^\infty m(I_j) \leq m^*(A) + \epsilon.$$
Set $G = \bigcup_{j=1}^\infty$. Then G is open, $G \supset A$ and
$$m(G) \leq \sum_{j=1}^\infty m(I_j) \leq m^*(A) + \epsilon.$$

(ii): For each j choose G_j open such that $G_j \supset A$ and $m(G_j) \leq m^*(A) + 1/j$. Put $H = \bigcap_{j=1}^\infty G_j$. Then H is a $\mathcal{G}_\delta$ containing A and $m(H) \leq m(G_j) \leq m^*(A) + 1/j$ for each j implies $m(H) \leq m^*(A)$. But $m^*(A) \leq m(H)$ so equality must hold.

From Theorem 3 we can obtain some topological-type characterizations of Lebesgue measurability.

Corollary 4 *Let $A \subseteq \mathbf{R}^n$. The following are equivalent:*

(i) $A \in \mathcal{M}_n$.

(ii) *For every $\epsilon > 0$ there exists open $G \supset A$ such that $m^*(G \backslash A) < \epsilon$.*

(iii) *For every $\epsilon > 0$ there exists closed $F \subset A$ such that $m^*(A \backslash F) < \epsilon$.*

(iv) *There exists a $\mathcal{G}_\delta$ $H \supset A$ such that $m^*(H \backslash A) = 0$.*

(v) *There exists an $\mathcal{F}_\sigma$ $K \subset A$ such that $m^*(A \backslash K) = 0$.*

(vi) *There exists a Borel set $B \in \mathcal{B}(\mathbf{R}^n)$ and an m-null set Z such that $A = B \cup Z$.*

Proof: (i)$\Rightarrow$(ii): First suppose $m^*(A) < \infty$. By Theorem 3 there exists open $G \supset A$ with $m(G) < m(A) + \epsilon$. By Proposition 2.2.3, $m(G \backslash A) < \epsilon$.

Now suppose A is measurable. Let $E_k = \{x : \|x\| \leq k\}$ and set $A_k = A \cap E_k$ so A_k is measurable and $m(A_k) < \infty$. By the part above, for each k there exists open $G_k \supset A_k$ with $m(G_k \backslash A_k) < \epsilon/2^k$. Put $G = \bigcup_{k=1}^\infty G_k$. Then G is open and $G \supset A$. Since $G \backslash A \subseteq \bigcup_{k=1}^\infty (G_k \backslash A_k)$,
$$m(G \backslash A) \leq \sum_{k=1}^\infty m(G_k \backslash A_k) < \sum_{k=1}^\infty \epsilon/2^k = \epsilon.$$

(ii)⇒(iv): For each k there exists open $G_k \supset A$ with $m^*(G_k \backslash A) < 1/k$. Set $H = \bigcap_{k=1}^{\infty} G_k$. Then H is a $\mathcal{G}_\delta$ and $H \supset A$. Since

$$m^*(H \backslash A) \leq m^*(G_k \backslash A) < 1/k$$

for each k, $m^*(H \backslash A) = 0$.

(iv)⇒(i): $A = H \backslash (H \backslash A)$ and H is measurable since it is a Borel set and $H \backslash A$ is measurable since it is m-null.

(i)⇒(iii): A measurable implies A^c measurable so there exists open $G \supset A^c$ with $m(G \backslash A^c) < \epsilon$. Then $F = G^c$ is closed and $F \subset A$ with $m(A \backslash F) = m(A \cap G) < \epsilon$.

(iii)⇒(v): For each k there exists closed $F_k \subset A$ such that $m^*(A \backslash F_k) < 1/k$. Put $K = \bigcup_{k=1}^{\infty} F_k$ so K is an $\mathcal{F}_\sigma$ contained in A with $m^*(A \backslash K) \leq m^*(A \backslash F_k) < 1/k$ for each k so $m^*(A \backslash K) = 0$.

(v)⇒(vi): Put $B = K$ and $Z = A \backslash K$.

(vi)⇒(i): This is clear since both Borel sets and m-null sets are Lebesgue measurable.

It follows from (vi) that Lebesgue measure is the completion of the measure m restricted to the Borel sets $\mathcal{B}(\mathbf{R}^n)$ [Exercise 1]. We show later that Lebesgue measure restricted to $\mathcal{B}(\mathbf{R})$ is not a complete measure (Example 9).

The properties in (ii) and (iii) for Lebesgue measure are called regularity. We now give the formal definitions of regularity; some of the basic properties of regular measures are given in §2.7.

Let S be a Hausdorff topological space and $\mathcal{B}(S)$ the Borel subsets of S. Let $\sum$ be a σ-algebra containing $\mathcal{B}(S)$ and $\mu : \sum \to [0, \infty]$ finitely additive.

Definition 5 $E \in \sum$ *is* inner regular *(with respect to μ) if*

$$\mu(E) = \sup\{\mu(K) : K \subset E, K \text{ compact}\};$$

$E \in \sum$ *is* outer regular *(with respect to μ) if $\mu(E) = \inf\{\mu(G) : G \supset E, G \text{ open}\}$. E is* regular *(with respect to μ) if E is both inner and outer regular.*

A *Borel measure* is a measure defined on the Borel sets which is finite on compact sets. A Borel measure is regular if every Borel set is outer regular and every open set is inner regular. Lebesgue measure restricted to the Borel sets is regular (Exer. 10).

Uniqueness of Lebesgue Measure:

We next consider the uniqueness of Lebesgue measure. A measure μ on $\mathcal{B}(\mathbf{R}^n)$ is said to be *translation invariant* if $\mu(B) = \mu(h + B)$ for every Borel set B and $h \in \mathbf{R}^n$. Lebesgue measure restricted to $\mathcal{B}(\mathbf{R}^n)$ is translation invariant (Exercises 2 and 3). We show that translation invariance along with regularity characterizes Lebesgue measure.

2.5. LEBESGUE MEASURE

Theorem 6 *If μ is a translation invariant regular measure on $\mathcal{B}(\mathbf{R}^n)$, then $\mu = cm$ for some positive constant c.*

Proof: Let $I = [0,1] \times \cdots \times [0,1]$ be the unit "cube" in $\mathbf{R}^n$ and set $c = \mu(I)$. I is the pairwise disjoint union of 2^{nk} bricks of side length 2^{-k} for any $k \in \mathbf{N}$, and since by translation invariance each of these bricks has the same μ-measure,

$$\mu(I) = 2^{nk}\mu(B) = cm(I) = 2^{nk}cm(B)$$

for any brick B with side length 2^{-k} so $\mu(B) = cm(B)$ for any such brick B. By Remark 2 $\mu(G) = cm(G)$ for every open set $G \subset \mathbf{R}^n$, and the same equality must hold for any Borel set by regularity of μ and m.

It is shown in 2.7.7 that every Borel measure on $\mathbf{R}^n$ is regular so the regularity assumption in Theorem 6 is redundant.

Lebesgue Measure in R:

We consider Lebesgue measure in $\mathbf{R}$ in more detail.

Since any singleton in $\mathbf{R}$ (or $\mathbf{R}^n$) obviously has Lebesgue measure 0, any countable subset of $\mathbf{R}$ (of $\mathbf{R}^n$) has Lebesgue measure 0. However, an uncountable set can have Lebesgue measure 0. We give an example of such a set.

Example 7 (Cantor Set) Let $I = [0,1]$. It is convenient to describe the complement of the Cantor set in I. Let E_0^1 be the open interval $(1/3, 2/3)$, the middle one-third of I. At the zeroth stage the interval E_0^1 is removed from I leaving two closed subintervals $[0, 1/3]$ and $[1/3, 1]$. At the first stage the two open middle thirds E_1^1 and E_1^2 are removed from these closed intervals leaving four closed subintervals. At the second stage the four open middle thirds $E_2^1, E_2^2, E_2^3, E_2^4$ are removed from these closed subintervals [see the figure below]. The construction is continued; at the kth stage 2^k open intervals E_k^i, $i = 1, \cdots, 2^k$, are removed. The Cantor set K is the subset remaining after these open subintervals have been removed, i.e., $K = I \setminus \bigcup_{k=0}^{\infty} \bigcup_{i=1}^{2^k} E_k^i$. Since each E_k^i is an open interval, K is obviously a closed set. K obviously contains the endpoints $1/3, 2/3, 1/9, \cdots$, and it may appear that these are all the points in K. However, we will show that K is, in fact, uncountable. First we calculate the Lebesgue measure of K. Note each E_k^i has length $1/3^{k+1}$ and since the $\{E_k^i\}$ are pairwise disjoint,

$$m(K) = 1 - \sum_{k=0}^{\infty} \sum_{i=1}^{2^k} 1/3^{k+1} = 1 - 1/3 \sum_{k=0}^{\infty} (2/3)^k = 1 - 1/3(1/(1-2/3)) = 0.$$

$E_2^1 \quad E_1^1 \quad E_2^2 \qquad\qquad E_0^1 \qquad\qquad E_2^3 \quad E_1^3 \quad E_2^4$

$\vdash\!(\ )\!\!-\!\!(\quad)\!(\ )\!\!-\!\!(\!\!\text{————}\!\!)\!(\quad)\!(\ )\!\!-\!\!(\quad)\!(\ )\!\dashv$

$0 \quad \frac{1}{27}\,\frac{2}{27}\,\frac{1}{9} \qquad \frac{2}{9}\,\frac{7}{27}\,\frac{8}{27}\,\frac{1}{3} \qquad\qquad \frac{2}{3}\,\frac{19}{27}\,\frac{20}{27}\,\frac{7}{9} \qquad \frac{8}{9}\,\frac{25}{27}\,\frac{26}{27}\quad 1$

We next observe some of the topological properties of K. First K is nowhere dense, i.e., K has no interior points. Note the "distance" between any two adjacent open intervals making up $A_N = \bigcup_{k=0}^{N} \bigcup_{i=1}^{2^k} E_k^i$ is $1/3^{N+1}$. Thus, if K were to contain an open interval (a, b), then for $1/3^{N+1} < b - a$, the distance between some two open intervals in A_N would be greater than $1/3^{N+1}$.

Next, K is perfect, i.e., every point of K is a limit point of K. Let $x \in K$ and $\epsilon > 0$. Choose N such that $1/3^{N+1} < \epsilon$. Then $S(x, \epsilon) = (x - \epsilon, x + \epsilon)$ must intersect at least one of the open intervals in A_N so $S(x, \epsilon)$ must contain a point of K distinct from x (namely, one of the endpoints of the open intervals). Thus, x is a limit point of K. [It actually follows from the fact that K is a perfect subset of $\mathbf{R}$ that K is uncountable ([R1] 2.43), but we will establish this below by a different technique.]

Note that the point $x \in I$ belongs to K if and only if in one of its ternary expansions, $x = .a_1 a_2 \cdots$ (base 3), the digit 1 does not occur. It follows from this observation that K is uncountable ([DeS] 1.8), and that $\mathbf{R}$ and K have the same cardinality.

Remark 8 Note that for any α, $0 < \alpha < 1$, by altering the size of the open intervals E_k^i to $(1 - \alpha)(1/3^{k+1})$, we may obtain a nowhere dense, perfect set of measure α which we refer to as an α-*Cantor set*. In particular, this shows that a set can have a positive Lebesgue measure and not contain an interval. These Cantor-like sets of positive Lebesgue measure can be used to construct sets of the second category which have Lebesgue measure 0 [see Exercise 7].

In §1.3 we showed the existence of a subset of $\mathbf{R}$ which was not Lebesgue measurable [Exercise 2]. The existence of such non-measurable sets in $\mathbf{R}^n$, $n > 1$, is addressed in Exer. 3.9.4. We now show that there are Lebesgue measurable sets in $\mathbf{R}$ which are not Borel sets.

Example 9 Let $0 < \epsilon < 1$ and K_ϵ an ϵ-Cantor set. Let K be the Cantor set and

$$\{E_k^i : i = 1, \cdots, 2^k; k = 0, 1, \cdots\}(\{A_k^i : k = 1, \cdots, 2^k; i = 0, 1, \cdots\})$$

the open subintervals of $[0, 1]\backslash K([0, 1]\backslash K_\epsilon)$. Let f be an increasing function which maps each A_k^i linearly onto E_k^i (see the graph below). $[0, 1]\backslash K_\epsilon$ is dense in $[0, 1]$ so we can extend f to $[0, 1]$ by setting $f(0) = 0$ and $f(x) = \sup\{f(y) : y < x, y \in [0, 1]\backslash K_\epsilon\}$. Then f is increasing, and f must be a continuous map of $[0, 1]$ onto itself since f cannot have any jumps (a jump in f would correspond to K containing an interval). Now K_ϵ must contain a subset P which is not Lebesgue measurable (Remark 1.3.2), and f carries P onto some subset P' of K which must be Lebesgue measurable since K has Lebesgue measure 0. By Exercise 11, P' cannot be a Borel set. This also shows that *Lebesgue measure restricted to the Borel sets is not a complete measure*.

2.5. LEBESGUE MEASURE

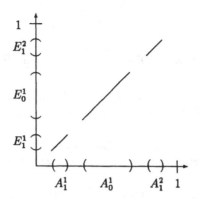

Thus, each of the containments $\mathcal{B}(\mathbf{R}) \subset \mathcal{M}(\mathbf{R}) \subset \mathcal{P}(\mathbf{R})$ is proper. (A cardinality argument can be used to establish that $\mathcal{B}(\mathbf{R}) \neq \mathcal{M}(\mathbf{R})$; see [Ra] 5.3. However, $\mathcal{M}(\mathbf{R})$ and $\mathcal{P}(\mathbf{R})$ have the same cardinality (Exercise 8).) The existence of Lebesgue measurable sets in $\mathbf{R}^2(\mathbf{R}^n)$ which are not Borel sets is considered in Exercise 3.9.9.

Another interesting result pertaining to Lebesgue measure due to H. Steinhaus asserts that if E is a Lebesgue measurable subset of $\mathbf{R}^n$ of positive Lebesgue measure, then 0 is an interior point of $E - E$ (see [AB], 15.12 or Exer. 3.11.9).

Remark 10 Lebesgue measure can be extended to a translation invariant measure on a σ-algebra which properly contains $\mathcal{M}$, but such extensions do not seem to be useful (see [KO]).

Exercise 1. Show $(\mathbf{R}^n, \mathcal{M}_n, m)$ is the completion of $(\mathbf{R}^n, \mathcal{B}(\mathbf{R}^n), m)$.

Exercise 2. Show $m^*(E) = m^*(E+h)$ for any $h \in \mathbf{R}^n$, $E \subset \mathbf{R}^n$. Show E is Lebesgue measurable if and only if $E + h$ is Lebesgue measurable for every $h \in \mathbf{R}^n$ and in this case $m(E + h) = m(E)$. [Hint: $E \cap (a + A) = a + (E - a) \cap A$, $t + A^c = (t + A)^c$.]

Exercise 3. Show $h + \mathcal{B}(\mathbf{R}^n) = \mathcal{B}(\mathbf{R}^n)$ for any $h \in \mathbf{R}^n$.

Exercise 4. Let $a > 0$. Show $m^*(aE) = a^n m^*(E)$ for $E \subset \mathbf{R}^n$. Show E is Lebesgue measurable if and only if aE is Lebesgue measurable for every $a > 0$.

Exercise 5. Show every countable subset of $\mathbf{R}^n$ is a Borel set.

Exercise 6. Show $E \subset \mathbf{R}^n$ is (Lebesgue) measurable if and only if $E \cap G$ is measurable for every (bounded) open G if and only if $E \cap F$ is measurable for every (bounded) closed F.

Exercise 7. Construct a set of category II in **R** which has Lebesgue measure 0.

Exercise 8. Show $\mathcal{M}(\mathbf{R})$ and $\mathcal{P}(\mathbf{R})$ have the same cardinality.

Exercise 9. Define a measure $\mu_n : \mathcal{M} \to \mathbf{R}^*$ by $\mu_n(E) = m(E \cap [n, \infty])$. Show $\mu_n(E) \downarrow \mu(E)$ for each $E \in \mathcal{M}$ but μ is not a measure.

Exercise 10. Show H being closed in Corollary 4 (iii) can be replaced by H being compact. Show every Lebesgue measurable set is regular.

Exercise 11. If $f : \mathbf{R}^n \to \mathbf{R}^m$ is continuous, show the inverse image of a Borel set is a Borel set. Note $\mathbf{R}^n$ and $\mathbf{R}^m$ can be replaced by topological spaces.

2.6 Lebesgue-Stieltjes Measures

In this section we consider the Lebesgue-Stieltjes measures introduced in Example 2.2.9. Let $f : \mathbf{R} \to \mathbf{R}$ be increasing, let S be the semi-algebra of half-closed intervals $[a, b)$, $a \leq b$ (Example 2.1.8). Define μ_f on S by

$$\mu_f[a, b) = f(b) - f(a), \ a \leq b. \tag{2.1}$$

Then μ_f is finitely additive on S (Example 2.2.9) and has a finitely additive extension to the algebra generated by S (Exer. 2.2.2). If f is left continuous, then μ_f is countably additive on S and by Theorem 2.4.8 has a unique countably additive extension to a complete, countably additive measure defined on a σ-algebra, $\mathcal{M}(\mu_f^*)$, which contains the Borel sets, $\mathcal{B} = \mathcal{B}(\mathbf{R})$, of $\mathbf{R}$ and is finite on compact subsets of $\mathbf{R}$. The σ-algebra $\mathcal{M}(\mu_f^*)$ can be very different from the σ-algebra of Lebesgue measurable sets (Exer. 3). We show conversely that if μ is any countable additive measure defined on the Borel sets of $\mathbf{R}$ and which is finite on compact sets, then μ is a Lebesgue-Stieltjes measure (restricted to $\mathcal{B}$).

Let $\mu : \mathcal{B} \to [0, \infty]$ be a Borel measure. Set

$$f(t) = \begin{cases} \mu[0, t) & t > 0 \\ 0 & t = 0 \\ -\mu[t, 0) & t < 0 \end{cases} \tag{2.2}$$

f is called the (cumulative) *distribution function* of μ.

The distribution function f of μ is increasing since if $0 < t < s$, then

$$f(s) - f(t) = \mu[0, s) - \mu[0, t) = \mu[t, s) \geq 0$$

and similarly if $t < s < 0$ or $t < 0 < s$. Also, the distribution function is left continuous. For suppose $t_k \downarrow 0$. If $t > 0$, then $f(t - t_k) - f(t) = \mu[t - t_k, 0) \downarrow 0$ since μ is finite on bounded intervals (2.2.5). Therefore, f is left continuous for $t > 0$ and, similarly, f is left continuous for $t \leq 0$.

Clearly, the Lebesgue-Stieltjes measure induced by the distribution function f is exactly μ. Since two Lebesgue-Stieltjes measures μ_f and μ_g are equal if and only if $f - g$ is a constant (Exer. 4), there is a one-one correspondence between Borel measures on $\mathbf{R}$ and the class of increasing, left-continuous functions on $\mathbf{R}$ which are normalized by requiring that they vanish at the origin.

We consider the analogue for Lebesgue-Stieltjes measures of the regularity results for Lebesgue measure given in Theorem 2.5.3 and Corollary 2.5.4. For this we require the following lemma.

Lemma 1 *For $E \subseteq \mathbf{R}$, $\mu_f^*(E) = \inf\{\sum_{i=1}^{\infty} \mu_f(a_i, b_i) : E \subseteq \bigcup_{i=1}^{\infty} (a_i, b_i)\}$.*

Proof: Denote the quantity on the right hand side of the equality above by $\nu(E)$. Suppose $E \subseteq \bigcup_{i=1}^{\infty}(a_i, b_i)$ and set $(a_i, b_i) = I_i$. Each I_i is a countable pairwise disjoint union of half-closed intervals of the form $[\alpha, \beta)$ so $\mu_f^*(E) \leq \nu(E)$.

Let $\epsilon > 0$. Assume $\mu_f^*(E) < \infty$. There exist $\{[a_i, b_i)\}$ such that $E \subseteq \bigcup_{i=1}^{\infty}[a_i, b_i)$ and
$$\sum_{i=1}^{\infty} \mu_f[a_i, b_i) \leq \mu_f^*(E) + \epsilon.$$
For each i there exists $\delta_i > 0$ such that f is continuous at $a_i - \delta_i < b_i$ and
$$f(a_i - \delta_i) - f(a_i) < \epsilon/2^i.$$
Then $E \subseteq \bigcup_{i=1}^{\infty}(a_i - \delta_i, b_i)$ and
$$\begin{aligned}\nu(E) &\leq \sum_{i=1}^{\infty} \mu_f(a_i - \delta_i, b_i) = \sum_{i=1}^{\infty} \mu_f[a_i - \delta_i, b_i) \\ &\leq \sum_{i=1}^{\infty}(\mu_f[a_i, b_i) + \mu_f(a_i - \delta_i, a_i)) \leq \mu_f^*(E) + 2\epsilon\end{aligned}$$
(Exer. 6). Hence, $\nu(E) \leq \mu_f^*(E)$.

A similar argument works if $\mu_f^*(E) = \infty$.

Employing the arguments as in Theorem 2.5.3 and Corollary 2.5.4, we obtain the analogues of those results for Lebesgue measure.

Theorem 2 *Let $E \subseteq \mathbf{R}$.*

(i) *For every $\epsilon > 0$ there exists open $G \supset E$ such that $\mu_f(G) \leq \mu_f^*(E) + \epsilon$.*

(ii) *There exists a $\mathcal{G}_\delta$ $H \supset E$ such that $\mu_f(H) = \mu_f^*(E)$.*

Corollary 3 *Let $E \subset \mathbf{R}$. The following are equivalent:*

(i) *E is μ_f^* measurable.*

(ii) *For every $\epsilon > 0$ there exists open $G \supset E$ such that $\mu_f^*(G \backslash E) < \epsilon$.*

(iii) *For every $\epsilon > 0$ there exists closed $F \subset E$ such that $\mu_f^*(E \backslash F) < \epsilon$.*

(iv) *There exists a $\mathcal{G}_\delta$ $H \supset E$ such that $\mu_f^*(H \backslash E) = 0$.*

(v) *There exists an $\mathcal{F}_\sigma$ $K \subset E$ such that $\mu_f^*(E \backslash K) = 0$.*

(vi) *There exists a Borel set B and a μ_f^* null set Z such that $E = B \cup Z$.*

2.6. LEBESGUE-STIELTJES MEASURES

Exercise 1. Show $(\mathbf{R}, \mathcal{M}(\mu_f^*), \mu_f)$ is the completion of $(\mathbf{R}, \mathcal{B}(\mathbf{R}), \mu_f)$.

Exercise 2. Show the map $f \to \mu_f$ which associates an increasing left continuous, normalized function with its Lebesgue-Stieltjes measure is additive and positive homogeneous.

Exercise 3. Let $f(x)$ equal the greatest integer less than x. Compute $\mathcal{M}(\mu_f^*)$.

Exercise 4. Show $\mu_f = \mu_g$ if and only if $f - g =$ constant.

Exercise 5. Show a Lebesgue-Stieltjes measure restricted to the Borel sets is regular.

Exercise 6. Let f be increasing and left continuous. For $a \in \mathbf{R}$ show $\mu_f(\{a\}) = f(a^+) - f(a)$ and f is continuous if and only if $\mu_f(\{a\}) = 0$.

Exercise 7. Describe μ_f for:

(a) $f(t) = \begin{cases} t+1 & t > 0 \\ 0 & t \le 0 \end{cases}$

(b) $f(t) = \begin{cases} t+1 & t > 0 \\ t & t \le 0 \end{cases}$

(c) $f(t) = \begin{cases} 0 & t \le 0 \\ n+t & n \le t < n+1 \end{cases} \quad n = 0, 1, \ldots$

2.6.1 Hewitt-Yosida Decomposition for Lebesgue-Stieltjes Measures

In this section we consider Lebesgue-Stieltjes measures on the interval $[0, 1)$ as discussed in §2.6 for measures on $\mathbf{R}$. We give examples of purely finitely set functions and also give an example of the Hewitt-Yosida decomposition of §2.4.1 for such Lebesgue-Stieltjes set functions.

Let $S = [0, 1)$ and let $\mathcal{S}$ be the semi-algebra of all subsets of S of the form $[a, b)$, $0 \le a < b \le 1$ and $\mathcal{A}$ the algebra generated by $\mathcal{S}$.

We give a simple example of a purely finitely additive set function on $\mathcal{A}$. Fix $0 < t \le 1$. Define $\lambda_t : \mathcal{A} \to \mathbf{R}$ by $\lambda_t(A) = 1$ if $\exists \delta > 0$ such that $[t - \delta, t) \subseteq A$ and $\lambda_t(A) = 0$ otherwise. Then λ_t is finitely additive (Exer. 2.2.8). Also, λ_t is purely finitely additive; for if $0 \le \nu \le \lambda_t$ and ν is countably additive on $\mathcal{A}$, then $\nu[t, 1) = 0$ and if $0 \le a < t - \delta$ for some $\delta > 0$, then $\nu[a, t - \delta) = 0$. Thus, if

$$I_1 = [0, t/2), I_2 = [t/2, 3t/4), \dots,$$

then $[0, t) = \bigcup\limits_{j=1}^{\infty} I_j$ so $\nu[0, t) = \sum\limits_{j} \nu(I_j) = 0$ and $\nu = 0$.

We next claim that the λ_t, $0 < t \le 1$, are the only purely finitely additive set functions which take on the values 0 and 1.

Theorem 1 *Let $\mu : \mathcal{A} \to \mathbf{R}$ be bounded, finitely additive and have range $\{0, 1\}$. Then $\exists t \in [0, 1)$ such that $\mu = \delta_t$ or $\mu = \lambda_t$.*

Proof: If $0 = a_0 < a_1 < \dots < a_n = 1$ is a partition of $[0, 1)$, then exactly one of the subintervals $[a_{j-1}, a_j)(j = 1, \dots, n)$ has μ-measure 1 while the remaining subintervals have μ-measure 0. Thus, starting with 0, 1/2, 1, then 0, 1/4, 1/2, 3/4, 1, etc., we obtain a nested sequence of intervals $[a_i, b_i)$ with $\mu[a_i, b_i) = 1$, $b_i - a_i = 1/2^i$. Let $t = \lim a_i = \lim b_i$. If $b_i > t$ $\forall i$ and $\mu[t, b_i) = 1 \forall i$, then $\mu = \delta_t$. On the other hand, if $[t, b_{i_0}) = \emptyset$ or $\mu[t, b_{i_0}) = 0$ for some i_0, then

$$\mu[a_i, b_i) = \mu[a_i, t) + \mu[t, b_i) = \mu[a_i, t) = 1$$

for all $i \ge i_0$ and $\mu = \lambda_t$.

Theorem 2 *Let $\mu : \mathcal{A} \to \mathbf{R}$ be bounded, finitely additive. Then $\exists \{t_i\} \subseteq [0, 1)$, $a_i \in \mathbf{R}$ with $\sum\limits_{i} |a_i| < \infty$ and $\nu : \mathcal{A} \to \mathbf{R}$ countably additive such that $\mu = \nu + \sum\limits_{i} a_i \lambda_{t_i}$.*

Proof: By the Jordan Decomposition (§2.2.1), we may assume $\mu \ge 0$. Let f be the distribution function of μ as in §2.6, $f(t) = \mu[0, t)$, $0 < t < 1$, and $f(0) = 0$. Then f is increasing with

$$\lim_{t \to 1^-} f(t) = f(1^-) \le \mu[0, 1) < \infty.$$

2.6. LEBESGUE-STIELTJES MEASURE

Let $\{t_i\}$ be the countable set for which the inequality $f(t^-) < f(t)$ holds and set $a_i = f(t_i) - f(t_i^-)$. Then

$$a_i = \lim_{\epsilon \to 0^+} \mu[t_i - \epsilon, \, t_i) \text{ and } \sum_i a_i \leq \mu[0, \, 1) < \infty.$$

Put

$$\nu = \mu - \sum_i a_i \lambda_{t_i}.$$

Then ν is non-negative and finitely additive, and we claim that ν is countably additive. To see this, let g be the distributive function of ν, $g(t) = \nu[0, \, t)$ for $0 < t \leq 1$ and $g(0) = 0$. Then g is left continuous so as in Example 2.2.9, ν is countably additive.

Since the measure $\sum a_i \nu_{t_i}$ in Theorem 1 is purely finitely additive (Exer. 1), the decomposition of μ given in Theorem 2 is exactly the Hewitt-Yosida Decomposition of Theorem 2.4.1.2.

Exercise 1. Show the measure $\sum a_i \lambda_{t_i}$ of Theorem 2 is purely finitely additive.

Exercise 2. Show there is a 1-1 correspondence between the finitely additive real-valued set functions μ defined on $\mathcal{A}$ and the real-valued functions $f : [0, \, 1] \to \mathbf{R}$ which satisfy $f(0) = 0$. (See Exer. 2.2.10.) Show μ is bounded if and only if f is of bounded variation.

2.7 Regular Measures

In this section we consider the properties of Lebesgue measure given in Corollary 2.5.4. We give a slightly more general definition of regularity than that given in §2.5. Let S be a Hausdorff topological space and $\mathcal{B}(S)$ the class of Borel sets of S. Let $\mathcal{A}$ be an algebra of subsets of S and $\mu : \mathcal{A} \to [0, \infty]$ finitely additive.

Definition 1 $E \in \mathcal{A}$ is inner regular *(with respect to μ) if*

$$\mu(E) = \sup\{\mu(K) : K \subset E, K \in \mathcal{A}, K \text{ compact}\}$$

and E outer regular if

$$\mu(E) = \inf\{\mu(U) : U \supset E, U \in \mathcal{A}, U \text{ open}\};$$

E is regular if E is both inner and outer regular.

A *Borel measure* on S is a measure defined on $\mathcal{B}(S)$ which is finite on compact sets; a Borel measure is *regular* if every Borel set is outer regular and every open set is inner regular.

Proposition 2 *Let μ be a regular Borel measure. Then every σ-finite Borel set B is inner regular.*

Proof: First assume $\mu(B) < \infty$. Let $\epsilon > 0$. Pick V open such that $V \supset B$ and $\mu(V) < \mu(B) + \epsilon$. Pick W open such that $V \backslash B \subset W \subset V$ and

$$\mu(W) < \mu(V \backslash B) + \epsilon = \mu(V) - \mu(B) + \epsilon.$$

Pick K compact such that $K \subset V$ and $\mu(V) < \mu(K) + \epsilon$. Set $K' = K \backslash W$. Then K' is compact, $K' \subset B$ and

$$0 \leq \mu(B) - \mu(K') = \mu(B \backslash K') \leq \mu(V \backslash K') = \mu((V \backslash K) \cup W) \leq \mu(V) - \mu(K) + \mu(W) < 3\epsilon.$$

If $\mu(B) = \infty$, let $B = \bigcup_{k=1}^{\infty} B_k$, $B_k \uparrow$, $\mu(B_k) < \infty$. Let $r > 0$. Choose N such that $\mu(B_N) > r$. Choose K compact such that $K \subset B_N$ and $\mu(K) > r$. Hence,

$$\mu(B) = \infty = \sup\{\mu(K) : K \text{ compact}, K \subset B\}.$$

A subset $E \subset S$ is called *σ-compact* if E is the countable union of compact subsets of S. For example, $\mathbf{R}^n$ is σ-compact. If μ is a Borel measure, then any σ-compact subset is σ-finite. From Proposition 2, we have

Corollary 3 *If μ is a regular Borel measure which is σ-finite, then every Borel set is regular. If S is σ-compact, then every Borel set is regular.*

2.7. REGULAR MEASURES

Proposition 4 *Let μ be a Borel measure on S. If G is σ-compact, then G is inner regular.*

Proof: Let $G = \bigcup_{j=1}^{\infty} K_j$, where each K_j is compact. If $K'_j = \bigcup_{i=1}^{j} K_i$, then K'_j is compact and $K'_j \uparrow G$ so $\mu(K'_j) \uparrow \mu(G)$.

Lemma 5 *Let S be such that every open set is σ-compact. If μ is a finite Borel measure on S, then for each $B \in \mathcal{B}(S)$*

(i) $\mu(B) = \inf\{\mu(V) : V \text{ open}, V \supset B\}$ *and*

(ii) $\mu(B) = \sup\{\mu(F) : F \text{ closed}, F \subset B\}$.

Proof: Let $\sum$ be all $B \in \mathcal{B}(S)$ satisfying (i) and (ii). By hypothesis and Proposition 4 every open set satisfies (i) and (ii) so if we show that $\sum$ is a σ-algebra, then $\sum = \mathcal{B}(S)$ and the proof is complete.

Let $E \in \sum$ and $\epsilon > 0$. There exists closed F, open V such that $F \subset E \subset V$ and $\mu(V) - \mu(F) = \mu(V \setminus F) < \epsilon$. Then $F^c \supset E^c \supset V^c$, F^c is open, V^c is closed and

$$\mu(F^c \setminus V^c) = \mu(V \setminus F) < \epsilon$$

so $E^c \in \sum$.

Let $\{E_k\} \subset \sum$ and set $E = \bigcup_{k=1}^{\infty} E_k$. For each k there exists closed F_k, open V_k such that $F_k \subset E_k \subset V_k$ and $\mu(V_k \setminus F_k) < \epsilon/2^k$. Set $V = \bigcup_{k=1}^{\infty} V_k$, $F = \bigcup_{k=1}^{\infty} F_k$ so V is open and $F \subset E \subset V$. Moreover, $\mu(V \setminus F) \leq \sum_{k=1}^{\infty} \mu(V_k \setminus F_k) < \epsilon$. Now $F'_k = \bigcup_{j=1}^{k} F_j$ is closed,

$$V \setminus F = V \setminus \bigcup_{k=1}^{\infty} F'_k = \bigcap_{k=1}^{\infty} (V \setminus F'_k)$$

and $V \setminus F'_k \downarrow V \setminus F$ so $\mu(V \setminus F) = \lim \mu(V \setminus F'_k) < \epsilon$. Hence, there exists k such that $\mu(V \setminus F'_k) < \epsilon$ and $E \in \sum$.

Theorem 6 *Let S be such that every open set is σ-compact and μ a Borel measure on S. Then every Borel set is regular; in particular, μ is regular.*

Proof: Every open set is inner regular by hypothesis and Proposition 4.

If μ is finite, every Borel set is outer regular by Lemma 5(i). Assume μ is infinite and let $S = \bigcup_{k=1}^{\infty} U_k$ where each U_k is open and $\mu(U_k) < \infty$. For each k set $\mu_k(B) = \mu(B \cap U_k)$ for $B \in \mathcal{B}(S)$. Hence, each μ_k is a finite Borel measure on S. Let $B \in \mathcal{B}(S)$ and $\epsilon > 0$. By Lemma 5, for each k there exists open $V_k \supset B$ such that

$\mu_k(V_k) < \mu_k(B) + \epsilon/2^k$ so $\mu(U_k \cap V_k \setminus B) < \epsilon/2^k$. Put $V = \bigcup_{k=1}^{\infty}(U_k \cap V_k)$ so V is open, $V \supset B$ and

$$\mu(V \setminus B) \leq \sum_{k=1}^{\infty} \mu(U_k \cap V_k \setminus B) < \epsilon$$

and B is outer regular. Hence, μ is regular.

By Corollary 3, every Borel set is inner regular.

Remark 7 Note $\mathbf{R}^n$ satisfies the hypothesis of Theorem 6 [take squares with centers having rational coordinates and rational side lengths] so every Borel measure on $\mathbf{R}^n$ is regular.

If $\mu : \mathcal{A} \to \mathbf{R}$ is a finitely additive, we say that μ is regular if $|\mu|$ is regular. From Theorem 6, we have

Corollary 8 Let $\mu : \mathcal{B}(S) \to \mathbf{R}$ be finitely additive. The following are equivalent:

(i) μ is regular

(ii) μ^+ and μ^- are regular

(iii) If $A \in \mathcal{B}(S)$ and $\epsilon > 0$, there exist compact $K \subset A$ and open $V \supset A$ such that $|\mu(B)| < \epsilon$ for every $B \in \mathcal{B}(S)$ with $B \subset V \setminus K$.

Finally, we have an interesting result of Alexanderoff on regular, finitely additive set functions.

Theorem 9 (Alexanderoff) Let $\mu : \mathcal{A} \to \mathbf{R}$ be bounded, finitely additive. If every element of $\mathcal{A}$ is both inner and outer regular, then μ is countably additive.

Proof: It suffices to show that $|\mu|$ is countably additive. Let $\{A_j\} \in \mathcal{A}$ be pairwise disjoint and such that $A = \bigcup_{j=1}^{\infty} A_j \in \mathcal{A}$. Let $\epsilon > 0$. There exists compact $F \subset A$, $F \in \mathcal{A}$, such that $|\mu|(A \setminus F) < \epsilon$. For each j there exists open $G_j \in \mathcal{A}$, $G_j \supset A_j$ such that $|\mu|(G_j \setminus A_j) < \epsilon/2^j$. Since $\bigcup_{j=1}^{\infty} G_j \supset F$, there exists n such that $\bigcup_{j=1}^{n} G_j \supset F$. Then

$$\sum_{j=1}^{\infty} |\mu|(A_j) \geq \sum_{j=1}^{\infty} |\mu|(G_j) - \epsilon \geq \sum_{j=1}^{n} |\mu|(G_j) - \epsilon \geq |\mu|(F) - \epsilon \geq \mu(A) - 2\epsilon.$$

Hence, $\sum_{j=1}^{\infty} |\mu|(A_j) \geq |\mu|(A)$.

For each n,

$$|\mu|(A) \geq |\mu|(\bigcup_{j=1}^{n} A_j) = \sum_{j=1}^{n} |\mu|(A_j)$$

so

$$|\mu|(A) \geq \sum_{j=1}^{\infty} |\mu|(A_j),$$

2.7. REGULAR MEASURES

and it follows that $|\mu|$ is countably additive.

Exercise 1. Let μ be as in Theorem 9. Show μ has unique, regular, countably additive extension to the σ-algebra generated by $\mathcal{A}$.

Exercise 2. If the measure μ is defined on a σ-algebra, show the union of a sequence of outer regular sets is outer regular and the union an increasing sequence of inner regular sets is inner regular.

Exercise 3. If μ is a finite measure on a σ-algebra, show the intersection of a sequence of inner regular sets is inner regular and the intersection a decreasing sequence of outer regular sets is outer regular.

2.8 The Nikodym Convergence and Boundedness Theorems

In this section we establish two remarkable results for signed measures which are due to Nikodym ([N1], [N2]). Our proofs are based on a theorem concerning infinite matrices due to Antosik and J. Mikusinski ([AS]) which we now establish. We begin with a simple lemma.

Lemma 1 *Let $x_{ij} \geq 0$ and $\epsilon_{ij} > 0$ for $i, j \in \mathbf{N}$. If $\lim_i x_{ij} = 0$ for each j and $\lim_j x_{ij} = 0$ for each i, then there is a subsequence $\{m_i\}$ of positive integers such that $x_{m_i m_j} < \epsilon_{ij}$ for $i \neq j$.*

Proof: Put $m_1 = 1$. There exists $m_2 > m_1$ such that $x_{m_1 m} < \epsilon_{12}$ and $x_{m m_1} < \epsilon_{21}$ for $m \geq m_2$. Then there exists $m_3 > m_2$ such that $x_{m_1 m} < \epsilon_{13}$, $x_{m_2 m} < \epsilon_{23}$, $x_{m m_1} < \epsilon_{31}$ and $x_{m m_2} < \epsilon_{32}$ for $m \geq m_3$. Continue.

Theorem 2 (Antosik-Mikusinski) *Let $x_{ij} \in \mathbf{R}$ for $i, j \in \mathbf{N}$. Suppose*

(I) $\lim_i x_{ij} = x_j$ *exists for each j.*

(II) *For each subsequence $\{m_j\}$ there is a subsequence $\{n_j\}$ of $\{m_j\}$ such that the sequence $\{\sum_{j=1}^{\infty} x_{i n_j}\}$ converges.*

Then $\lim_i x_{ij} = x_j$ uniformly for $j \in \mathbf{N}$. Also, $\lim_j x_{ij} = 0$ uniformly for $i \in \mathbf{N}$ and $\lim_i x_{ii} = 0$.

Proof: If the conclusion fails, there is a $\delta > 0$ and a subsequence $\{k_i\}$ such that $\sup_j |x_{k_i, j} - x_j| > \delta$. For notational convenience, assume $k_i = i$. Set $i_1 = 1$ and pick j_1 such that $|x_{i_1 j_1} - x_{j_1}| > \delta$. By (I) there exists $i_2 > i_1$, with $|x_{i_1 j_1} - x_{i_2 j_1}| > \delta$ and $|x_{ij} - x_j| < \delta$ for $i \geq i_2$ and $1 \leq j \leq j_1$. Now pick j_2 such that $|x_{i_2 j_2} - x_{j_2}| > \delta$ and note that $j_2 > j_1$. Continuing by induction produces two increasing sequences $\{i_k\}$ and $\{j_k\}$ such that $|x_{i_k j_k} - x_{i_{k+1} j_k}| > \delta$. Set $z_{k\ell} = x_{i_k j_\ell} - x_{i_{k+1} j_\ell}$ and note

$$|z_{kk}| > \delta. \qquad (2.1)$$

Consider the infinite matrix $M = [z_{k\ell}]$. By (I) the columns of this matrix converge to 0. By (II) the rows of the matrix $[x_{ij}]$ converge to 0 so the same holds for the matrix M. By Lemma 1 there is a subsequence $\{m_k\}$ such that $|z_{m_k m_\ell}| < 1/2^{k+\ell}$ for $k \neq \ell$. By (II) there is a subsequence $\{n_k\}$ of $\{m_k\}$ such that

$$\lim_k \sum_{\ell=1}^{\infty} z_{n_k n_\ell} = 0. \qquad (2.2)$$

2.8. THE NIKODYM CONVERGENCE AND BOUNDEDNESS THEOREMS

Then

$$|z_{n_k n_k}| \leq |\textstyle\sum_{\ell=1}^{\infty} z_{n_k n_\ell}| + \left|\textstyle\sum_{\substack{\ell=1 \\ \ell \neq k}}^{\infty} z_{n_k n_\ell}\right| \leq |\textstyle\sum_{\ell=1}^{\infty} z_{n_k n_\ell}| + \textstyle\sum_{\ell \neq k} |z_{n_k n_\ell}| \quad (2.3)$$

$$\leq |\textstyle\sum_{\ell=1}^{\infty} z_{n_k n_\ell}| + \textstyle\sum_{\ell=1}^{\infty} 1/2^{k+\ell} = |\textstyle\sum_{\ell=1}^{\infty} z_{n_k n_\ell}| + 1/2^k.$$

Now the first term on the right hand side of (3) goes to 0 as $k \to \infty$ by (2) and since the second term obviously goes to 0, $\lim_k z_{n_k n_k} = 0$ contradicting (1).

The uniform convergence of the limit, $\lim_i x_{ij} = x_i$, and the fact that $\lim_j x_{ij} = 0$ for each i implies that the double limit $\lim_{i,j} x_{ij}$ exists and is equal to 0 so, in particular, $\lim_j x_{ij} = 0$ uniformly for $i \in \mathbb{N}$ and $\lim_i x_{ii} = 0$.

We will use this matrix theorem of Antosik and Mikusinski below in the proof of the Nikodym Convergence Theorem. Note the theorem gives a sufficient condition for the diagonal of an infinite matrix to converge to 0 so it is sometimes referred to as a Diagonal Theorem.

We now establish the Nikodym Convergence Theorem. Let Σ be a σ-algebra of subsets of S and $\mu_i : \Sigma \to \mathbf{R}$, $i \in \mathbf{N}$, a sequence of (finite) signed measures. Nikodym's Convergence Theorem then asserts that the pointwise limit of a sequence of finite signed measures is a signed measure. For this we first require a definition and a preliminary lemma.

Definition 3 *The sequence of signed measures, $\{\mu_i\}$, is said to be* uniformly countably additive *if the series $\{\sum_{j=1}^{\infty} \mu_i(E_j)\}_i$ converge uniformly for every pairwise disjoint sequence $\{E_j\} \subset \Sigma$.*

Lemma 4 *The following are equivalent:*

(i) $\{\mu_i\}$ *is uniformly countably additive.*

(ii) *For each decreasing sequence $\{E_j\}$ from Σ with $E_j \downarrow \emptyset$, $\lim_j \mu_i(E_j) = 0$ uniformly for $i \in \mathbf{N}$.*

(iii) *For each pairwise disjoint sequence $\{E_j\}$ from Σ, $\lim_j \mu_i(E_j) = 0$ uniformly for $i \in \mathbf{N}$.*

Proof: The equivalence of (i) and (ii) is exactly like the proof of Proposition 2.2.5. Certainly (i) implies (iii).

Suppose (iii) holds but (ii) fails to hold. Then we may assume (by passing to a subsequence if necessary) that there exist a $\delta > 0$ and a decreasing sequence $F_j \downarrow \emptyset$ with $|\mu_j(F_j)| > \delta$ for every j. There exists k_1 such that $|\mu_1(F_{k_1})| < \delta/2$. Then there exists $k_2 > k_1$ such that $|\mu_{k_1}(F_{k_2})| < \delta/2$. Continuing by induction produces an increasing sequence $\{k_j\}$ such that $|\mu_{k_j}(F_{k_j+1})| < \delta/2$. If $E_j = F_{k_j} \setminus F_{k_j+1}$, then $\{E_j\}$ is a pairwise disjoint sequence from $\sum$ with

$$|\mu_{k_j}(E_j)| \geq |\mu_{k_j}(F_{k_j})| - |\mu_{k_j}(F_{k_j+1})| > \delta/2$$

which contradicts (iii).

Theorem 5 (Nikodym Convergence Theorem) *Suppose* $\lim_i \mu_i(E) = \mu(E)$ *exists for every* $E \in \sum$, *then*

(i) $\{\mu_i\}$ *is uniformly countably additive and*

(ii) μ *is countably additive (i.e., μ is a signed measure).*

Proof: (i): Let $\{E_j\}$ be a pairwise disjoint sequence from $\sum$. Consider the matrix $M = [\mu_i(E_j)]$. The columns of M converge by hypothesis and if $\{m_j\}$ is any increasing sequence of positive integers,

$$\lim_i \sum_{j=1}^\infty \mu_i(E_{m_j}) = \lim_i \mu_i(\bigcup_{j=1}^\infty E_{m_j})$$

exists by hypothesis. Hence, M satisfies conditions (I) and (II) of Theorem 2, and $\lim_j \mu_i(E_j) = 0$ converges uniformly for $i \in \mathbf{N}$. By Lemma 4, $\{\mu_i\}$ is uniformly countably additive.

(ii): If $\{E_j\}$ is a decreasing sequence from $\sum$ with $E_j \downarrow \emptyset$, then by (i) and Lemma 4 $\lim_j \mu_i(E_j) = 0$ uniformly for $i \in \mathbf{N}$. Hence,

$$\lim_j \mu(E_j) = \lim_j \lim_i \mu_i(E_j) = \lim_i \lim_j \mu_i(E_j) = 0$$

and μ is countably additive by Proposition 2.2.5.

Remark 6 The σ-algebra assumption in Theorem 5 is important; the result is false for countably additive set functions defined on algebras (see Exer. 4 or [Sw3]).

We next turn to the Nikodym Boundedness Theorem. For this we require a technical lemma.

Lemma 7 *Let $\mathcal{A}$ be an algebra of subsets of S and $\nu_i : \mathcal{A} \to \mathbf{R}$ bounded and finitely additive for each $i \in \mathbf{N}$. Then $\{\nu_i(E) : i \in \mathbf{N}, E \in \mathcal{A}\}$ is bounded if and only if $\{\nu_i(E_i) : i \in \mathbf{N}\}$ is bounded for each pairwise disjoint $\{E_i\} \subset \mathcal{A}$.*

Proof: Suppose $\sup\{|\nu_i(E)| : i \in \mathbf{N}, E \in \mathcal{A}\} = \infty$. Note that for each $M > 0$ there is a partition (E, F) of S and an integer i such that $\min\{|\nu_i(E)|, |\nu_i(F)|\} > M$. [This follows since

$$|\nu_i(E)| > M + \sup\{|\nu_j(S)| : j \in \mathbf{N}\}$$

2.8. THE NIKODYM CONVERGENCE AND BOUNDEDNESS THEOREMS 69

implies $|\nu_i(S\backslash E)| \geq |\nu_i(E)| - |\nu_i(S)| > M$.] Hence, there exist i_1 and a partition (E_1, F_1) of S such that $\min\{|\nu_{i_1}(E_1)|, |\nu_{i_1}(F_1)|\} > 1$. Now either

$$\sup\{|\nu_i(A \cap E_1)| : i \in \mathbf{N}, A \in \mathcal{A}\} = \infty$$

or

$$\sup\{|\nu_i(A \cap F_1)| : i \in \mathbf{N}, A \in \mathcal{A}\} = \infty.$$

Pick whichever of E_1 or F_1 satisfies this condition and label it B_1 and set $A_1 = S\backslash B_1$. Now treat B_1 as S above to obtain a partition (A_2, B_2) of B_1 and an $i_2 > i_1$ satisfying $|\nu_{i_2}(A_2)| > 2$ and $\sup\{|\nu_i(A \cap B_2)| : i \in \mathbf{N}, A \in \mathcal{A}\} = \infty$. Proceeding by induction produces a subsequence $\{i_j\}$ and a pairwise disjoint sequence $\{A_j\}$ such that $|\mu_{i_j}(A_j)| > j$. This establishes the sufficiency; the necessity is clear.

Theorem 8 (Nikodym Boundedness Theorem) *If $\{\mu_i(E) : i \in \mathbf{N}\}$ is bounded for each $E \in \sum$, then $\{\mu_i(E) : i \in \mathbf{N}, E \in \sum\}$ is bounded, i.e., $\{\mu_i\}$ is uniformly bounded on $\sum$ when $\{\mu_i\}$ is pointwise bounded on $\sum$.*

Proof: Let $\{E_j\}$ be a pairwise disjoint sequence from $\sum$ and let $t_i \to 0$. Then $\lim_i t_i\mu_i(E) = 0$ for each $E \in \sum$. Hence, by the Nikodym Convergence Theorem $\{t_i\mu_i\}$ is uniformly countably additive, and $\lim_i t_i\mu_i(E_i) = 0$ by Lemma 4. By Exercise 1, $\{\mu_i(E_i)\}$ is bounded, and the result follows from Lemma 7.

Remark 9 As in the Nikodym Convergence Theorem, the σ-algebra assumption in Theorem 8 is important; see Exercise 3. The result is also valid for bounded, finitely additive set functions defined on σ-algebras (see Exercise 2).

Despite the examples which show that neither the Nikodym Convergence or Boundedness Theorems are valid for algebras, there are versions of both results for set functions defined on domains which are not σ-algebras. For references to such results and a discussion of the history of these two important theorems of measure theory see [DU].

Exercise 1. Show $\{s_i\} \subset \mathbf{R}$ is bounded if and only if $\lim s_i t_i = 0$ for every sequence $\{t_i\}$ with $\lim t_i = 0$.

Exercise 2. Use Drewnowski's Lemma to show that Theorem 8 is valid for bounded, finitely additive set functions defined on a σ-algebra.

Exercise 3. Let $\mathcal{A}$ be the algebra of subsets of $\mathbf{N}$ which are either finite or have finite complements. Define $\delta_n : \mathcal{A} \to \mathbf{R}$ by $\delta_n(E) = 1$ if $n \in E$ and 0 otherwise. Define $\mu_n : \mathcal{A} \to \mathbf{R}$ by

$$\mu_n(E) = n(\delta_{n+1}(E) - \delta_n(E))$$

if E is finite and

$$\mu_n(E) = -n(\delta_{n+1}(E) - \delta_n(E))$$

otherwise. Show each μ_n is bounded, countably additive and $\{\mu_n\}$ is pointwise bounded on $\mathcal{A}$ but is not uniformly bounded on $\mathcal{A}$.

Exercise 4. Let $S = [0,1)$ and $\mathcal{A}$ the algebra generated by intervals of the form $[a,b)$, $0 \leq a \leq b \leq 1$. Define μ_n on $\mathcal{A}$ by $\mu_n(A) = nm(A \cap [0,1/n))$. Show μ_n is countably additive, $\lim \mu_n(A) = \mu(A)$ exists for every $A \in \mathcal{A}$ and μ is not countably additive.

Chapter 3

Integration

3.1 Measurable Functions

We now begin our study of the Lebesgue integral by introducing the class of functions which will be considered. Recall that in section 1.3 when we were describing Lebesgue's definition of the integral a necessary condition for a bounded function $f : [a, b] \to \mathbf{R}$ to be integrable was that the set $\{t \in [a, b] : \alpha < f(t) < \beta\}$ had to be (Lebesgue) measurable for each α, β. The functions which satisfy this condition are called measurable functions; we now introduce and study this class of functions.

Let $\sum$ be a σ-algebra of subsets of a set S.

Definition 1 *An extended real-valued function* $f : S \to \mathbf{R}^*$ *is* $\sum$*-measurable, or simply measurable if* $\sum$ *is understood, if* $\{t \in S : f(t) < a\} \in \sum$ *for every* $a \in \mathbf{R}$.

Proposition 2 *Let* $f : S \to \mathbf{R}^*$. *The following are equivalent*

(i) *f is measurable,*

(ii) $\{t : f(t) \geq a\} \in \sum$ *for every* $a \in \mathbf{R}$,

(iii) $\{t : f(t) > a\} \in \sum$ *for every* $a \in \mathbf{R}$,

(iv) $\{t : f(t) \leq a\} \in \sum$ *for every* $a \in \mathbf{R}$,

(v) *the same as (i), (ii), (iii) or (iv) except with "for every $a \in \mathbf{R}$" replaced by "for every a in any dense subset of $\mathbf{R}$".*

Proof: (i)$\Rightarrow$(ii): $\{t : f(t) \geq a\} = S \setminus \{t : f(t) < a\}$.
(ii)$\Rightarrow$(iii): $\{t : f(t) > a\} = \bigcup_{n=1}^{\infty} \{t : f(t) \geq a + 1/n\}$.
(iii)$\Rightarrow$(iv): $\{t : f(t) \leq a\} = S \setminus \{t : f(t) > a\}$.
(iv)$\Rightarrow$(i): $\{t : f(t) < a\} = \bigcup_{n=1}^{\infty} \{t : f(t) \leq a - 1/n\}$.

For (v) suppose $D \subset \mathbf{R}$ is dense and $\{t : f(t) \leq d\} \in \Sigma$ for every $d \in D$. Let $a \in \mathbf{R}$. Pick $\{d_n\} \subset D$ such that $d_n \uparrow a$, $d_n \neq a$. Then

$$\{t : f(t) < a\} = \bigcup_{n=1}^{\infty} \{t : f(t) \leq d_n\} \in \Sigma$$

so (iv) holds. The other cases are similar.

Other characterizations of measurable functions are given in Exercise 1.

Corollary 3 *If $f : S \to \mathbf{R}^*$ is measurable, then $\{t : f(t) = a\} \in \Sigma$ for every $a \in \mathbf{R}^*$.*

Proof: For $a \in \mathbf{R}$,

$$\{t : f(t) = a\} = \bigcap_{n=1}^{\infty} \{t : a - 1/n < f(t) < a + 1/n\}.$$

For $a = \infty$, $\{t : f(t) = \infty\} = \bigcap_{n=1}^{\infty} \{t : f(t) > n\}$, and the case $a = -\infty$ is similar.

Example 4 The converse of Corollary 3 is false. Let P be a subset of $(0, 1)$ which is not Lebesgue measurable and set $Q = (0, 1) \setminus P$. Define $f : (0, 1) \to \mathbf{R}$ by $f(t) = t$ if $t \in Q$ and $f(t) = -t$ if $t \in P$. Then f is 1-1 so $\{t : f(t) = a\}$ is Lebesgue measurable for every $a \in \mathbf{R}^*$, but f is not Lebesgue measure.

Concerning the algebraic properties of measurable functions, we have

Proposition 5 *Let $f_1, f_2 : S \to \mathbf{R}^*$ be measurable. Assume $f_1 + f_2$ and $f_1 f_2$ are defined on S. Then*

$$f_1 + f_2, a f_1, f_1 f_2, f_1 \vee f_2, f_1 \wedge f_2$$

are measurable functions.

Proof: First, $f_1 + f_2$ is measurable. If $a \in \mathbf{R}$ and $f_1(t) + f_2(t) < a$, then $f_1(t) < a - f_2(t)$ and, therefore, there is a rational r such that $f_1(t) < r < a - f_2(t)$. Hence,

$$\{t : f_1(t) + f_2(t) < a\} = \bigcup_{r \in \mathbf{Q}} \{t : f_1(t) < r\} \cap \{t : f_2(t) < a - r\}.$$

That $a f_1$ is measurable for $a \in \mathbf{R}$ is easily checked.

To show $f_1 f_2$ is measurable, first note f_1^2 is measurable since

$$\{t : f_1^2(t) > a\} = \{t : f_1(t) > \sqrt{a}\} \cup \{t : f_1(t) < -\sqrt{a}\}$$

for $a \geq 0$ and $\{t : f_1^2(t) > a\} = S$ for $a < 0$. But $f_1 f_2 = [(f_1 + f_2)^2 - f_1^2 - f_2^2]/2$.

Since

$$\{t : f_1 \vee f_2(t) > a\} = \{t : f_1(t) > a\} \cup \{t : f_2(t) > a\}, f_1 \vee f_2$$

is measurable and, similarly, $f_1 \wedge f_2$.

3.1. MEASURABLE FUNCTIONS

Definition 6 *Let $f : S \to \mathbf{R}^*$. Define $f^+ = f \vee 0$, $f^- = (-f) \vee 0$. Hence, $f = f^+ - f^-$ and $|f| = f^+ + f^-$.*

Corollary 7 *$f : S \to \mathbf{R}^*$ is measurable if and only if f^+ and f^- are measurable. If f is measurable, then $|f|$ is measurable.*

The converse of the second statement in Corollary 7 is false (Exercise 6). Concerning sequences of measurable functions, we have

Proposition 8 *Let $f_k : S \to \mathbf{R}^*$ be measurable for each $k \in \mathbf{N}$. Then*

$$f = \sup\{f_k : k \in \mathbf{N}\}, g = \inf\{f_k : k \in \mathbf{N}\}, \overline{\lim} f_k, \underline{\lim} f_k$$

are measurable.

Proof: For $a \in \mathbf{R}$, $\{t : f(t) > a\} = \bigcup_{k=1}^{\infty} \{t : f_k(t) > a\}$ so f is measurable. Similarly, g is measurable. The other two statements follow immediately.

Corollary 9 *If $f_k : S \to \mathbf{R}^*$ is measurable for each $k \in \mathbf{N}$ and if $\{f_k\}$ converges pointwise to the function $f : S \to \mathbf{R}^*$, then f is measurable.*

We next consider the measurability of compositions.

Definition 10 *A function $f : \mathbf{R}^n \to \mathbf{R}^*$ is called a Borel function if*

$$\{t : f(t) < a\} \in \mathcal{B}(\mathbf{R}^n)$$

for every $a \in \mathbf{R}$, i.e., if f is $\mathcal{B}(\mathbf{R}^n)$-measurable.

For example, any continuous real-valued function on $\mathbf{R}^n$ is a Borel function (Exer. 2.5.11). Any Borel function is obviously Lebesgue measurable.

Proposition 11 *Let $f : S \to \mathbf{R}$ be measurable and $g : \mathbf{R} \to \mathbf{R}^*$ be a Borel function. Then $g \circ f$ is measurable.*

Proof: For $a \in \mathbf{R}$, $(g \circ f)^{-1}(a, \infty) = f^{-1}(g^{-1}(a, \infty))$ so the result follows from Exercise 1.

It is not, in general, true that the composition of measurable functions is measurable.

Example 12 Let K be the Cantor set in $[0,1]$ and let K_ϵ be an ϵ-Cantor set with $0 < \epsilon < 1$. Let $f : [0,1] \to [0,1]$ be the continuous function with maps the open intervals in the complement of K_ϵ onto the open intervals in the complement of K constructed in Example 2.5.9. Let P be a subset of K_ϵ which is not Lebesgue measurable and $P' = f(P)$. Since K has measure 0, P' is measurable. Observe that $C_{P'} \circ f = C_P$ is not measurable while $C_{P'}$ is measurable and f is even continuous.

Almost Everywhere:

Definition 13 *Let μ be a measure on $\sum$. A statement about the points in a subset $E \subset S$ is said to hold μ-almost everywhere [$\mu-a.e.$] in E if the statement is true for all of the points of E except possibly for the points in a subset of E of μ measure 0. For example, if $f : S \to \mathbf{R}^*$, to say that $f = 0$ $\mu-a.e.$ means that $\mu\{t : f(t) \neq 0\} = 0$.*

Proposition 14 *Let μ be a complete measure and let $f, g : S \to \mathbf{R}^*$. If f is $\sum$-measurable and $f = g$ $\mu-a.e.$ in S, then g is $\sum$-measurable.*

Proof: Let $Z = \{t : f(t) \neq g(t)\}$. Then $\mu(Z) = 0$. For $a \in \mathbf{R}$,

$$\{t : g(t) < a\} = \{t : f(t) < a\} \cup \{t \in Z : g(t) < a\} \setminus \{t \in Z : f(t) \geq a\}$$

so g is $\sum$-measurable.

Corollary 15 *Let μ be a complete measure and $f_k, f : S \to \mathbf{R}^*$ for $k \in \mathbf{N}$. If each f_k is $\sum$-measurable and $\{f_k\}$ converges pointwise to f $\mu-a.e.$, then f is $\sum$-measurable.*

Proof: $g = \overline{\lim} f_k$ is $\sum$-measurable (Proposition 7) and $f = g$ $\mu - a.e.$ so the result follows from Proposition 14.

Without the completeness assumption on the measure μ, Proposition 14 and Corollary 15 may fail (Exercise 4).

Finally, we consider an important result on a.e. convergence which is due to Egoroff. Let $f_k, f : S \to \mathbf{R}$ be $\sum$-measurable functions.

Definition 16 *The sequence $\{f_k\}$ converges μ-almost uniformly to f if for every $\epsilon > 0$ there exists $E \in \sum$ such that $\mu(S \setminus E) < \epsilon$ and $f_k \to f$ uniformly on E.*

Clearly, if $f_k \to f$ uniformly on S, then $\{f_k\}$ converges μ-almost uniformly for any measure μ, but the converse does not hold (Exercise 11). If $\{f_k\}$ converges μ-almost uniformly to f, then $\{f_k\}$ converges $\mu-a.e.$ to f (Exercise 13), but the converse is false (Example 19). However, if μ is a finite measure, $\mu-a.e.$ convergence does imply μ-almost uniform convergence; this is Egoroff's Theorem which we now prove.

For $\sigma > 0$, set $E_k(\sigma) = \{t : |f_k(t) - f(t)| \geq \sigma\}$.

Proposition 17 (i) $\{f_k\}$ converges μ-almost uniformly to $f \Leftrightarrow$ for every

$$\sigma > 0 \quad \lim_n \mu(\bigcup_{m \geq n} E_m(\sigma)) = 0.$$

(ii) $\{f_k\}$ converges $\mu-a.e.$ to $f \Leftrightarrow$ for every $\sigma > 0$, $\mu(\bigcap_{n=1}^{\infty} \bigcup_{m=n}^{\infty} E_m(\sigma)) = 0$.

3.1. MEASURABLE FUNCTIONS

Proof: (i): $\Rightarrow$: Let $\epsilon > 0$. There exists $E \in \Sigma$ such that $\mu(S \setminus E) < \epsilon$ and $f_k \to f$ uniformly on E. Thus, there exists N such that $E \subset \bigcap_{m=N}^{\infty} E_m(\sigma)^c$ so $E^c \supset \bigcup_{m=N}^{\infty} E_m(\sigma)$. For $n \geq N$,
$$\mu(\bigcup_{m=n}^{\infty} E_m(\sigma)) \leq \mu(E^c) < \epsilon.$$

$\Leftarrow$: For every p there exists N_p such that $\mu(\bigcup_{m \geq N_p} E_m(1/p)) < \epsilon/2^p$. Set

$$F = \bigcup_{p=1}^{\infty} \bigcup_{m=N_p}^{\infty} E_m(1/p).$$

Then $\mu(F) < \epsilon$ and $f_k \to f$ uniformly on $E = F^c = \bigcap_{p=1}^{\infty} \bigcap_{m=N_p}^{\infty} E_m(1/p)^c$.

(ii): Let $A = \{t : f_k(t) \to f(t)\}$. Then

$$A = \bigcap_{\sigma > 0} \bigcup_{n=1}^{\infty} \bigcap_{m=n}^{\infty} E_m(\sigma) = \bigcap_{p=1}^{\infty} \bigcup_{n=1}^{\infty} \bigcap_{m=n}^{\infty} E_m(1/p). \tag{1}$$

Since $f_k \to f$ μ-a.e. if and only if $\mu(A^c) = 0$, (ii) follows from (1).

Theorem 18 (Egoroff) *If μ is finite and $f_k \to f$ $\mu - a.e.$, then $f_k \to f$ μ-almost uniformly.*

Proof: Set $A_n = \bigcup_{m=n}^{\infty} E_m(\sigma)$ and note $A_n \downarrow$. Since μ is finite, $\lim \mu(A_n) = \mu\left(\bigcap_{n=1}^{\infty} A_n\right)$ by 2.2.6 so the result follows from Proposition 17.

Example 19 The finiteness requirement in Egoroff's Theorem cannot be dropped. Consider $f_k = C_{[k,\infty)}$ in **R** with Lebesgue measure. $\{f_k\}$ converges pointwise to 0 but for every k $f_k(t) = 1$ for t in a set with infinite Lebesgue measure.

Example 20 The conclusion in Egoroff's Theorem cannot be improved to read, "$f_k \to f$ uniformly on a set E with $\mu(S \setminus E) = 0$". Let $S = (0, 1)$ with Lebesgue measure and $f_k = C_{(0,1/k]}$. Then $\{f_k\}$ converges pointwise to 0. Suppose $E \subset S$ has $m(E) = 1$. Then for every k there exists $t_k \in E \cap (0, 1/k)$ so $f_k(t_k) = 1$ and $\{f_k\}$ does not converge uniformly to 0 on E.

For an interesting extension of Egoroff's Theorem, see [Ba2].

Exercise 1. Let $f : S \to \mathbf{R}^*$. Show f is Σ-measurable if and only if $f^{-1}(G) \in \Sigma$ for every open $G \subset \mathbf{R}$ and $f^{-1}(\infty) \in \Sigma$, $f^{-1}(-\infty) \in \Sigma$ if and only if $f^{-1}(B) \in \Sigma$ for every Borel set $B \subset \mathbf{R}$ and $f^{-1}(\infty) \in \Sigma$, $f^{-1}(-\infty) \in \Sigma$.

Exercise 2. Let $E \subset S$. Show C_E is Σ-measurable if and only if $E \in \Sigma$.

Exercise 3. Let $f : S \to \mathbf{R}^*$ and $a > 0$. If f is measurable, show $|f|^a$ is measurable (agree $|\infty|^a = \infty$). Show $1/f$ is measurable.

Exercise 4. Show that Proposition 14 (Corollary 15) is false if completeness is dropped. [Let B be a Borel set of measure 0 and $Z \subset B$ a Lebesgue measurable subset which is not a Borel set (Example 12). Define $g : \mathbf{R} \to \mathbf{R}$ by $g(t) = 1$ if $t \in \mathbf{R}\backslash B$, $g(t) = 2$ if $t \in B\backslash Z$ and $g(t) = 3$ if $t \in Z$.]

Exercise 5. Let $f_k : S \to \mathbf{R}^*$ be measurable. Show $\{t : \lim f_k(t) \text{ exists}\} \in \sum$.

Exercise 6. Give an example where $|f|$ is measurable but f is not.

Exercise 7. If $f : S \to \mathbf{R}$ is measurable, show $t \to sign\ f(t)$ is measurable.

Exercise 8. Show if $f : \mathbf{R} \to \mathbf{R}$ is continuous m–a.e., then f is Lebesgue (Borel) measurable.

Exercise 9. Show if $f : \mathbf{R} \to \mathbf{R}$ is differentiable on $\mathbf{R}$, then f' is a Borel function.

Exercise 10. If $f : \mathbf{R}^n \to \mathbf{R}$ is Lebesgue measurable and $a \in \mathbf{R}^n$ show that $x \to f(x + a)$ is Lebesgue measurable.

Exercise 11. Given an example where $\{f_k\}$ converges almost uniformly but not uniformly.

Exercise 12. Let $f, g : \mathbf{R} \to \mathbf{R}$ be continuous and $f = g$ m-a.e. Show $f = g$.

Exercise 13. Show that if $\{f_k\}$ converges μ-almost uniformly to f, then $f_k \to f$ μ–a.e. Hint: Use Proposition 17.

Exercise 14. If μ is counting measure on S, show $\{f_k\}$ converges μ-almost uniformly to f if and only if $f_k \to f$ uniformly on S.

Exercise 15. Suppose μ is finite and $\{f_k\}$ is a sequence of measurable functions which converges μ–a.e. to the measurable function f. Show there is a sequence, $\{E_j\}$, from $\sum$ such that $\mu(S\backslash \bigcup_{j=1}^{\infty} E_j) = 0$ and $f_k \to f$ uniformly on each E_j.

3.1. MEASURABLE FUNCTIONS

Exercise 16. Show that the image of a Lebesgue measurable set under a continuous function needn't be Lebesgue measurable. What about inverse images?

Exercise 17. Let $f : \mathbf{R} \to \mathbf{R}$ be Lebesgue measurable. Show there exists a set with positive Lebesgue measure on which f is bounded.

Exercise 18. Show $\sup \{f_a : a \in A\}$ ($\inf \{f_a : a \in A\}$) needn't be measurable when each f_a is measurable and A is uncountable (compare Proposition 8).

3.1.1 Approximation of Measurable Functions

In this section we show that measurable functions can be approximated by "nice" functions.

Let Σ be a σ-algebra of subsets of S and μ a measure on Σ.

Definition 1 *A function $f : S \to \mathbf{R}$ is Σ-simple, or simple if Σ is understood, if f is Σ-measurable and the range of f is finite.*

If $f : S \to \mathbf{R}$ is simple and $\mathcal{R}f = \{a_1, ..., a_n\}$ with $a_i \neq a_j$ for $i \neq j$, then $A_i = f^{-1}(a_i) \in \Sigma$, $A_i \cap A_j = \emptyset$ if $i \neq j$, and $f = \sum_{i=1}^{n} a_i C_{A_i}$; this is called the *standard representation* of f. Thus, a simple function is just a linear combination of characteristic functions of measurable sets.

We show that any measurable function can be approximated by simple functions.

Theorem 2 *Let $f : S \to \mathbf{R}^*$ be non-negative and measurable. Then there exists a sequence of non-negative simple functions, $\{\varphi_n\}$, such that $\varphi_n(t) \uparrow f(t) \forall t \in S$. If f is bounded, the convergence is uniform on S.*

Proof: For each n and $t \in S$ set

$$\varphi_n(t) = \begin{cases} (i-1)/2^n & \text{if } (i-1)/2^n \leq f(t) < i/2^n \\ n & \text{if } f(t) \geq n \end{cases} \quad i = 1, \cdots, n2^n$$

Each φ_n is non-negative, Σ-simple and $\varphi_n(t) \leq f(t)$ for all t. Also, $\varphi_{n+1}(t) \geq \varphi_n(t)$ since if $(i-1)/2^n \leq f(t) < i/2^n$, then $(2i-2)/2^{n+1} \leq f(t) < 2i/2^{n+1}$ so $\varphi_{n+1}(t) \geq (2i-2)/2^{n+1} = \varphi_n(t)$.

If $t \in S$ and $n > f(t)$, then

$$0 \leq f(t) - \varphi_n(t) < 1/2^n \tag{3.1}$$

so $\varphi_n(t) \uparrow f(t)$. The last statement follows from (1).

Corollary 3 *Let $f : S \to \mathbf{R}^*$ be measurable. Then there exists a sequence of simple functions, $\{\varphi_n\}$, such that $\{\varphi_n\}$ converges to f pointwise on S with $|\varphi_n(t)| \leq |f(t)|$ for all $t \in S$. Moreover, if f is bounded, the convergence is uniform on S.*

Proof: Apply Theorem 2 to f^+ and f^-.

We next consider the approximation of measurable functions by continuous functions. Let S be a topological space and Σ a σ-algebra of subsets of S which contains the Borel sets, $\mathcal{B}(S)$. We have an important approximation theorem due to Lusin.

3.1. MEASURABLE FUNCTIONS

Theorem 4 (Lusin) *Let μ be a measure on Σ such that for every $E \in \Sigma$ and $\epsilon > 0$ there exists an open $U \supset E$ such that $\mu(U \backslash E) < \epsilon$ and let $f : S \to \mathbf{R}$ be Σ-measurable. Given $\epsilon > 0$ there exists $H \in \Sigma$ such that $\mu(H) < \epsilon$ and $f|_{S \backslash H}$ is continuous.*

Proof: Pick a countable family of open subsets of $\mathbf{R}$, $\{V_j\}$, such that any open set in $\mathbf{R}$ is a union of elements of $\{V_j\}$. For each j, pick an open set U_j in S such that $U_j \supset f^{-1}(V_j)$ and $\mu(U_j \backslash f^{-1}(V_j)) < \epsilon/2^j$. Put $H = \bigcup_{j=1}^{\infty} (U_j \backslash f^{-1}(V_j))$. Then

$$\mu(H) < \sum_{j=1}^{\infty} \epsilon/2^j = \epsilon.$$

Set $g = f|_{S \backslash H}$.

We first claim that $g^{-1}(V_j) = U_j \cap H^c$. Clearly $g^{-1}(V_j) \subset U_j \cap H^c$. On the other hand,

$$\begin{aligned} U_j \cap H^c &\subset U_j \cap [S \backslash (U_j \backslash f^{-1}(V_j))] \\ &= U_j \cap S \cap [U_j \backslash f^{-1}(V_j)]^c \\ &= U_j \cap S \cap [(U_j)^c \cup f^{-1}(V_j)] \\ &= S \cap f^{-1}(V_j) = f^{-1}(V_j). \end{aligned}$$

Intersecting with H^c gives the desired containment and establishes the claim.

We next claim that g is continuous on H^c. Let V be open in $\mathbf{R}$. There exists $M \subset \mathbf{N}$ such that $V = \bigcup_{j \in M} V_j$. From above, $g^{-1}(V) = \bigcup_{j \in M} g^{-1}(V_j) = H^c \cap (\bigcup_{j \in M} U_j)$ so $g^{-1}(V)$ is open in H^c.

Remark 5 Note that both Lebesgue and Lebesgue-Stieltjes measures satisfy the assumptions on μ (2.5.4 and 2.6.3). If μ is also inner regular (2.5.5), then H can be taken to be a closed set. This proof of Lusin's Theorem is from [Fe].

By using the Tietze Extension Theorem, we can establish a more convenient form of Lusin's Theorem. The Tietze Extension Theorem asserts that a bounded continuous function defined on a closed subset of a normal topological space has a continuous extension to the whole space which has the same bound as the original function. A proof for topological spaces can be found in [Si], for metric spaces in [Di], and for $\mathbf{R}^n$ in [Ba].

Corollary 6 *Let μ and f be as in Theorem 4 and assume further that μ is finite, S is normal and every set in Σ is inner regular. Then for every $\epsilon > 0$ there exists a continuous function $g : S \to \mathbf{R}$ such that $\mu\{t : f(t) \neq g(t)\} < \epsilon$. If f is bounded by M, g can be chosen to be bounded by M.*

Proof: Let H be as in Theorem 4 but with ϵ replaced by $\epsilon/2$. Choose $K \subset H^c$ compact such that $\mu(H^c \backslash K) < \epsilon/2$. Then $\mu(S \backslash K) < \epsilon$ and $f|_K$ is continuous and, hence, bounded. By the Tietze Extension Theorem, $f|_K$ can be extended to a continuous function g on S with the same bound as $f|_K$.

Note Corollary 6 is applicable to Lebesgue measurable subsets of $\mathbf{R}^n$ with finite measure.

Theorem 7 *Let the assumptions be as in Corollary 6. Then there exists a sequence of continuous functions $\{f_k\}$ on S such that $f_k \to f$ μ–a.e. Moreover, if f is bounded by M, the $\{f_k\}$ can be chosen to be bounded by M.*

Proof: By Corollary 6, for each k there exist a compact K_k such that $\mu(K_k^c) < 1/2^k$ and a continuous function f_k on S such that $f_k|_{K_k} = f|_{K_k}$.

Let $Z = \overline{\lim} K_k^c = \bigcap_{j=1}^{\infty} \bigcup_{k=j}^{\infty} K_k^c$ so for every j,

$$\mu(Z) \leq \mu(\bigcup_{k=j}^{\infty} K_k^c) \leq \sum_{k=j}^{\infty} 1/2^k = 1/2^{j-1}$$

and $\mu(Z) = 0$.

If $t \in Z^c$, $t \notin \bigcup_{k=j}^{\infty} K_k^c$ for some j so $t \in K_k$ for $k \geq j$ and $f_k(t) = f(t)$ for $k \geq j$. Hence, $f_k(t) \to f(t)$ or $f_k \to f$ μ–a.e.

The last statement follows from Corollary 5.

The proof of Theorem 7 and Corollary 3.1.15 also gives the following result.

Corollary 8 *Let μ be a complete measure on $\sum$. Suppose $f : S \to \mathbf{R}$ is such that for every $\epsilon > 0$ there exists a continuous function $g : S \to \mathbf{R}$ such that*

$$\mu\{t : f(t) \neq g(t)\} < \epsilon.$$

Then f is measurable.

Corollaries 6 and 8 give a characterization of measurable functions (for certain measures) which on $\mathbf{R}$ is due to Lusin. In their treatment of measure and integration Bourbaki take this characterization as the definition of measurability for a function ([Bb]).

Exercise 1. Let $f : S \to \mathbf{R}$ be measurable. Show there exists a sequence, $\{\varphi_k\}$, of countably valued, measurable functions such that $\varphi_k \to f$ uniformly on S with $|\varphi_k| \leq |f|$. [Hint: Consider the proof of Theorem 2.]

Exercise 2. If $\{t : f(t) \neq 0\}$ has σ-finite μ measure, show the sequence $\{\varphi_k\}$ in Theorem 2 and Corollary 3 can be chosen such that $\{t : \varphi_k(t) \neq 0\}$ has finite μ measure for every k.

Exercise 3. Let $f : \mathbf{R} \to \mathbf{R}^*$ be Lebesgue measurable. Show there exists a Borel function $g : \mathbf{R} \to \mathbf{R}^*$ such that $f = g$ m–a.e.

3.2 The Lebesgue Integral

In this section we define the Lebesgue integral with respect to an arbitrary measure. In Lebesgue's original construction of the integral, he considered a bounded function f defined on an interval $[a, b]$ and approximated the function from above and below by simple functions ψ and φ, defined the integral of the simple functions ψ and φ in a natural way, and then used these two integrals to approximate the integral of f (§1.3, equation (2)). We follow basically the same procedure as Lebesgue, except that we will consider unbounded functions directly so we cannot consider upper approximations by simple functions. It will be seen that this leads to no difficulties because we will restrict the functions which we will consider to the class of measurable functions (Remark 5).

Let Σ be a σ-algebra of subsets of S and let μ be a measure on Σ. If φ is a non-negative, Σ-simple function and $\varphi = \sum_{k=1}^{n} a_k C_{A_k}$ is the standard representation of φ, we define the *integral of φ* with respect to μ to be $\int_S \varphi d\mu = \sum_{k=1}^{n} a_k \mu(A_k)$ [here we are using the convention that $0 \cdot \infty = 0$]. If $E \in \Sigma$, we define the integral of φ over E (with respect to μ) to be

$$\int_E \varphi d\mu = \int_S C_E \varphi d\mu (= \sum_{k=1}^{n} a_k \mu(E \cap A_k)).$$

We say that φ is μ-*integrable over* E if $\int_E \varphi d\mu < \infty$.

Proposition 1 *Let $\varphi : S \to \mathbf{R}$ be non-negative, simple with*

$$\varphi = \sum_{i=1}^{m} b_i C_{B_i}, \ B_i \cap B_j = \emptyset$$

for $i \neq j$. Then

$$\int_S \varphi d\mu = \sum_{i=1}^{m} b_i \mu(B_i).$$

Proof: Let $\varphi = \sum_{j=1}^{n} a_j C_{A_j}$ be the standard representation of φ. Then $A_j = \bigcup_{b_i = a_j} B_i$ so

$$\int_S \varphi d\mu = \sum_{j=1}^{n} a_j \mu(A_j) = \sum_{j=1}^{n} a_j \sum_{b_i = a_j} \mu(B_i) = \sum_{i=1}^{m} b_i \mu(B_i).$$

Remark 2 Note that only the finite additivity of μ was used in the proof of Proposition 1; the non-negativity of the simple function was used only to insure that there were not arithmetic problems of the form $\infty - \infty$ encountered. Thus, if φ is a simple function and μ is a finitely additive set function on an algebra $\mathcal{A}$ which takes on only values in **R**, the integral $\nu(E) = \int_E \varphi d\mu$ is well-defined and $\nu : \mathcal{A} \to \mathbf{R}$ is finitely additive. Moreover,

$$\left|\int_E \varphi d\mu\right| \leq \int_E |\varphi| d|\mu| \leq \sup\{|\varphi(t)| : t \in E\} |\mu|(E) \text{ for } E \in \mathcal{A},$$

where $|\mu|$ is the total variation of μ [Proposition 2.2.1.7].

Proposition 3 *Let φ, ψ be non-negative, simple.*

(i) *If $t \geq 0$, then $\int_S t\varphi d\mu = t \int_S \varphi d\mu$.*

(ii) *$\int_S (\varphi + \psi) d\mu = \int_S \varphi d\mu + \int_S \psi d\mu$.*

(iii) *If $\varphi \leq \psi$, then $\int_S \varphi d\mu \leq \int_S \psi d\mu$.*

(iv) *The set function $E \to \int_E \varphi d\mu$ is a measure on Σ.*

Proof: (i) is immediate. For (ii) let

$$\varphi = \sum_{i=1}^m a_i C_{A_i}, \psi = \sum_{j=1}^n b_j C_{B_j}$$

be the standard representations for φ and ψ. Set $E_{ij} = A_i \cap B_j$ so

$$\varphi = \sum_{i,j} a_i C_{E_{ij}}, \psi = \sum_{i,j} b_j C_{E_{ij}}$$

and by Proposition 1,

$$\int_S (\varphi + \psi) d\mu = \sum_{i,j} (a_i + b_j) \mu(E_{ij}) = \int_S \varphi d\mu + \int_S \psi d\mu.$$

If $\varphi \leq \psi$, then $a_i \leq b_j$ when $E_{ij} \neq \emptyset$ so

$$\int_S \varphi d\mu = \sum_{i,j} a_i \mu(E_{ij}) \leq \sum_{i,j} b_j \mu(E_{ij}) = \int_S \psi d\mu$$

and (iii) holds.

For (iv), if $\{E_j\} \subset \Sigma$ is a pairwise disjoint sequence from Σ with union E, then

$$\int_E \varphi d\mu = \sum_{i=1}^m a_i \mu(A_i \cap E) = \sum_{i=1}^m a_i \sum_{j=1}^\infty \mu(A_i \cap E_j) = \sum_{j=1}^\infty \sum_{i=1}^m a_i \mu(A_i \cap E_j) = \sum_{j=1}^\infty \int_{E_j} \varphi d\mu.$$

We now define the integral of a non-negative, measurable function.

Definition 4 *If $f : S \to [0, \infty]$ is Σ-measurable, we define the integral of f with respect to μ to be*

$$\int_S f d\mu = \sup\{\int_S \varphi d\mu : \varphi \text{ simple}, 0 \leq \varphi \leq f\}.$$

If $E \in \Sigma$, we define the integral of f over E (with respect to μ) to be

$$\int_E f d\mu = \int_S C_E f d\mu.$$

3.2. THE LEBESGUE INTEGRAL

Every non-negative measurable function has an integral but it may be infinite. Note from Proposition 3 (iii), the definition of the integral for simple functions given in Definition 4 agrees with the previous definition.

Remark 5 Note the integral defined above is analogous to a lower integral in the Riemann theory of integration. There is no need to go through a "lower integral-upper integral" procedure because we have restricted our considerations to measurable functions. Indeed, if $f : S \to \mathbf{R}$ is bounded and μ is a finite, complete measure, then

$$\sup\{\int_S \varphi d\mu : \varphi \leq f, \text{ simple}\} = \inf\{\int_S \psi d\mu : f \leq \psi, \psi \text{ simple}\} \quad (3.1)$$

holds if and only if f is measurable. We indicate a proof of this result in Appendix I at the end of this section.

From Proposition 3, we have

Proposition 6 *Let f, g be non-negative, measurable.*

(i) *If $0 \leq f \leq g$, then $\int_S f d\mu \leq \int_S g d\mu$.*

(ii) *If $t \geq 0$, then $\int_S t f d\mu = t \int_S f d\mu$.*

One of the most important properties of the Lebesgue integral is the ease with which it handles limits. We now establish one of the most important results in this direction, the Monotone Convergence Theorem (MCT).

Theorem 7 *(MCT). Let $\{f_k\}$ be a sequence of non-negative, measurable functions such that $f_k(t) \uparrow$ for every $t \in S$. If $f(t) = \lim f_k(t)$, then $\int_S f d\mu = \lim \int_S f_k d\mu$.*

Proof: Since $0 \leq f_k \leq f_{k+1} \leq f$, from Proposition 6 $\{\int_S f_k d\mu\}$ is increasing and $\lim \int_S f_k d\mu \leq \int_S f d\mu$.

For the reverse inequality, fix $0 < a < 1$ and let φ be an arbitrary simple function with $0 \leq \varphi \leq f$. Set $E_k = \{t : f_k(t) \geq a\varphi(t)\}$. Since $\{f_k\}$ is increasing, $\{E_k\}$ is increasing, and since $\{f_k\}$ converges pointwise to f, $\bigcup_{k=1}^{\infty} E_k = S$. Then

$$\int_S f_k d\mu \geq \int_{E_k} f_k d\mu \geq a \int_{E_k} \varphi d\mu.$$

By Propositions 6, 3(iv) and 2.2.5,

$$\lim \int_S f_k d\mu \geq a \lim \int_{E_k} \varphi d\mu = a \int_S \varphi d\mu.$$

Letting a approach 1 gives

$$\lim \int_S f_k d\mu \geq \int_S \varphi d\mu,$$

and since φ is arbitrary,

$$\lim \int_S f_k d\mu \geq \int_S f d\mu.$$

Corollary 8 *Let f_k be non-negative, measurable. Then*

$$\sum_{k=1}^{\infty} \int_S f_k d\mu = \int_S \sum_{k=1}^{\infty} f_k d\mu.$$

Proof: Let $\{\varphi_j\}$ ($\{\psi_j\}$) be a sequence of non-negative simple functions such that $\varphi_j \uparrow f_1$ ($\psi_j \uparrow f_2$) [Theorem 3.1.2]. Then $(\varphi_j + \psi_j) \uparrow (f_1 + f_2)$ so by the MCT and Proposition 3(ii),

$$\lim \int_S (\varphi_j + \psi_j) d\mu = \lim \int_S \varphi_j d\mu + \lim \int_S \psi_j d\mu = \int_S (f_1 + f_2) d\mu = \int_S f_1 d\mu + \int_S f_2 d\mu.$$

By induction and the MCT,

$$\int_S \sum_{k=1}^{\infty} f_k d\mu = \lim_n \int_S \sum_{k=1}^{n} f_k d\mu = \lim_n \sum_{k=1}^{n} \int_S f_k d\mu = \sum_{k=1}^{\infty} \int_S f_k d\mu.$$

Another important property of the Lebesgue integral is that as a set function it defines a countably additive measure.

Theorem 9 *If f is non-negative and measurable, then the set function $E \to \int_E f d\mu$ is a measure on Σ.*

Proof: Let $\{\varphi_k\}$ be a sequence of non-negative, simple functions such that $\varphi_k \uparrow f$ (Theorem 3.1.2). By the MCT,

$$\lim \int_E \varphi_k d\mu = \int_E f d\mu$$

for each $E \in \Sigma$. The set function $E \to \int_E \varphi_k d\mu$ is a measure by Proposition 3(iv) so the result follows from Exercise 2.2.13.

Proposition 10 *Let f be non-negative and measurable. Then $\int_S f d\mu = 0 \Leftrightarrow f = 0$, μ-a.e. in S.*

Proof: $\Leftarrow$: The result is clear for simple functions and, therefore, follows immediately from the definition of the integral.

$\Rightarrow$: Set $A_k = \{t : f(t) \geq 1/k\}$. Then $A = \{t : f(t) > 0\} = \bigcup_{k=1}^{\infty} A_k$. Therefore, if $\mu(A) > 0$, then $\mu(A_k) > 0$ for some k and

$$\int_S f d\mu \geq \int_{A_k} f d\mu \geq \mu(A_k)/k > 0.$$

If $f : S \to \mathbf{R}^*$ is measurable, we say that f has a μ-integral over $E \in \Sigma$ if one of the integrals $\int_S f^+ d\mu$, $\int_S f^- d\mu$ is finite, and if this is the case, the μ-*integral* of f over E is then defined to be

$$\int_E f d\mu = \int_E f^+ d\mu - \int_E f^- d\mu.$$

3.2. THE LEBESGUE INTEGRAL

If $\int_E f d\mu$ is finite, we say that f is μ-integrable over E.

When it is necessary to indicate a variable of integration, we sometimes write the integral as
$$\int_E f d\mu = \int_E f(s) d\mu(s).$$
This is a particularly useful notation when the integrand f depends on a parameter.

When the measure μ is Lebesgue measure, we say the function f is Lebesgue integrable and call $\int_E f dm$ the Lebesgue integral of f. We sometimes denote the Lebesgue integral by
$$\int_E f dm = \int_E f(t) dm(t) = \int_E f(t) dt,$$
and if $E = [a, b]$ is an interval, we often write
$$\int_E f dm = \int_a^b f dm = \int_a^b f(t) dt.$$

If $\varphi : S \to \mathbf{R}$ is a simple function, then φ has a μ-integral over E if and only if one of the sets $\{t \in E : \varphi(t) > 0\}$, $\{t \in E : \varphi(t) < 0\}$ has finite μ-measure; in this case, if
$$\varphi_k = \sum_{k=1}^n a_k C_{A_k},$$
then
$$\int_E \varphi d\mu = \sum_{k=1}^n a_k \mu(A_k \cap E)$$
so φ is μ-integrable over E if and only if $\mu(A_k \cap E) < \infty$ for $a_k \neq 0$.

One of the interesting properties of the Lebesgue integral is that it is an absolute integral for measurable functions.

Proposition 11 *Let $f : S \to \mathbf{R}^*$ be measurable. Then f is μ-integrable $\Leftrightarrow$ $|f|$ is μ-integrable.*

In this case, $|\int_S f d\mu| \leq \int_S |f| d\mu$.

Proof: The first part is immediate from the inequalities $f^+, f^- \leq |f| = f^+ + f^-$, Proposition 6 and Corollary 8.

Since
$$\int_S f d\mu = \int_S f^+ d\mu - \int_S f^- d\mu \leq \int_S f^+ d\mu + \int_S f^- d\mu = \int_S |f| d\mu$$
and
$$-\int_S f d\mu = \int_S f^- d\mu - \int_S f^+ d\mu \leq \int_S f^- d\mu + \int_S f^+ d\mu = \int_S |f| d\mu,$$
the last inequality follows.

Corollary 12 Let $f, g : S \to \mathbf{R}^*$ be measurable with $|f| \leq g$ μ-a.e. If g is μ-integrable, then f is μ-integrable.

Proof: Since $\int_S |f|\, d\mu \leq \int_S g\, d\mu$ [Exercise 12], the result follows from Proposition 11.

In particular, if f is μ-integrable over $E \in \Sigma$, then f is μ-integrable over $F \in \Sigma$ when $F \subset E$.

The integral is a linear functional over the class of integrable functions.

Theorem 13 If f, g are μ-integrable, then $af + bg$ is μ-integrable for any $a, b \in \mathbf{R}$ and
$$\int_S (af + bg)\, d\mu = a \int_S f\, d\mu + b \int_S g\, d\mu.$$

Proof: Since $|af + bg| \leq |a|\,|f| + |b|\,|g|$, the first statement follows from Proposition 11 and Corollaries 8 and 12. It is easily checked that
$$\int_S af\, d\mu = a \int_S f\, d\mu.$$

Let $h = f + g$. Then $h = h^+ - h^- = f^+ - f^- + g^+ - g^-$ so $h^+ + f^- + g^- = f^+ + g^+ + h^-$ and from Corollary 8
$$\int_S h^+\, d\mu + \int_S f^-\, d\mu + \int_S g^-\, d\mu = \int_S f^+\, d\mu + \int_S g^+\, d\mu + \int_S h^-\, d\mu$$
which gives
$$\int_S f\, d\mu + \int_S g\, d\mu = \int_S (f + g)\, d\mu.$$

Concerning the "size" or "growth" of an integrable function, we have

Proposition 14 Let $f : S \to \mathbf{R}^*$ be μ-integrable. Then

(i) For every $a > 0$, $E_a = \{t : |f(t)| \geq a\}$ has finite μ-measure,

(ii) f is finite μ-a.e.

(iii) $\{t : f(t) \neq 0\}$ has σ-finite μ-measure.

Proof: (i): $\infty > \int_S |f|\, d\mu \geq a\mu(E_a)$.
(ii): If $Z = \{t : |f(t)| = \infty\}$, then for every $a > 0$
$$\int_S |f|\, d\mu \geq \int_Z |f|\, d\mu \geq a\mu(Z)$$
so $\mu(Z) = 0$.

3.2. THE LEBESGUE INTEGRAL

(iii) follows from (i) since

$$\{t : f(t) \neq 0\} = \bigcup_{k=1}^{\infty} \{t : |f(t)| \geq 1/k\}.$$

We next establish the other important convergence theorem for the Lebesgue integral, the Dominated Convergence Theorem (DCT). This result removes the very restrictive monotonicity requirement in the MCT. For this we use an important result due to Fatou.

Theorem 15 (Fatou) *Let $\{f_k\}$ be non-negative and measurable. Then*

$$\int_S \underline{\lim} f_k d\mu \leq \underline{\lim} \int_S f_k d\mu.$$

Proof: Set $h_k = \inf\{f_j : j \geq k\}$. Then $h_k \uparrow \underline{\lim} f_k$ so by the MCT

$$\int_S \underline{\lim} f_k d\mu = \lim \int_S h_k d\mu \leq \underline{\lim} \int_S f_k d\mu$$

since $f_k \geq h_k$.

Concerning the hypothesis and conclusion, see Exercises 14 and 15.

Theorem 16 (DCT) *Let f_k, f, g be measurable with g μ-integrable and such that $|f_k| \leq g$ μ-a.e. If $f_k \to f$ μ-a.e., then $\{f_k\}$ and f are μ-integrable with*

$$\int_S f d\mu = \lim \int_S f_k d\mu.$$

Moreover, $\lim \int_S |f_k - f| d\mu = 0$.

Proof: Since $|f_k| \leq g$ μ-a.e. and $|f| \leq g$ μ-a.e., these functions are μ-integrable. Now $g - f_k \geq 0$ μ-a.e. so by Fatou's Theorem

$$\int_S \underline{\lim}(g - f_k) d\mu = \int_S (g - f) d\mu = \int_S g d\mu - \int_S f d\mu$$

$$\leq \underline{\lim} \int_S (g - f_k) d\mu = \int_S g d\mu - \overline{\lim} \int_S f_k d\mu.$$

Hence,

$$\int_S f d\mu \geq \overline{\lim} \int_S f_k d\mu.$$

Similarly, $f_k + g \geq 0$ μ-a.e. so

$$\int_S \underline{\lim}(f_k + g) d\mu = \int_S f d\mu + \int_S g d\mu \leq \underline{\lim} \int_S (f_k + g) d\mu = \underline{\lim} \int_S f_k d\mu + \int_S g d\mu$$

and
$$\lim \int_S f_k d\mu \geq \int_S f d\mu.$$

Thus, $\lim \int_S f_k d\mu = \int_S f d\mu$.

Since $|f_k - f| \leq 2g$ μ-a.e., the last statement follows immediately from the first.

We cannot expect the conclusion of the DCT to hold without some condition on the $\{f_k\}$. For example, take $f_k = kC_{[0,1/k]}$ and Lebesgue measure on $[0,1]$. On the other hand, the domination condition in the DCT is sufficient but is not necessary. For example, take $f_k(t) = C_{[k-1/2,k+1/2]}(t)/t$. Then $f_k \to 0$ and

$$\int_0^\infty f_k dm = \ell n((k+1/2)/(k-1/2)) \to 0$$

[this computation is justified in the next section where it is shown that the Lebesgue integral generalizes the Riemann integral], but there is no Lebesgue integrable function g dominating the sequence $\{f_k\}$ [if $g \geq f_k$, then $g(t) \geq 1/t$ for $t > 1$].

In Theorem 9 we showed that the integral of a non-negative measurable function defined a measure. Exercise 26 extends this results to the integral of an arbitrary function having a μ-integral. We next consider another important property of the integral as a set function.

Theorem 17 *Let f be μ-integrable. Then $\lim_{\mu(E) \to 0} \int_E f d\mu = 0$.*

Proof: First, suppose $f \geq 0$. Set $f_k = f \wedge k$. Then $0 \leq f_k \uparrow f$ so by the MCT,

$$\lim \int_S f_k d\mu = \int_S f d\mu.$$

If $\epsilon > 0$ is given, choose k such that $\int_S (f - f_k) d\mu < \epsilon/2$ and $0 < \delta < \epsilon/2k$. If $\mu(E) < \delta$, then

$$\int_E f d\mu \leq \int_S (f - f_k) d\mu + \int_E f_k d\mu \leq \epsilon/2 + k \cdot \epsilon/2k = \epsilon.$$

If f is μ-integrable, then $|\int_E f d\mu| \leq \int_E |f| d\mu$ so the general result follows from the first part.

Finally, we have

Proposition 18 *Suppose that f has a μ-integral over S. If $\int_E f d\mu = 0$ for every $E \in \Sigma$, then $f = 0$ μ-a.e. [so f is μ-integrable with $\int_S f d\mu = 0$ by Exercise 12].*

Proof: If $E = \{t : f(t) \geq 0\}$, then $\int_E f d\mu = 0$ so $f = 0$ μ-a.e. in E by Proposition 10. Similarly, $f = 0$ μ-a.e. in $F = \{t : f(t) < 0\}$.

3.2. THE LEBESGUE INTEGRAL

Remark 19 We have chosen to basically follow Lebesgue's original construction of the integral. We began with a measure and then constructed the integral from the measure. It is possible to follow the reverse order; this approach to the integral is due to Daniell. In the Daniell approach one begins with an "elementary integral" I, a linear functional defined on a vector space of functions, Z, defined on some subset S, which is positive in the sense that $I(f) \geq 0$ whenever $f \geq 0$, $f \in Z$, and which satisfies a mild sequential continuity condition [for example, the Riemann integral on the class of continuous functions]. The functional I is then extended to a positive linear functional J defined on a vector space X of functions on S which contains Z. There is a natural measure associated with the extension $J : \Sigma = \{E : C_E \in X\}$ is a σ-algebra of subsets of S and $\mu(E) = J(C_E)$ is a measure on Σ. Under appropriate assumptions, $J(f) = \int_S f d\mu$. For treatments of the Daniell integral, see [Roy] or [Ta].

Appendix I: We give a proof of the statement in Remark 5.

First, suppose that f is measurable and the range of f is contained in $[\ell, L]$. Let $\epsilon > 0$ and $\ell = y_0 < y_1 \cdots < y_n = L$ be a partition of $[\ell, L]$ with

$$\max\{\ell_{i+1} - \ell_i : i = 0, 1, \ldots, n-1\} < \epsilon.$$

Set

$$E_i = \{t : \ell_{i-1} \leq f(t) < \ell_i\}, i = 1, \ldots, n,$$

and

$$\psi = \sum_{i=1}^n \ell_i C_{E_i}, \varphi = \sum_{i=1}^n \ell_{i-1} C_{E_i}.$$

Then $\varphi \leq f \leq \psi$ and

$$\int_S (\psi - \varphi) d\mu \leq \sum_{i=1}^n (\ell_i - \ell_{i-1}) \mu(E_i) < \epsilon \mu(S)$$

so (1) (in Remark 5) holds.

Suppose (1) holds. For each k there exist simple functions φ_k and ψ_k such that $\varphi_k \leq f \leq \psi_k$ and $\int_S (\psi_k - \varphi_k) d\mu < 1/k$. Set $\varphi = \sup \varphi_k$, $\psi = \inf \psi_k$. Then φ and ψ are measurable with $\varphi \leq \psi$ and

$$\int_S (\psi - \varphi) d\mu \leq \int_S (\psi_k - \varphi_k) d\mu < 1/k$$

for each k so $\int_S (\psi - \varphi) d\mu = 0$ and $\psi = \varphi$ μ-a.e. by Proposition 10. Hence $\psi = f = \varphi$ μ-a.e. and f must be measurable.

Appendix II: Countable Additivity of Lebesgue measure on $\mathbf{R}^n$.

We show, using the MCT, that Lebesgue measure m_n on the semi-ring $\mathcal{S}_n$ is countably additive (Example 2.2.10). The proof is by induction on n. For $n = 1$, this follows from Example 2.2.9. Assume the result is true for n. Let

$$B_i = A_i \times [a_i, b_i) \in \mathcal{S}_{n+1}$$

with $B_i \in \mathcal{S}_{n+1}$ pairwise disjoint and with union $B = A \times [a,b) \in \mathcal{S}_{n+1}$, $A \in \mathcal{S}_n$. Fix $x \in \mathbf{R}^n$. Then
$$\sum_{i=1}^{k} C_{A_i}(x) C_{[a_i, b_i)} \uparrow C_A(x) C_{[a,b)}$$
as $k \to \infty$ so by the MCT for the measure m, we obtain
$$\sum_{i=1}^{k} C_{A_i}(x)(b_i - a_i) \uparrow C_A(x)(b-a).$$

We now apply the MCT to the measure m_n, which is countably additive by the induction hypothesis, to obtain
$$\sum_{i=1}^{k} m_n(A_i)(b_i - a_i) = \sum_{i=1}^{k} m_{n+1}(B_i) \uparrow m_n(A)(b-a) = m_{n+1}(B).$$

Hence, m_{n+1} is countably additive.

Note that for the proof above we only required the MCT for simple functions. This can be obtained by proving Propositions 1 and 3 and then using the method of proof in Theorem 7 (MCT).

As noted earlier in Example 2.2.10 a geometric proof of this result which does not use integration theory can be given, but the proof is surprisingly difficult.

Appendix III: As promised following Theorem 2.2.1.5 we give an example of a real-valued, finitely additive set function defined on a σ-algebra which is not bounded. For our construction we first require a lemma which utilizes integration with respect to a finitely additive set function as described in Remark 2.

Lemma 20 *Let $\mathcal{A}$, $\mathcal{B}$ be algebras of subsets of S with $\mathcal{A} \subset \mathcal{B}$ and let $\alpha : \mathcal{A} \to \mathbf{R}$ be finitely additive. If $B \in \mathcal{B} \setminus \mathcal{A}$ and $b \in \mathbf{R}$, there exists $\beta : \mathcal{B} \to \mathbf{R}$, finitely additive, such that β is an extension of α with $\beta(B) = b$.*

Proof: Let $\mathcal{S}(\mathcal{A})$ [$\mathcal{S}(\mathcal{B})$] be the vector space of all $\mathcal{A}$-simple [$\mathcal{B}$-simple] functions. Then α induces a linear functional $\hat{\alpha} : \mathcal{S}(\mathcal{A}) \to \mathbf{R}$ via integration with respect to α, i.e., $\hat{\alpha}(f) = \int_S f d\alpha$ [Remark 2]. The linear functional $\hat{\alpha}$ has a linear extension, $\hat{\beta}$, to $\mathcal{S}(\mathcal{B})$ such that $\hat{\beta}(C_B) = b$. Then $\beta(E) = \hat{\beta}(C_E)$, $E \in \mathcal{B}$, defines the desired finitely additive extension of α.

We now present our example ([Gi]).

Example 21 *Let $\{E_k\}_{k=0}^{\infty}$ be a pairwise disjoint sequence of intervals such that $\bigcup_{k=0}^{\infty} E_k = \mathbf{R}$. Let $\mathcal{A}_k$ be the algebra generated by $\{E_0, E_1, \ldots, E_k\}$ so*
$$\mathcal{A}_0 \subset \mathcal{A}_1 \subset \mathcal{A}_2 \subset \ldots \subset \mathcal{M},$$

3.2. THE LEBESGUE INTEGRAL

the Lebesgue measurable subsets of $\mathbf{R}$. Set $\alpha_0 = 0$ on $\mathcal{A}_0$; let α_1 be a finitely additive extension of α_0 to $\mathcal{A}_1$ such that $\alpha_1(E_1) = 1$ [Lemma 20], and by induction there exists a sequence $\{\alpha_k\}$ of finitely additive set functions defined on $\{\mathcal{A}_k\}$ such that α_{k+1} extends α_k and $\alpha_k(E_k) = k$. Now $\mathcal{A} = \bigcup_{k=0}^{\infty} \mathcal{A}_k$ is an algebra and $\alpha = \bigcup_{k=0}^{\infty} \alpha_k$ is finitely additive on $\mathcal{A}$. By Lemma 20, there is a real-valued finitely additive extension of α, μ, to $\mathcal{M}$, and since $\mu(E_k) = k$ for every k, μ is unbounded.

Exercise 1. If μ is counting measure on S, show $f : S \to \mathbf{R}$ is μ-integrable if and only if $A = \{t : f(t) \neq 0\}$ is countable and $\sum_{t \in A} |f(t)| < \infty$. In this case, show $\int_S f\,d\mu = \sum_{t \in A} f(t)$.

Exercise 2. Let $\{f_k\}$ be μ-integrable and $\sum_{k=1}^{\infty} \int_S |f_k|\,d\mu < \infty$. Show the series $\sum f_k$ converges pointwise (absolutely) μ-a.e. to a μ-integrable function f with

$$\int_S f\,d\mu = \sum_{k=1}^{\infty} \int_S f_k\,d\mu.$$

Exercise 3. Let f_k, f be non-negative, μ-integrable with $f_k \to f$ μ-a.e. Suppose

$$\int_S f_k\,d\mu \to \int_S f\,d\mu.$$

Show $\int_E f_k\,d\mu \to \int_E f\,d\mu$ for every $E \in \Sigma$. [Hint: Apply Fatou's Theorem to E and $S \setminus E$.]

Exercise 4. Let f be μ-integrable. Suppose there exists $k > 0$ such that

$$\left| \int_E f\,d\mu / \mu(E) \right| \leq k$$

for every $0 < \mu(E) < \infty$. Show $|f| \leq k$ μ-a.e.

Exercise 5. Let μ be finite. Assume $\{f_k\}$ are μ-integrable and $f_k \to f$ uniformly on S. Show f is μ-integrable and $\int_S |f_k - f|\,d\mu \to 0$. Can finiteness be dropped?

Exercise 6. Let f, g be non-negative, measurable. Assume f is μ-integrable and set $\nu(E) = \int_E f\,d\mu$ for $E \in \Sigma$. Show g is ν-integrable if and only if fg is μ-integrable and in this case $\int_E g\,d\nu = \int_E fg\,d\mu$ for $E \in \Sigma$. [Hint: First consider simple functions g.]

Exercise 7. Let $f : [a, b] \to \mathbf{R}$ be m-integrable. If $\int_a^b f(t) t^k\,dm(t) = 0$ for $k = 0, 1, 2, \ldots$, show $f = 0$ m-a.e.

Exercise 8. Let $f : \mathbf{R} \to \mathbf{R}$ be m-integrable. For $a \in \mathbf{R}$ set $f_a(t) = f(t+a)$. Show f_a is m-integrable and $\int_{\mathbf{R}} f\,dm = \int_{\mathbf{R}} f_a dm$. [See Exercise 2.5.2.] What about $\int_{\mathbf{R}} f(at)dm(t)$? [Exercise 2.5.4.]

Exercise 9. (Chebychev Inequality). Let f be μ-integrable and $c > 0$. Show $\mu\{t : |f(t)| \geq c\} \leq \int_S |f|\,d\mu/c$.

Exercise 10. Let f be μ-integrable. Show
$$\lim_{c \to \infty} \mu\{t : |f(t)| \geq c\} = 0$$
and
$$\lim_{c \to \infty} \int_{\{t:|f(t)|\geq c\}} |f|\,d\mu = 0.$$

Exercise 11. Let $\{f_k\}$, g be μ-integrable with $|f_k| \leq g$ μ-a.e. Show
$$\lim_{c \to \infty} \int_{\{t:|f_k(t)|\geq c\}} |f_k|\,d\mu = 0$$
uniformly for $k \in \mathbf{N}$.

Exercise 12. Show that if $\mu(E) = 0$, then any measurable function f is μ-integrable over E with $\int_E f\,d\mu = 0$.

Exercise 13. Show the analogues of the MCT and DCT do not hold for the Riemann integral.

Exercise 14. Show that strict inequality can occur in Fatou's Theorem. [Consider $f_k = C_{(0,2)}$ for k odd, $f_k = C_{(1,3)}$ for k even.]

Exercise 15. Show the non-negativity assumption cannot be dropped in Fatou's Theorem.

Exercise 16 (Bounded Convergence Theorem; BCT). Let μ be a finite measure and f_k, f be measurable. Assume $\exists M > 0$ such that $|f_k| \leq M$ μ-a.e. If $f_k \to f$ μ-a.e., show $\int_S f\,d\mu = \lim_k \int_S f_k d\mu$.

Exercise 17. Let $f : S \to \mathbf{R}^*$ be non-negative and μ σ-finite. Show that f is measurable if and only if $f \wedge g$ is μ-integrable for every μ-integrable g.

3.2. THE LEBESGUE INTEGRAL

Exercise 18. Let $\{E_i\} \subset \Sigma$ be pairwise disjoint and $E = \bigcup_{i=1}^{\infty} E_i$. Let f be measurable and μ-integrable over each E_i. Show that f is μ-integrable over E if and only if

$$\sum_{i=1}^{\infty} \int_{E_i} |f|\, d\mu < \infty.$$

Show this last condition cannot be replaced by $\left|\sum_{i=1}^{\infty} \int_{E_i} f\, d\mu\right| < \infty$.

Exercise 19. Suppose $f : \mathbf{R} \to \mathbf{R}$ is uniformly continuous on $\mathbf{R}$ and m-integrable over $\mathbf{R}$. Show f vanishes at ∞ and is bounded. Can uniform continuity be replaced by continuity?

Exercise 20. If f is bounded and measurable and g is μ-integrable, show fg is μ-integrable.

Exercise 21. Let μ be finite and f measurable. If fg is μ-integrable for every μ-integrable g, show there exists $M \geq 0$ such that $|f| \leq M$ μ-a.e. [Such functions are called μ-essentially bounded.]

Exercise 22. Let f be non-negative and measurable and set

$$E_k = \{t : k \leq f(t) < k+1\} \text{ for } k = 0, 1, \ldots.$$

If f is μ-integrable, show $\sum_{k=0}^{\infty} k\mu(E_k) < \infty$. Show

$$\sum_{k=0}^{\infty} (k+1)\mu(E_k) < \infty$$

implies that f is μ-integrable.

Exercise 23. If f is μ-integrable and $F_k = \{t : |f(t)| \geq k\}$, show $\lim k\mu(F_k) = 0$.

Exercise 24. Show that $E \subset \mathbf{R}^n$ has Lebesgue measure 0 if and only if there exists a sequence of m-integrable functions $\{f_k\}$ such that

$$\sum_{k=1}^{\infty} \int_{\mathbf{R}^n} |f_k|\, dm < \infty$$

and

$$\sum_{k=1}^{\infty} |f_k(x)| = \infty,$$

for every $x \in E$.

Exercise 25. Show the measurability assumption in Proposition 11 and Corollary 12 is important.

Exercise 26. Let $f : S \to \mathbf{R}^*$ have a μ-integral (possibly infinite). Show $\nu(E) = \int_E f d\mu$ defines a signed measure on $\sum$. Give a Hahn Decomposition for ν. Show
$$\nu^+(E) = \int_E f^+ d\mu, \ \nu^-(E) = \int_E f^- d\mu$$
and
$$|\nu|(E) = \int_E |f| d\mu.$$

Exercise 27. Let f_k be non-negative, μ-integrable and suppose $f_k(t) \downarrow f(t)$ for $t \in S$. Show $f = 0$ μ-a.e. if and only if $\lim \int_S f_k d\mu = 0$.

Exercise 28. If μ is a regular measure and f is non-negative and μ-integrable, show that $\nu = \int f d\mu$ is a regular measure.

Exercise 29. Let f_k be non-negative and measurable. If $f_k \to f$ pointwise and $f_k \leq f$, show
$$\int_S f_k d\mu \to \int_S f d\mu.$$

Exercise 30. Let $f : [0,1] \to \mathbf{R}$ be Lebesgue integrable. Show $t \to t^k f(t)$ is Lebesgue integrable for each $k \in \mathbf{N}$ and $\int_0^1 t^k f(t) dt \to 0$.

Exercise 31. Let ν be a signed measure on $\sum$ with $\nu = \nu^+ - \nu^-$ its Jordan decomposition. If $f : S \to \mathbf{R}^*$ is $\sum$-measurable, say that f is ν-integrable if and only if f is both ν^+ and ν^- integrable and define
$$\int f d\nu = \int f d\nu^+ - \int f d\nu^-.$$
Show f is ν-integrable if and only if f is $|\nu|$-integrable and in this case
$$\left| \int f d\nu \right| \leq \int |f| d|\nu|.$$
Show $|\nu|(E) = \sup\{|\int_E f d\nu| : |f| \leq 1\}$.

Exercise 32. Let μ be a finite measure and f_k non-negative and μ-integrable with $f_k \to 0$ μ-a.e. Show $\int_S f_k d\mu \to 0$ if and only if for every $\varepsilon > 0$ there exists $\delta > 0$ such that $\mu(E) < \delta$ implies $\int_E f_k d\mu < \varepsilon$ for all k.

3.2. THE LEBESGUE INTEGRAL

Exercise 33. Show μ is σ-finite if and only if $\exists$ a μ-integrable function f with $f(t) > 0$ for all $t \in S$.

Exercise 34. If $f(t) > 0$ for all $t \in E$, $\mu(E) > 0$ and f is μ-integrable over E, show $\int_E f d\mu > 0$.

Exercise 35. Let $f : \mathbf{R}^n \to \mathbf{R}^*$ be Lebesgue integrable. If $\int_K f dm = 0$ for every compact (open) K, show $f = 0$ m-a.e.

Exercise 36. Find $\lim \int_0^1 \frac{1+kt}{(1+t)^k} dt$.

Exercise 37. Let $\{f_k\}$ be measurable and $f_k \to f$ μ-a.e. Let g_k, g be μ-integrable, $g_k \to g$ μ-a.e. and $\int_S g_k d\mu \to \int_S g d\mu$. If $|f_k| \le g_k$ μ-a.e., show f is μ-integrable and $\int_S f_k d\mu \to \int_S f d\mu$.

Exercise 38. Let $f : [a, b] \to \mathbf{R}^*$ be Lebesgue integrable and $\int_a^x f dm = 0$ for every $a \le x \le b$. Show $f = 0$ m-a.e.

3.3 The Riemann and Lebesgue Integrals

Let $f : [a,b] \to \mathbf{R}$ be bounded. If $\pi = \{a = x_0 < x_1 < \ldots < x_n = b\}$ is a partition of $[a,b]$, set $\delta_i = [x_{i-1}, x_i]$, $m_i = \inf\{f(t) : t \in \delta_i\}$, $M_i = \sup\{f(t) : t \in \delta_i\}$ and $\mu(\pi) = \max\{x_i - x_{i-1}\}$. The *upper* (*lower*) *sum* of f with respect to π is

$$U(f, \pi) = \sum_{i=1}^{n} M_i(x_i - x_{i-1})$$

$[(L(f, \pi) = \sum_{i=1}^{n} m_i(x_i - x_{i-1})]$, and the *upper* (*lower*) *integral* of f is

$$\overline{\int}_a^b f = \inf\{U(f, \pi) : \pi \text{ is a partition of } [a,b]\}$$

$[\underline{\int}_a^b f = \sup\{(L(f, \pi) : \pi \text{ a partition of } [a,b]\}]$. The function f is *Riemann integrable* [over $[a,b]$] if $\overline{\int}_a^b f = \underline{\int}_a^b f$, and the Riemann integral of f is defined to be the common value; in order to distinguish the Riemann integral from the Lebesgue integral, we denote the Riemann integral by $\mathcal{R}\int_a^b f$.

We now show that any Riemann integrable function is Lebesgue integrable and the two integrals agree in this case. Our proof also gives a necessary and sufficient condition for a function to be Riemann integrable.

Theorem 1 *Let $f : [a,b] \to \mathbf{R}$ be bounded.*

(i) *If f is Riemann integrable over $[a,b]$, then f is Lebesgue integrable over $[a,b]$ and $\mathcal{R}\int_a^b f = \int_a^b f\, dm$.*

(ii) *f is Riemann integrable over $[a,b]$ if and only if f is continuous m-a.e. in $[a,b]$.*

Proof: Choose a sequence of partitions, $\{\pi_k\}$, such that $\pi_1 \subset \pi_2 \subset \cdots$, $\mu(\pi_k) \to 0$ and $\lim L(f, \pi_k) = \underline{\int}_a^b f$, $\lim U(f, \pi_k) = \overline{\int}_a^b f$ [this can be done by choosing $\{P_k\}$ to satisfy the last three conditions and then setting $\pi_k = \bigcup_{j=1}^{k} P_j$]. If π_k is given by $\{a = x_0 < x_1 < \ldots < x_n = b\}$, let $m_i = \inf\{f(t) : x_{i-1} \leq t \leq x_i\}$, $M_i = \sup\{f(t) : x_{i-1} \leq t \leq x_i\}$ and define simple functions ℓ_k and u_k by $\ell_k = \sum_{i=1}^{n} m_i C_{[x_{i-1}, x_i)}$, $u_k = \sum_{i=1}^{n} M_i C_{[x_{i-1}, x_i)}$ on $[a,b)$ and $\ell_k(b) = u_k(b) = f(b)$ so $\int_a^b \ell_k dm = L(f, \pi_k)$, $\int_a^b u_k dm = U(f, \pi_k)$. Since $\pi_{k+1} \supset \pi_k$, we have

$$\ell_1(x) \leq \ell_2(x) \leq \ldots \leq f(x) \leq \ldots \leq u_2(x) \leq u_1(x) \text{ for } a \leq x \leq b. \quad (3.1)$$

3.3. THE RIEMANN AND LEBESGUE INTEGRALS

Set $\ell(x) = \lim \ell_k(x)$, $u(x) = \lim u_k(x)$ and note $u(x) \geq \ell(x)$. By the MCT,

$$\int_a^b \ell \, dm = \lim \int_a^b \ell_k \, dm = \underline{\int_a^b} f, \quad \int_a^b u \, dm = \lim \int_a^b u_k \, dm = \overline{\int_a^b} f. \quad (3.2)$$

For the proof of (i), assume that f is Riemann integrable. From (2) we obtain $\int_a^b \ell \, dm = \mathcal{R} \int_a^b f = \int_a^b u \, dm$ so $\int_a^b (u - \ell) \, dm = 0$ and since $u \geq \ell$, $u = \ell$ m-a.e. so from (1), $u = \ell = f$ m-a.e. Hence, f is Lebesgue integrable and $\int_a^b f \, dm = \mathcal{R} \int_a^b f$.

For the proof of (ii), let $C = \bigcup_{k=1}^\infty \pi_k$ so C is countable and $m(C) = 0$. If $x \notin C$, then f is continuous at x if and only if $u(x) - \ell(x) = \lim(u_k(x) - \ell_k(x)) = 0$.

Hence, if f is continuous m-a.e., then from (1), $u = \ell = f$ m-a.e. and from (2) $\int_a^b \ell \, dm = \underline{\int_a^b} f = \int_a^b u \, dm = \overline{\int_a^b} f$ so f is Riemann integrable.

If, conversely, f is Riemann integrable, then from (2) $\int_a^b u \, dm = \int_a^b \ell \, dm$ so $u = \ell$ m-a.e. and f is continuous m-a.e.

This theorem now allows us to compute the Lebesgue integral for a large class of functions, and we will freely use properties of the Riemann integral to calculate Lebesgue integrals.

We can use Theorem 1 to establish the analogue of the BCT for the Riemann integral.

Corollary 2 (Arzela) *Let f_k, $f : [a, b] \to \mathbf{R}$ be Riemann integrable with $f_k \to f$ pointwise on $[a, b]$. If there exists $M > 0$ such that $|f_k(t)| \leq M$ for all k, $t \in [a, b]$, then*

$$\lim \mathcal{R} \int_a^b f_k = \mathcal{R} \int_a^b f.$$

Proof: Theorem 1 and the BCT.

Note that we must assume the Riemann integrability of the limit function f in Corollary 2 [Exercise 1]. Our proof of Arzela's Theorem, of course, depends on measure theory and the Lebesgue integral; for an elementary proof not using such machinery see [Lew].

Improper Riemann Integrals:

If $f : [a, \infty) \to \mathbf{R}$ is Riemann integrable over $[a, b]$ for each $b > a$, the improper Riemann integral of f over $[a, \infty)$ is defined to be

$$\lim_{b \to \infty} \mathcal{R} \int_a^b f = \mathcal{R} \int_a^\infty f,$$

provided the limit exists and is finite. [This integral is also called the *Cauchy-Riemann integral*.] If $\mathcal{R} \int_a^\infty |f|$ exists, it follows from Theorem 1 and the MCT and DCT that f is Lebesgue integrable over $[a, \infty)$ and $\int_a^\infty f \, dm = \mathcal{R} \int_a^\infty f$. However, a function can be Cauchy-Riemann integrable over $[a, \infty)$ and not be Lebesgue integrable. For example, consider $\mathcal{R} \int_1^\infty \frac{\sin x}{x} dx$. Integrating by parts, gives

$$\int_1^b \frac{\sin x}{x} dx = \cos 1 - \frac{\cos b}{b} + \int_1^b \frac{\cos x}{x^2} dx$$

and since $|\cos x|/x^2 \le 1/x^2$, $\mathcal{R}\int_1^\infty \frac{\sin x}{x}dx$ exists. On the other hand,

$$\int_\pi^{n\pi} \left|\frac{\sin x}{x}\right| dx = \sum_{k=1}^{n-1} \int_{k\pi}^{(k+1)\pi} \left|\frac{\sin x}{x}\right| dx \ge \sum_{k=1}^{n-1} \frac{1}{(k+1)\pi} \int_{k\pi}^{(k+1)\pi} |\sin x|\, dx = \sum_{k=1}^{n-1} \frac{2}{(k+1)\pi}$$

so $\frac{\sin x}{x}$ is not Lebesgue integrable over $[1, \infty)$. Similar remarks apply to improper Riemann integrals for unbounded functions on bounded intervals.

Defects in the Riemann Integral:

I. The first defect is the lack of good convergence theorems. For example, there exists a uniformly bounded sequence of Riemann integrable functions which converge pointwise to a function which is not Riemann integrable (Exercise 1.3.1).

II. Closely related to the lack of strong convergence theorems is the incompleteness of the space of Riemann integrable functions. Let $\mathcal{R}[a,b]$ be the vector space of Riemann integrable functions on $[a,b]$ and define a semi-metric d on $\mathcal{R}[a,b]$ by $d(f,g) = \int_a^b |f-g|$. Convergence in this semi-metric is called convergence in mean and will be studied for the Lebesgue integral in §3.5. We show that d is not complete; in §3.5 we show that the analogous space for the Lebesgue integral is complete.

Let $H \subset [0,1]$ be a 1/2-Cantor set. Let $\{I_k\}$ be the open intervals making up $[0,1]\backslash H$ and set

$$J_n = \bigcup_{k=1}^n I_k, \varphi_n = C_{J_n}.$$

Then each φ_n is Riemann integrable and $\varphi_n \to C_{H^c} = \varphi$ pointwise. Since $|\varphi_n - \varphi| \le 1$, by the BCT

$$\int_a^b |\varphi_n - \varphi|\, dm = 0$$

so $\{\varphi_n\}$ is Cauchy in $\mathcal{R}[a,b]$ with respect to d.

We show that $\{\varphi_n\}$ has no limit in $(\mathcal{R}[a,b], d)$. For if $\varphi_n \to f \in \mathcal{R}[a,b]$ with respect to d, then $f = \varphi$ m-a.e. For any $t_0 \in [0,1]$ and $\epsilon > 0$, $(t_0 - \epsilon, t_0 + \epsilon)\backslash H$ contains an interval so $(t_0 - \epsilon, t_0 + \epsilon)$ must contain points t such that $f(t) = 1$. Hence, $\overline{\lim}_{t \to t_0} f(t) = 1$ for every $t_0 \in [0,1]$. But $f = \varphi$ m-a.e. and $m(H) = 1/2$ so $f = 0$ on a set with positive measure. That is, f is discontinuous on a set with positive measure and, therefore, is not Riemann integrable.

III. Finally, the Fundamental Theorem of Calculus (*FTC*) in its full generality fails for the Riemann integral. The desired form of the FTC would be: if $f : [a,b] \to \mathbf{R}$ is differentiable everywhere in $[a,b]$, then the derivative f' is integrable and $\int_a^b f' = f(b) - f(a)$. To obtain the FTC for the Riemann integral it is necessary to assume that the derivative f' is Riemann integrable, i.e., the Riemann integral cannot integrate arbitrary derivatives [the Lebesgue integral also suffers this same defect; Example 4.3.1].

It is easy to give examples of functions with unbounded derivatives [see, for example, Example 4.3.1]; however, we give an example of a bounded derivative which is not Riemann integrable. Let $H \subset [0,1]$ be a 1/2-Cantor set. Let (a,b) be one of the open

3.3. THE RIEMANN AND LEBESGUE INTEGRALS

intervals making up $[0,1]\backslash H$. Define f on (a,b) by setting $f(t) = (t-a)^2 \sin(1/(t-a))$ on (a,α) where $\alpha < (a+b)/2$ is such that $f'(\alpha) = 0$, letting f be the constant $f(\alpha)$ on $[\alpha, (a+b)/2]$ and defining f on $[(a+b)/2, b]$ by reflection in the line $t = (a+b)/2$. [A sketch is helpful.] We extend f to $[0,1]$ by setting $f(t) = 0$ for $t \in H$. Then f' exists everywhere in $[0,1]\backslash H$ and $|f'(t)| \leq 1$ for $t \in H^c$.

We claim that $f'(x) = 0$ for $x \in H$. Let $\epsilon > 0$ and suppose $|x - t| < \epsilon$. If $t \in H$, then $(f(t) - f(x))/(t - x) = 0$. If $t \notin H$, then t belongs to some open interval (a,b) making up $[0,1]\backslash H$. Suppose a is the endpoint nearest x. Then

$$|(f(t) - f(x))/(t-x)| = |f(t)/(t-x)| \leq |f(t)/(t-a)| \leq |t-a|^2 / |t-a| < \epsilon$$

Hence, $f'(x) = 0$.

Thus, f' exists everywhere and is bounded. But, f' is discontinuous on H and $m(H) = 1/2$ so f' is not Riemann integrable.

Exercise 1. Show the Riemann integrability of the limit function f cannot be dropped in Corollary 2.

Exercise 2. Let $K \subset [0,1]$ be the Cantor set. Is C_K Riemann integrable?

Exercise 3. Let $0 < \epsilon < 1$ and K_ϵ an ϵ-Cantor set. Is C_{K_ϵ} Riemann integrable?

Exercise 4. Show the following functions are Lebesgue measurable and determine whether they are Lebesgue integrable.

(a) $f(t) = 1/t^p$, $0 < t \leq 1$, $p > 0$,

(b) $f(t) = (-1)^k/k$ for $t \in [k-1, k)$ and $f(t) = 0$ otherwise,

(c) $f(t) = (-1)^k/2^k$ for $t \in [k-1, k)$ and $f(t) = 0$ otherwise,

(d) $f(t) = 1/t$ for $0 < t < 1$, $-1/\sqrt{t-1}$ for $1 < t < 2$ and $f(t) = 0$ otherwise,

(e) $f(t) = 1/t^p$, $1 \leq t < \infty$, $p > 0$.

Exercise 5. Let $\mathcal{A}$ be all subsets of $[a,b]$ such that C_A is Riemann integrable (Exer. 2.1.8). Define μ on $\mathcal{A}$ by $\mu(A) = \int_a^b C_A$. Show μ is countably additive.

3.4 Integrals Depending on a Parameter

Let (S, Σ, μ) be a measure space and $I \neq \emptyset$. If $f : S \times I \to \mathbf{R}^*$, we write $f(\cdot, t)$ $[f(s, \cdot)]$ for the function $f(\cdot, t)(s) = f(s, t)$ $[f(s, \cdot)(t) = f(s, t)]$. If $f(\cdot, t)$ is μ-integrable for every $t \in I$, we say that the integral $F(t) = \int_S f(s, t) d\mu(s)$ depends on the parameter $t \in I$. In this section we study properties which the function F inherits from f.

First, we consider continuity.

Theorem 1 *Let I be a metric space and $f : S \times I \to \mathbf{R}$. Assume*

(i) *$f(\cdot, t)$ is μ-integrable for every $t \in I$.*

(ii) *$f(s, \cdot)$ is continuous at $t_0 \in I$ for each $s \in S$.*

(iii) *There exists a μ-integrable function $g : S \to \mathbf{R}$ such that $|f(s, t)| \leq g(s)$ for all $s \in S$, $t \in I$. Then $F(t) = \int_S f(s, t) d\mu(s)$ is continuous at t_0.*

Proof: Let $\{t_k\}$ be a sequence from I converging to t_0. Then $f(s, t_k) \to f(s, t_0)$ for every $s \in S$ by (ii). By (iii) $|f(\cdot, t_k)| \leq g$ for every k so the DCT implies $F(t_k) \to F(t_0)$ and F is continuous at t_0.

As an example we consider the Gamma function.

Example 2 (Gamma Function) The Gamma function is defined by

$$\Gamma(x) = \int_0^\infty t^{x-1} e^{-t} dt$$

for $x > 0$. First, we observe that the integral exists. For $0 < t \leq 1$, $t^{x-1} e^{-t} \leq t^{x-1}$ and the function $t \to t^{x-1}$ is integrable over $[0, 1]$ for $x > 0$ (Exer. 3.3.4) so $\int_0^1 t^{x-1} e^{-t} dt$ is finite. For $t > 1$, $t^{x-1} e^{-t} = [t^{x+1} e^{-t}] t^{-2}$ and the function $t \to t^{x+1} e^{-t}$ is bounded since $\lim_{t \to \infty} t^{x+1} e^{-t} = 0$ so $\int_1^\infty t^{x-1} e^{-t} dt < \infty$ (Exer. 3.3.4). Hence, $\int_0^\infty t^{x-1} e^{-t} dt < \infty$.

Next, we show that Γ is continuous for $x > 0$. Let $x_0 > 0$. If $0 \leq t \leq 1$ and $x_0/2 < x$, then $t^x \leq t^{x_0/2}$ and $e^{-t} \leq 1$ for $t \geq 0$. Therefore, $t^{x-1} e^{-t} \leq t^{(x_0/2)-1}$ so by Theorem 1, $x \to \int_0^1 t^{x-1} e^{-t} dt$ is continuous at x_0. For $t > 1$ and $x_0/2 \leq x \leq 2x_0$, there is a B such that $t^{x-1} e^{-t} \leq B t^{-2}$ so by Theorem 1

$$x \to \int_1^\infty t^{x-1} e^{-t} dt$$

is continuous at x_0. Hence, Γ is continuous at x_0.

Next, we consider differentiability.

Theorem 3 (Leibniz) *Let I be an interval in $\mathbf{R}$ and $f : S \times I \to \mathbf{R}$. Assume*

3.4. INTEGRALS DEPENDING ON A PARAMETER

(i) $f(\cdot, t)$ is μ-integrable for every $t \in I$.

(ii) $\frac{\partial f}{\partial t}(s, t)$ exists for every $s \in S$, $t \in I$.

(iii) There exists a μ-integrable function $g : S \to \mathbf{R}$ such that
$$\left|\frac{\partial f}{\partial t}(s, t)\right| \leq g(s) \text{ for } s \in S, t \in I.$$

If $F(t) = \int_S f(s,t) d\mu(s)$, then F is differentiable on I with $F'(t) = \int_S \frac{\partial f}{\partial t}(s,t) d\mu(s)$.

Proof: Fix $t \in I$ and let $t_k \to t$, $t_k \in I$, $t_k \neq t$. Then for each $s \in S$,
$$\lim_k \frac{f(s, t_k) - f(s, t)}{t_k - t} = \frac{\partial f}{\partial t}(s, t) \text{ and } \frac{f(\cdot, t_k) - f(\cdot, t)}{t_k - t}$$
is μ-integrable for each k. By the Mean Value Theorem, for each s, k there is a $z_{s,k}$ between t_k and t such that
$$\frac{f(s, t_k) - f(s, t)}{t_k - t} = \frac{\partial f}{\partial t}(s, z_{s,k}) \text{ so } \left|\frac{f(s, t_k) - f(s, t)}{t_k - t}\right| \leq g(s)$$
for every $s \in S$ by (iii). Hence, the DCT implies
$$\lim_k \int_S \frac{f(s, t_k) - f(s, t)}{t_k - t} d\mu(s) = \int_S \frac{\partial f}{\partial t}(s, t) d\mu(s) = F'(t).$$

As an example of how Theorems 1 and 3 can be used, we evaluate the integral $\int_0^\infty e^{-x^2} dx$.

Example 4 (Euler) Since $0 \leq e^{-x^2} \leq e^{-x}$ for $x \geq 1$, e^{-x^2} is Lebesgue integrable over $[0, \infty)$. For $t \geq 0$, set $f(t) = (\int_0^t e^{-x^2} dx)^2$ and $g(t) = \int_0^1 e^{-t^2(x^2+1)}/(x^2 + 1) dx$. From standard results on Riemann integration, f is differentiable on $[0, \infty)$ with
$$f'(t) = 2e^{-t^2} \int_0^t e^{-x^2} dx.$$

Since
$$\left|\frac{\partial}{\partial t}\left(e^{-t^2(x^2+1)}/(x^2+1)\right)\right| = \left|-2(te^{-t^2})e^{-t^2 x^2}\right| \leq B$$
for $t \geq 0$, $0 \leq x \leq 1$, Theorem 2 implies that g is differentiable on $[0, \infty)$ with
$$g'(t) = -2te^{-t^2} \int_0^1 e^{-t^2 x^2} dx.$$

Setting $u = tx$ for $t > 0$ in this last integral gives
$$g'(t) = -2e^{-t^2} \int_0^t e^{-u^2} du.$$

Hence, $f'(t) + g'(t) = 0$ for $t > 0$ and $f(t) + g(t) = c$ for $t > 0$. Now f is continuous on $[0,\infty)$ and since $\left|e^{-t^2(x^2+1)}/(x^2+1)\right| \leq 1/(x^2+1)$ for $t \geq 0$, $0 \leq x \leq 1$, g is also continuous on $[0,\infty)$ by Theorem 1. Also, by the DCT,

$$\lim_{t\to\infty} \int_0^1 e^{-t^2(x^2+1)}/(x^2+1)dx = 0.$$

Thus, $f + g = c$ on $[0,\infty)$ and

$$\begin{aligned} f(0) + g(0) &= \int_0^1 \tfrac{1}{1+x^2}dx = \pi/4 \\ &= \lim_{t\to\infty}(f(t) + g(t)) = \left(\int_0^\infty e^{-x^2}dx\right)^2 \end{aligned}$$

so $\int_0^\infty e^{-x^2}dx = \sqrt{\pi}/2$.

Exercise 1. For $x > 0$ show $\Gamma(x+1) = x\Gamma(x)$. If x is a positive integer show $\Gamma(x+1) = x!$. Show $\Gamma(1/2) = \sqrt{\pi}$.

Exercise 2. Show $F(x) = \int_1^\infty \frac{\sin t}{x^2+t^2}dt$ is continuous for $x \in \mathbf{R}$.

Exercise 3. Show $\int_0^\infty x^{2n}e^{-x^2}dx = (2n)!\sqrt{\pi}/(2^{2n}n!2)$. Hint: For $n = 0$ this is Example 4.

Exercise 4. Show $\int_0^\infty e^{-tx^2}dx = \sqrt{(\pi/t)}/2$ for $t > 0$.

Exercise 5. Show the function $F(x) = \int_0^\infty e^{-xt}/(1+t)dt$ is differentiable for $x > 0$.

Exercise 6. Show the Gamma function is differentiable.

Exercise 7. Let $F(t) = \int_0^\infty e^{-tx}\frac{\sin x}{x}dx$ for $t > 0$. Show $F(t) = \frac{\pi}{2} - \arctan t$. Hint: $F'(t) = -1/(1+t^2)$ so $F(t) = c - \arctan t$. Evaluate c by considering $\{F(n)\}$.

Exercise 8. Evaluate $F(x) = \int_0^1 (t^x - 1)/\ln t\, dt$ for $x > 0$. Hint: Find $F'(x)$ and note $F(x) \to 0$ as $x \to 0$.

3.5 Convergence in Mean

In this section we consider the approximation of integrable functions with respect to a natural semi-metric induced by the integral.

Let $(S, \sum, \mu)$ be a measure space. We denote by $L^1(\mu)$ the space of all functions $f : S \to \mathbf{R}$ which are μ-integrable. By 3.2.13 $L^1(\mu)$ is a vector space (under the usual operations of pointwise addition and scalar multiplication). We define a semi-metric d_1 on $L^1(\mu)$ by $d_1(f,g) = \int_S |f - g| \, d\mu$; note $d_1(f,g) = 0$ if and only if $f = g$ μ-a.e. For convenience we set $\|f\|_1 = \int_S |f| \, d\mu$; then $d_1(f,g) = \|f - g\|_1$. $\|\ \|_1$ is called the L^1-norm of f; we consider more general such norms in §6.1. Convergence in this semi-metric is called μ-mean convergence or convergence in μ-mean.

We often write $f_k \to f$ μ-mean when $\|f_k - f\|_1 \to 0$. If I is an interval in $\mathbf{R}^n$, we denote by $L^1(I)$ the space of real-valued integrable functions on I when I is equipped with the Lebesgue measure.

We compare mean convergence with other modes of convergence in §3.7.

One of the most important properties of $L^1(\mu)$ is its completeness with respect to the metric of mean convergence.

Theorem 1 (Riesz-Fischer) $L^1(\mu)$ is complete.

Proof: Let $\{f_k\}$ be Cauchy in $L^1(\mu)$. Then there exists an increasing sequence $\{n_k\}$ such that $\|f_{n_{k+1}} - f_{n_k}\|_1 < 1/2^k$. Since

$$\sum_{k=1}^{\infty} \int_S \left|f_{n_{k+1}} - f_{n_k}\right| d\mu < \infty,$$

the series

$$|f_{n_1}| + \sum_{k=1}^{\infty} \left|f_{n_{k+1}} - f_{n_k}\right|$$

converges μ-a.e. to a real-valued function which belongs to $L^1(\mu)$ (Exer. 3.2.2). Therefore, the series

$$f_1 + \sum_{k=1}^{\infty} (f_{n_{k+1}} - f_{n_k}) = \lim f_{n_k} = f$$

converges μ-a.e. But $\{f_{n_k}\}$ also converges in mean to f since given any $\epsilon > 0$, $\int_S \left|f_{n_k} - f_{n_j}\right| d\mu < \epsilon$ for large k, j, and by Fatou's Lemma, letting $j \to \infty$ gives $\int_S |f_{n_k} - f| \, d\mu \le \epsilon$ for large k. This means that $f \in L^1(\mu)$ and $f_{n_k} \to f$ μ-mean. Hence, $f_k \to f$ μ-mean.

The analogue of Theorem 1 for the Riemann integral is false (§3.3), and its failure is perhaps the most important reason for the overwhelming use of the Lebesgue integral.

We now consider some dense subspaces of $L^1(\mu)$:

Dense Subsets of $L^1(\mu)$:

Theorem 2 *The vector space of Σ-simple μ-integrable functions, $S(\Sigma)$, is dense in $L^1(\mu)$ (with respect to d_1). Moreover, given $f \in L^1(\mu)$ there exists a sequence of simple functions $\{\varphi_k\}$ in $L^1(\mu)$ such that $\varphi_k \to f$ pointwise and $\|\varphi_k - f\|_1 \to 0$; if $f \geq 0$, the $\{\varphi_k\}$ can be chosen such that $\varphi_k \uparrow f$.*

Proof: If $f \in L^1(\mu)$, pick a sequence of Σ-simple functions $\{\varphi_n\}$ such that $\varphi_n \to f$ pointwise with $|\varphi_n| \leq |f|$ (3.1.1.3). The DCT implies $d_1(\varphi_n, f) \to 0$. The last statement follows from 3.1.1.2.

For the next result assume that μ is a premeasure on a semi-ring S of subsets of S and that Σ is the σ-algebra of μ^*-measurable subsets of S (§2.4). Let μ denote the restriction of μ^* to Σ.

Theorem 3 *The vector space of S-simple μ-integrable functions is dense in $L^1(\mu)$.*

Proof: From Theorem 2 it suffices to show that for each $E \in \Sigma$ with $\mu(E) < \infty$ and each $\epsilon > 0$ there is an S-simple function φ such that $\|C_E - \varphi\|_1 < \epsilon$. Pick $\{A_j\} \subset S$ pairwise disjoint such that $\bigcup_{j=1}^{\infty} A_j \supset E$ and $\sum_{j=1}^{\infty} \mu(A_j) < \mu(E) + \epsilon/2$ (2.4.11). Set $B = \bigcup_{j=1}^{\infty} A_j$ and note $\mu(B) < \infty$ and $\sum_{j=1}^{\infty} C_{A_j} = C_B$. Choose N such that

$$0 \leq \int_S (C_B - \sum_{j=1}^{N} C_{A_j}) d\mu = \sum_{j=N+1}^{\infty} \mu(A_j) < \epsilon/2.$$

Then

$$\left\| C_E - \sum_{j=1}^{N} C_{A_j} \right\|_1 \leq \int_S |C_E - C_B| d\mu + \int_S \left| C_B - \sum_{j=1}^{N} C_{A_j} \right| d\mu$$
$$= \sum_{j=1}^{\infty} \mu(A_j) - \mu(E) + \epsilon/2 < \epsilon.$$

We next consider the approximation of integrable functions by continuous functions. A topological space S is *locally compact* if every point in S has a neighborhood with compact closure. For example, $\mathbf{R}^n$ is locally compact. For locally compact Hausdorff spaces, we have the important lemma of Urysohn.

Lemma 4 (Urysohn) *Let S be a locally compact Hausdorff space, $K \subset S$ compact and V open with $K \subset V$. Then there exists a continuous function $f : S \to [0,1]$ such that $f(t) = 1$ for $t \in K$ and $f(t) = 0$ for $t \notin V$.*

3.5. CONVERGENCE IN MEAN

See [Si] §28 for the proof in a general topological space. For metric spaces (such as $\mathbf{R}^n$), one can use the function $f(t) = \operatorname{dist}(t, V^c)/(\operatorname{dist}(t, K) + \operatorname{dist}(t, V^c))$.

We need a slight refinement of Urysohn's Lemma. If $f : S \to \mathbf{R}$ is a continuous function, the *support of f*, denoted by $\operatorname{spt}(f)$, is the closure of the set

$$\{t \in S : f(t) \neq 0\}.$$

The space of continuous functions on S with compact support is denoted by $C_c(S)$.

Lemma 5 *Let S be locally compact, Hausdorff. If K is compact and V is open with $K \subset V$, then there exists $f \in C_c(S)$ such that $f : S \to [0,1]$, $f(t) = 1$ for $t \in K$ and $f(t) = 0$ for $t \in V^c$.*

Proof: If $S = K$, trivial so assume $x \in S \backslash K$. For each $y \in K$ there is an open neighborhood $N_y \subset V$ of y with compact closure and an open neighborhood U_y of x such that $N_y \cap U_y = \emptyset$. Then $\{N_y : y \in K\}$ is an open cover of K and, therefore, has a finite subcover, $N_1, \ldots, N_j$ with $V \supset U = \bigcup_{i=1}^{j} N_i \supset K$ and since $\overline{N}_i$ is compact, $\overline{U} \supset K$ is compact. By Lemma 4 there is a continuous function $f : S \to [0,1]$ such that $f(t) = 1$ for $t \in K$ and $f(t) = 0$ for $t \in U^c$. Since $\operatorname{spt}(f) \subset \overline{U}$, $f \in C_c(S)$.

Theorem 6 *Let S be locally compact Hausdorff and let μ be a Borel measure on $\mathcal{B}(S)$ such that every Borel set is inner regular. Then $C_c(S)$ is dense in $L^1(\mu)$.*

Proof: By Theorem 2 it suffices to show that C_B for $B \in \mathcal{B}(S)$ and $\mu(B) < \infty$ can be approximated by a function in $C_c(S)$. Let $\epsilon > 0$. Since B and B^c are inner regular, there exist $K_1 \subset B$ compact with $\mu(B \backslash K_1) < \epsilon/2$ and $K_2 \subset B^c$ compact with $\mu(B^c \backslash K_2) < \epsilon/2$. By Lemma 4 there is a function $f \in C_c(S)$ such that $0 \leq f(t) \leq 1$ for $t \in S$, $f(t) = 1$ for $t \in K_1$ and $f(t) = 0$ for $t \in K_2$. Then

$$\int_S |f - C_B|\, d\mu = \int_{B \backslash K_1} |f - C_B|\, d\mu + \int_{B^c \backslash K_2} |f - C_B|\, d\mu \leq \mu(B \backslash K_1) + \mu(B^c \backslash K_2) < \epsilon.$$

Note that this theorem is applicable to Lebesgue and Lebesgue-Stieltjes measures (Exer. 2.5.10 and 2.6.5). See also Proposition 2.7.2.

Exercise 1. Show L^1 is not, in general, closed under pointwise products. [Hint: Consider t^a for $0 \leq t \leq 1$.]

Exercise 2. Show the polynomials are dense in $L^1[a, b]$. Generalize to $\mathbf{R}^n$.

Exercise 3. Show $L^1[a, b]$ is separable. Generalize to $\mathbf{R}^n$.

Exercise 4. Give an example of a non-regular measure for which $C_c(S)$ is not dense in $L^1(\mu)$. [Hint: Use $\mathbf{R}$ with Lebesgue measure and the discrete metric.]

Exercise 5. What is the completion of $\mathcal{R}[a,b]$?

Exercise 6. Let μ be a finite measure on the σ-algebra $\sum$. Define a semi-metric d on $\sum$ by
$$d(A,B) = \mu(A \triangle B) = \int_S |C_A - C_B|\, d\mu.$$
Show d is a complete semi-metric.

3.6 Convergence in Measure

We consider another type of convergence with respect to a measure. Let (S, Σ, μ) be a measure space and f_k, $f : S \to \mathbf{R}$ Σ-measurable functions. We say that $\{f_k\}$ *converges to f in μ-measure*, $f_k \to f$ μ-measure, if for every $\sigma > 0$,

$$\lim_k \mu\{t : |f_k(t) - f(t)| \geq \sigma\} = 0.$$

We compare the various modes of convergence in §3.7. We now develop some of the properties of convergence in μ-measure.

Proposition 1 *Let f_k, f, g_k, $g : S \to \mathbf{R}$ be Σ-measurable.*

(i) *If $f_k \to f$ μ-measure and $g_k \to g$ μ-measure, then $af_k + bg_k \to af + bg$ μ-measure for $a, b \in \mathbf{R}$.*

(ii) *If $f_k \to 0$ μ-measure and $g_k \to 0$ μ-measure, then $f_k g_k \to 0$ μ-measure.*

(iii) *If $\mu(S) < \infty$ and $f_k \to f$ μ-measure, then $f_k g \to fg$ μ-measure.*

(iv) *If $\mu(S) < \infty$, $f_k \to f$ μ-measure and $g_k \to g$ μ-measure, then $f_k g_k \to fg$ μ-measure.*

(v) *If $f_k \to f$ μ-measure and $f_k \to g$ μ-measure, then $f = g$ μ-a.e.*

Proof: (i): For $\sigma > 0$,

$$\mu\{t : |f_k(t) + g_k(t) - f(t) - g(t)| \geq \sigma\} \leq \mu\{t : |f_k(t) - f(t)| \geq \sigma/2\} + \mu\{t : |g_k(t) - g(t)| \geq \sigma/2\}$$

implies that $f_k + g_k \to f + g$ μ-measure. That $af_k \to af$ μ-measure is clear.

(ii) follows from

$$\{t : |f_k(t)g_k(t)| \geq \sigma\} \subset \{t : |f_k(t)| \geq \sqrt{\sigma}\} \cup \{t : |g_k(t)| \geq \sqrt{\sigma}\}.$$

(iii): Since $\mu(S) < \infty$, $\lim_k \mu\{t : |g(t)| \geq k\} = 0$ (2.2.4). Let $\epsilon > 0$. There exists N such that $\mu\{t : |g(t)| \geq N\} < \epsilon/2$. Now

$$\mu\{t : |f_k(t)g(t) - f(t)g(t)| \geq \sigma\} \leq \mu\{t : |f_k(t) - f(t)| \geq \sigma/N\} + \mu\{t : |g(t)| \geq N\}$$

so if k_0 is chosen such that $k \geq k_0$ implies $\mu\{t : |f_k(t) - f(t)| \geq \sigma/N\} < \epsilon/2$, we have for $k \geq k_0$ that $\mu\{t : |f_k(t)g(t) - f(t)g(t)| \geq \sigma\} < \epsilon$.

(iv) follows from (i), (ii), (iii) and the fact that

$$f_k g_k - fg = (f_k - f)(g_k - g) + f(g_k - g) + g(f_k - f).$$

(v): Since
$$\{t : f(t) \neq g(t)\} = \bigcup_{k=1}^{\infty} \{t : |f(t) - g(t)| \geq 1/k\},$$
it suffices to show that $\mu\{t : |f(t) - g(t)| \geq \sigma\} = 0$ for $\sigma > 0$. But
$$\{t : |f(t) - g(t)| \geq \sigma\} \subset \{t : |f_k(t) - f(t)| \geq \sigma/2\} \cup \{t : |f_k(t) - g(t)| \geq \sigma/2\}.$$

Example 2 The finiteness condition in (iii) and (iv) cannot be dropped. Let $S = \mathbf{R}$ and consider Lebesgue measure. Let $f_k(t) = 1/k$ and $g(t) = t$ for $t \in \mathbf{R}$. Then $f_k \to 0$ in m-measure but for $\sigma > 0$, $m\{t : |f_k(t)g(t)| \geq \sigma\} = \infty$.

Concerning convergence in measure and convergence a.e., we have an important result of F. Riesz.

Theorem 3 (F. Riesz) *Let $f_k \to f$ μ-measure. Then there is a subsequence $\{f_{n_k}\}$ such that $f_{n_k} \to f$ μ-a.e.*

Proof: For each k there exists n_k such that $j \geq n_k$ implies
$$\mu\{t : |f_j(t) - f(t)| \geq 1/2^k\} < 1/2^k.$$
Set
$$E_k = \{t : |f_{n_k}(t) - f(t)| \geq 1/2^k\}.$$
If $t \notin \bigcup_{k=j}^{\infty} E_k$, then $|f_{n_k}(t) - f(t)| < 1/2^k$ for $k \geq j$. Thus, if
$$t \notin \bigcap_{j=1}^{\infty} \bigcup_{k=j}^{\infty} E_k = \overline{\lim} E_k = E,$$
then $f_{n_k}(t) \to f(t)$, i.e., $f_{n_k} \to f$ pointwise on $S \backslash E$.
But
$$\mu(E) \leq \sum_{k=j}^{\infty} \mu(E_k) \leq \sum_{k=j}^{\infty} 1/2^k = 1/2^{j-1}$$
for every j so $\mu(E) = 0$.

The assertion that there is a subsequence which converges a.e. in Theorem 3 is important – the entire sequence may not converge a.e. as the following example shows.

Example 4 Let $S = [0,1]$ and $A_k^n = [(k-1)/2^n, k/2^n]$ for $k = 1, \ldots, 2^n$ and $n \in \mathbf{N}$. Consider the sequence $C_{A_1^1}, C_{A_2^1}, C_{A_1^2}, \ldots$ [see the sketch below].

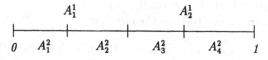

3.6. CONVERGENCE IN MEASURE

This sequence converges to 0 in m-measure since $m\{t : C_{A_k^n}(t) \geq \sigma\} \leq m(A_k^n) = 1/2^n$. However, this sequence doesn't converge to 0 at any point since it is 1 infinitely often at any point. [See Exercise 4.]

We can formulate a Cauchy-type condition for convergence in μ-measure. The sequence $\{f_k\}$ is said to be *Cauchy in μ-measure* if and only if for every $\sigma > 0$ and $\epsilon > 0$ there exists N such that $k, j \geq N$ implies

$$\mu\{t : |f_k(t) - f_j(t)| \geq \sigma\} < \epsilon.$$

We now establish a completeness type result for convergence in μ-measure which is also due to F. Riesz.

Lemma 5 *If $\{f_k\}$ is Cauchy in μ-measure and has a subsequence $\{f_{n_k}\}$ which converges in μ-measure to a measurable function f, then $f_k \to f$ μ-measure.*

Proof: Let $\sigma, \epsilon > 0$. There exists N such that $k, j \geq N$ implies

$$\mu\{t : |f_k(t) - f_j(t)| \geq \sigma/2\} < \epsilon/2.$$

Choose $n_k \geq N$ such that

$$\mu\{t : |f_{n_k}(t) - f(t)| \geq \sigma/2\} < \epsilon/2.$$

Then for $j \geq N$,

$$\mu\{t : |f_j(t) - f(t)| \geq \sigma\} \leq \mu\{t : |f_j(t) - f_{n_k}(t)| \geq \sigma/2\}$$
$$+ \mu\{t : |f_{n_k}(t) - f(t)| \geq \sigma/2\} < \epsilon.$$

Theorem 6 (F. Riesz) *Let $\{f_k\}$ be Cauchy in μ-measure. Then there exists a measurable function f such that $f_k \to f$ μ-measure.*

Proof: For each k there exists n_k such that $i, j \geq n_k$ implies

$$\mu\{t : |f_i(t) - f_j(t)| \geq 1/2^k\} < 1/2^k.$$

We may assume $n_k < n_{k+1}$. Let

$$E_k = \{t : |f_{n_{k+1}}(t) - f_{n_k}(t)| \geq 1/2^k\}$$

so $\mu(E_k) < 1/2^k$. If $F_k = \bigcup_{j=k}^{\infty} E_j$, then $\mu(F_k) < \sum_{j=k}^{\infty} 1/2^j = 1/2^{k-1}$ so if $A = \bigcap_{k=k}^{\infty} F_k = \overline{\lim} E_k$, then $\mu(A) = 0$.

We claim that $\{f_{n_k}\}$ converges pointwise on $S \setminus A$. If $t \in S \setminus A$, then $t \notin \bigcup_{j=k}^{\infty} E_j$ for some k so if $i, j \geq k$ with $i > j$,

$$\left|f_{n_i}(t) - f_{n_j}(t)\right| \leq \sum_{\ell=0}^{i-j-1} \left|f_{n_{j+\ell+1}}(t) - f_{n_{j+\ell}}(t)\right| \leq \sum_{\ell=0}^{i-j-1} 1/2^{j+\ell} < 1/2^{j-1}, \quad (3.1)$$

and $\{f_{n_k}(t)\}$ converges.

Now set $f(t) = \lim f_{n_k}(t)$ if $t \in S\backslash A$ and $f(t) = 0$ for $t \in A$. Then f is measurable since $f = C_{S\backslash A}\overline{\lim} f_{n_k}$.

Finally, we claim that $f_{n_k} \to f$ μ-measure. Let $\sigma, \epsilon > 0$. Choose N such that $\mu(F_N) < 1/2^{N-1} < \min(\sigma, \epsilon)$. If $j \geq N$, then

$$\{t : |f_{n_j}(t) - f(t)| \geq \sigma\} \subset \{t : |f_{n_j}(t) - f(t)| > 1/2^{N-1}\} \subset F_N$$

since if $t \notin F_N$ and $j \geq N$, passing to the limit in (1) as $i \to \infty$ gives $|f(t) - f_{n_j}(t)| \leq 1/2^{N-1}$. Thus, if $j \geq N$, $\mu\{t : |f_{n_j}(t) - f(t)| \geq \sigma\} \leq \mu(F_N) < \epsilon$.

Lemma 5 now gives the result.

Metric of Convergence in Measure:

We now describe a semi-metric which characterizes convergence in measure for finite measures. Assume henceforth that μ is a finite measure. We first require a lemma.

Lemma 7 *If* $a, b \in \mathbf{R}$, $\frac{|a+b|}{1+|a+b|} \leq \frac{|a|}{1+|a|} + \frac{|b|}{1+|b|}$.

Proof: Note the function $h(t) = t/(1+t)$ is increasing for $t > -1$. First, if a and b have the same signs, we may assume that $a, b \geq 0$ so

$$|a+b|/(1+|a+b|) = (a+b)/(1+a+b) \leq a/(1+a) + b/(1+b) = |a|/(1+|a|) + |b|/(1+|b|).$$

On the other hand, if a and b have different signs, we may assume that $|a| \geq |b|$. Then $|a+b| \leq |a|$ implies

$$|a+b|/(1+|a+b|) \leq |a|/(1+|a|) \leq |a|/(1+|a|) + |b|/(1+|b|).$$

Definition 8 *Let* $L^0(\mu)$ *be the vector space of all real valued measurable functions on* S. *For* $f, g \in L^0(\mu)$, *set*

$$d(f,g) = \int_S \frac{|f-g|}{1+|f-g|} d\mu$$

[note the integral exists since μ is finite and the integrand is bounded].

By Lemma 7, d defines a semi-metric on $L^0(\mu)$ which is translation invariant in the sense that $d(f,g) = d(f+h, g+h)$ for any $h \in L^0(\mu)$ and is such that $d(f,0) = 0$ if and only if $f = 0$ μ-a.e.

We show that convergence in the semi-metric d is exactly convergence in μ-measure.

Theorem 9 *Let* $f_k, f \in L^0(\mu)$.

(i) $f_k \to f$ *in* μ-*measure* $\Leftrightarrow d(f_k, f) \to 0$.

3.6. CONVERGENCE IN MEASURE

(ii) $\{f_k\}$ *is Cauchy in μ-measure* $\Leftrightarrow$ $\{f_k\}$ *is Cauchy with respect to d.*

Proof: (i): Let $\epsilon > 0$ and set $E_k = \{t : |f_k(t) - f(t)| \geq \epsilon\}$. Then
$$d(f_k, f) = \int_{E_k} \frac{|f_k - f|}{1+|f_k-f|} d\mu + \int_{S\setminus E_k} \frac{|f_k-f|}{1+|f_k-f|} d\mu \qquad (3.2)$$
$$\leq \mu(E_k) + \epsilon\mu(S\setminus E_k) \leq \mu(E_k) + \epsilon\mu(S).$$

Thus, if $f_k \to f$ μ-measure, (2) implies that $d(f_k, f) \to 0$.

Since $h(t) = t/(1+t)$ is increasing for $t > -1$,
$$d(f_k, f) \geq \int_{E_k} \frac{|f_k - f|}{1+|f_k-f|} d\mu \geq \frac{\epsilon}{1+\epsilon}\mu(E_k). \qquad (3.3)$$

Thus, if $d(f_k, f) \to 0$, (3) implies that $f_k \to f$ μ-measure.

(ii) follows from the inequalities (2) and (3) with f replaced by f_j.

Corollary 10 *d is a complete semi-metric on $L^0(\mu)$.*

Proof: Theorems 6 and 9.

If I is a bounded interval in $\mathbf{R}^n$, we denote by $\underline{L^0(I)}$ the space of all Lebesgue measurable functions on I, and we assume that the semi-metric on $L^0(I)$ is the semi-metric of convergence in Lebesgue measure.

There is a semi-metric which characterizes convergence in measure for infinite measures, but it is much more complicated than the semi-metric d; see [DS] III.2 for a description.

Exercise 1. If μ is counting measure on S, show convergence in μ-measure is exactly uniform convergence on S.

Exercise 2. Show that if $f_k \to f$ μ-measure and $f = g$ μ-a.e., then $f_k \to g$ μ-measure.

Exercise 3. If $f_k \to f$ μ-measure, show $\{f_k\}$ is Cauchy in μ-measure.

Exercise 4. Find a subsequence in Example 4 which converges m-a.e. to 0.

Exercise 5. Can the condition "$\mu(S) < \infty$" in Proposition 1 (iii) be replaced by "g is bounded"?

Exercise 6. Let $f : S \to \mathbf{R}$ be measurable. Show f is μ-integrable if and only if there exists a sequence of μ-integrable simple functions $\{\varphi_k\}$ such that

(i) $\varphi_k \to f$ μ-measure and

(ii) $\lim_{k,j} \int_S |\varphi_k - \varphi_j| d\mu = 0$.

3.7 Comparison of Modes of Convergence

In this section we pause to compare the various modes of convergence which have been introduced. Let (S, Σ, μ) be a measure space. We have considered the following types of convergence for sequences of measurable functions defined on S.

unif:	uniform convergence on S.
a. unif:	almost uniform convergence with respect to μ (Definition 3.1.16).
a.e.:	almost everywhere convergence with respect to μ (3.1.13).
mean:	convergence in μ-mean (§3.5).
meas:	convergence in μ-measure (§3.6).

We have the obvious implication that unif $\Rightarrow$ a. unif and from Exercise 3.1.13 we have a. unif $\Rightarrow$ a.e. It is also clear that a. unif $\Rightarrow$ meas (Exer. 1). There is one other general implication which we now establish.

Proposition 1 *If $f_k \to f$ μ-mean, then $f_k \to f$ μ-measure.*

Proof: Let $\sigma > 0$ and $E_k = \{t : |f_k(t) - f(t)| \geq \sigma\}$. Then

$$\int_S |f_k - f|\, d\mu \geq \int_{E_k} |f_k - f|\, d\mu \geq \sigma \mu(E_k) \text{ so } \mu(E_k) \to 0.$$

We now give examples to show that these are the only possible general implications which are valid. [Recall from Egoroff's Theorem a.e. $\Rightarrow$ a. unif for finite measures.]

2. a.e. $\not\Rightarrow$ a. unif: Take $f_k = C_{[k,\infty)}$ in $\mathbf{R}$ with Lebesgue measure.

3. a. unif $\not\Rightarrow$ unif: Take $f_k(t) = t^k$, $0 \leq t \leq 1$, and Lebesgue measure.

4. meas $\not\Rightarrow$ a.e.: Example 3.6.4.

5. meas $\not\Rightarrow$ a. unif: Example 3.6.4.

6. meas $\not\Rightarrow$ mean: Take $f_k = kC_{[0,1/k]}$ on $[0,1]$ with Lebesgue measure.

7. a.e. $\not\Rightarrow$ mean: Same as 6.

8. mean $\not\Rightarrow$ a.e.: Example 3.6.4.

9. a.e. $\not\Rightarrow$ meas: Same as 2.

10. a. unif $\not\Rightarrow$ mean: Same as 6.

11. unif $\not\Rightarrow$ mean: Take $f_k = C_{[0,k]}/k$ on $\mathbf{R}$ with Lebesgue measure.

3.7. COMPARISON OF MODES OF CONVERGENCE

12. mean $\not\Rightarrow$ a. unif: Example 3.6.4.

We can summarize the relationships above by means of the chart:

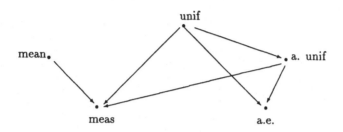

If μ is a finite measure, then by Egoroff's Theorem we have that a.e. $\Rightarrow$ a. unif (so also a.e. $\Rightarrow$ meas), and by Exercise 3.2.5, we also have unif $\Rightarrow$ mean. In this case, we have the chart:

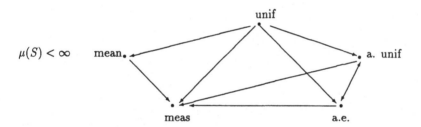

It is also possible to consider the case where a sequence of measurable function $\{f_k\}$ is dominated by a μ-integrable function g, i.e., $|f_k| \le g$ μ-a.e. In this case it follows from the DCT that a.e. $\Rightarrow$ mean and, similarly, meas $\Rightarrow$ mean (Exercise 2). We refer the reader to Munroe ([Mu] p. 237) for a complete discussion.

Exercise 1. Show that a. unif $\Rightarrow$ meas.

Exercise 2. Show that the DCT is valid if convergence μ-a.e. is replaced by convergence in μ-measure. [Hint: Theorem 3.6.3.]

Exercise 3. Show $L^0[a, b]$ is separable.

Exercise 4. Show a $\sum$-measurable function f is μ-integrable if and only if $\exists$ a sequence of μ-integrable $\sum$-simple functions $\{\varphi_n\}$ such that $\varphi_n \to f$ μ-measure and $\{\varphi_n\}$ is Cauchy in $L^1(\mu)$. Moreover, $\int_S f d\mu = \lim \int_S \varphi_n d\mu$.

3.8 Mikusinski's Characterization of the Lebesgue Integral

In this section we give an interesting characterization of the Lebesgue integral due to J. Mikusinski. The characterization is used later in the proof of Fubini's Theorem.

Let S be a semi-ring of subsets of S and let μ be a premeasure on S. Let μ also denote the countably additive extension of μ to the σ-algebra, $\mathcal{M}$, of μ^*-measurable subsets of S (§2.4).

For Mikusinski's description of the Lebesgue integral, we require two lemmas.

Lemma 1 *Let $E \in \mathcal{M}$ be μ-null and $\epsilon > 0$. There exists $\{J_k\} \subset S$ such that $\sum_{k=1}^{\infty} \mu(J_k) < \epsilon$ and $\sum_{k=1}^{\infty} C_{J_k}(t) = \infty$ for $t \in E$ [so t belongs to infinitely many J_k for $t \in E$].*

Proof: For each i, there is a sequence $\{I_{ij} : j \in \mathbf{N}\} \subset S$ covering E such that $\sum_{j=1}^{\infty} \mu(I_{ij}) < \epsilon/2^i$ (Definition 2.4.1). Now arrange the double sequence $\{I_{ij} : i, j \in \mathbf{N}\}$ into a sequence $\{J_k\}$. If $t \in E$, t belongs to infinitely many $\{J_k\}$ so $\{C_{J_k}(t)\}$ is 1 infinitely often and

$$\sum_{k=1}^{\infty} \mu(J_k) = \sum_{i=1}^{\infty} \sum_{j=1}^{\infty} \mu(I_{ij}) < \epsilon.$$

We say that a series $\sum a_k$ is *absolutely convergent to x* if the series is absolutely convergent and $\sum_{k=1}^{\infty} a_k = x$.

Lemma 2 *If $f : S \to \mathbf{R}^*$ is μ-integrable, there is a sequence of S-simple functions $\{\alpha_k\}$ such that the series $\sum_k \alpha_k$ is absolutely convergent to f μ-a.e. and*

$$\sum_{k=1}^{\infty} \int_S |\alpha_k| \, d\mu < \infty.$$

Proof: By Theorem 3.5.3 there exists a sequence of S-simple functions $\{\varphi_k\}$ such that $\|\varphi_k - f\|_1 \to 0$. Then $\varphi_k \to f$ μ-measure (Proposition 3.7.1) so by Theorem 3.6.3 there is a subsequence, which we continue to denote by $\{\varphi_k\}$, which converges μ-a.e. to f. We may additionally assume that $\|\varphi_{k+1} - \varphi_k\|_1 < 1/2^k$. Set $\varphi_0 = 0$ and $\alpha_k = \varphi_k - \varphi_{k-1}$ for $k \geq 1$. Then $\sum_{k=1}^n \alpha_k = \varphi_n \to f$ μ-a.e. or $\sum_{k=1}^{\infty} \alpha_k = f$ μ-a.e. Since $\sum_{k=1}^{\infty} \int_S |\alpha_k| \, d\mu < \infty$, the MCT implies that $\sum_{k=1}^{\infty} |\alpha_k|$ converges in $\mathbf{R}$ μ-a.e. so the series $\sum \alpha_k$ is absolutely convergent to f μ-a.e.

Theorem 3 *$f : S \to \mathbf{R}$ is μ-integrable $\Leftrightarrow$ there is a sequence of S-simple functions $\{\psi_k\}$ satisfying*

(1) $\sum_{k=1}^{\infty} \int_S |\psi_k| \, d\mu < \infty$

(2) $f(t) = \sum_{k=1}^{\infty} \psi(t)$ for any t for which $\sum_{k=1}^{\infty} |\psi_k(t)| < \infty$.

In this case,

$$\int_S f \, d\mu = \sum_{k=1}^{\infty} \int_S \psi_k \, d\mu.$$

Proof: $\Leftarrow$: The MCT implies that $\sum |\psi_k|$ is μ-integrable so the series $\sum |\psi_k|$ converges μ-a.e. in **R**. By hypothesis, $f(t) = \sum_{k=1}^{\infty} \psi_k(t)$ for such points. Since $|f| \leq \sum |\psi_k|$ μ-a.e., the DCT implies that f is μ-integrable and $\int_S f \, d\mu = \sum_{k=1}^{\infty} \int_S \psi_k \, d\mu$.

$\Rightarrow$: Let $\{\alpha_k\}$ be as in Lemma 2 and let E be a μ-null set where $\sum_{k=1}^{\infty} |\alpha_k(t)| < \infty$ and $f(t) = \sum_{k=1}^{\infty} \alpha_k(t)$ for $t \notin E$. Let $\{J_k\}$ be as in Lemma 1 (with respect to E) and set $\beta_k = C_{J_k}$. Define a sequence of S-simple functions, $\{\psi_k\}$, by:

$$\alpha_1, \beta_1, -\beta_1, \alpha_2, \beta_2, -\beta_2, \ldots \quad [\psi_{3k-2} = \alpha_k, \psi_{3k-1} = \beta_k, \psi_{3k} = -\beta_k].$$

If $t \in E$, the series $\sum |\psi_k(t)|$ diverges since $\beta_k(t) = 1$ infinitely often. If $\sum_{k=1}^{\infty} |\psi_k(t)| < \infty$, then $\sum_{k=1}^{\infty} \beta_k(t) < \infty$ [the series actually only has a finite number of non-zero terms] so $t \notin E$ and $\sum_{k=1}^{\infty} \psi_k(t) = \sum_{k=1}^{\infty} \alpha_k(t) = f(t)$. Moreover,

$$\sum_{k=1}^{\infty} \int_S |\psi_k| \, d\mu \leq \sum_{k=1}^{\infty} \int_S |\alpha_k| \, d\mu + 2 \sum_{k=1}^{\infty} \int_S |\beta_k| \, d\mu < \infty$$

by Lemmas 1 and 2.

Note, in particular, that if conditions (1) and (2) are satisfied, the function f is measurable and the series $\sum \psi_k$ converges to f μ-a.e.

One of the important features of Mikusinski's characterization of the integral is that null sets are not mentioned. We use this to good advantage in our proof of Fubini's Theorem given in §3.9.

The characterization of the integral in Theorem 3 was given for the Lebesgue integral in **R** by MacNeille ([Mac]) and for the Lebesgue integral in $\mathbf{R}^n$ by J. Mikusinski ([M1]). In $\mathbf{R}^n$ the characterization has an even simpler form; the S-simple functions in Theorem 3 can be replaced by scalar multiples of characteristic functions of half-closed intervals. In this case the value of the integral only depends on the value of the measure of such half-closed intervals so a definition of the Lebesgue integral can be based on the characterization which does not require the full development of properties of Lebesgue measure. Such a development of the integral is carried out in [M2] and [DM].

3.8. MIKUSINSKI'S CHARACTERIZATION – LEBESGUE INTEGRAL

Exercise 1. Assume μ is complete. Show that if $\{\psi_k\}$ is a sequence of μ-integrable functions satisfying (1) and (2), then f is measurable, μ-integrable and

$$\int_S f d\mu = \sum_{k=1}^{\infty} \int_S \psi_k d\mu.$$

3.9 Product Measures and Fubini's Theorem

Let $(S, \mathcal{S}, \mu)$ and $(T, \mathcal{T}, \nu)$ be measure spaces. If $A \subset S$, $B \subset T$, a set of the form $A \times B$ is called a *rectangle* in $S \times T$ and A and B are called its *sides*. If $A \in \mathcal{S}$, $B \in \mathcal{T}$, $A \times B$ is called a *measurable rectangle*. Let $\mathcal{R}$ be the family of all measurable rectangles in $S \times T$. Since

$$(A \times B) \cap (C \times D) = (A \cap C) \times (B \cap D)$$

and

$$S \times T \setminus A \times B = (A^c \times B) \cup (A \times B^c) \cup (A^c \times B^c),$$

$\mathcal{R}$ is a semi-algebra of subsets of $S \times T$.

We define the product of μ and ν, denoted by $\mu \times \nu$, on $\mathcal{R}$ by setting $\mu \times \nu(A \times B) = \mu(A)\nu(B)$ [here $0 \cdot \infty = 0$ as usual].

Theorem 1 $\mu \times \nu$ *is a premeasure on* $\mathcal{R}$.

Proof: Let $\{A_i \times B_i\} \subset \mathcal{R}$ be pairwise disjoint with $A \times B = \bigcup_{i=1}^{\infty} A_i \times B_i \in \mathcal{R}$. Then

$$C_{A \times B}(s, t) = \sum_{i=1}^{\infty} C_{A_i \times B_i}(s, t)$$

for $s \in S$, $t \in T$. Fix s and integrate with respect to ν to obtain from the MCT that

$$C_A(s)\nu(B) = \sum_{i=1}^{\infty} C_{A_i}(s)\nu(B_i).$$

Now integrate with respect to μ and apply the MCT to obtain

$$\mu(A)\nu(B) = \sum_{i=1}^{\infty} \mu(A_i)\nu(B_i)$$

as desired.

The premeasure $\mu \times \nu$ can now be extended to a complete measure on the σ-algebra of $(\mu \times \nu)^*$-measurable subsets of $S \times T$ (§2.4). We denote this measure by $\underline{\mu \times \nu}$ and call it the *product* of μ and ν [see, however, Example 2 below]. We refer to the elements of the σ-algebra of $(\mu \times \nu)^*$-measurable sets as $\underline{\mu \times \nu}$-*measurable* sets and functions which are measurable with respect to this σ-algebra as being $\mu \times \nu$-*measurable*. Note that if both μ and ν are finite (σ-finite), then $\mu \times \nu$ is finite (σ-finite). Thus, if μ and ν are σ-finite, $\mu \times \nu$ is the unique measure on the smallest σ-algebra containing $\mathcal{R}$, denoted by $\underline{\sigma(\mathcal{S} \times \mathcal{T})}$, which satisfies $\mu \times \nu(A \times B) = \mu(A)\nu(B)$ [Theorem 2.4.12].

If m is Lebesgue measure on $\mathbf{R}$, then $m \times m$ is a complete measure on a σ-algebra in $\mathbf{R}^2$ which contains all of the measurable rectangles and, hence, contains all of the open (Borel) sets in $\mathbf{R}^2$. Thus, $m \times m$ and m_2 agree on $\mathcal{B}(\mathbf{R}^2)$ and, hence, must be identical (Theorem 2.4.21 and Corollary 2.5.4 (vi)).

3.9. PRODUCT MEASURES AND FUBINI'S THEOREM

Example 2 (Bartle) In general the product measure may not be the only countably additive extension of $\mu \times \nu$ to the σ-algebra generated by $\mathcal{R}$.

Let $\mu : \mathcal{P}(\mathbf{R}) \to \mathbf{R}^*$ be defined by $\mu(A) = \infty$ if A is uncountable and $\mu(A) = 0$ otherwise. Let $\sum$ be the σ-algebra generated by $\mathcal{P}(\mathbf{R}) \times \mathcal{P}(\mathbf{R})$. Let P_T (P_S) be the projection of $\mathbf{R} \times \mathbf{R}$ onto the first (second) coordinate; $P_S(s,t) = s$ ($P_T(s,t) = t$).

Define π on $\sum$ by $\pi(E) = 0$ if $E = G \cup H$, where $P_S(G)$ and $P_T(H)$ are countable, and $\pi(E) = \infty$ otherwise. Then π is a measure on $\sum$ such that $\pi(A \times B) = \mu(A)\mu(B) = \mu \times \mu(A \times B)$.

Define ρ on $\sum$ by $\rho(E) = 0$ if $E = G \cup H \cup K$, where $P_S(G)$, $P_T(H)$ and the projection of K onto the diagonal $\{(x,x) : x \in \mathbf{R}\}$ are countable, and $\rho(E) = \infty$ otherwise. Then ρ is a measure on $\sum$ such that $\rho(A \times B) = \mu(A)\mu(B)$.

If $E = \{(x,y) : x + y = 0\}$, then $E \in \sum$ but $\rho(E) = 0$ while $\pi(E) = \infty$.

Many other pathologies can occur for product measures; see, for example, [Ga] or [DG].

We now consider the evaluation of integrals with respect to a product measure. As is the case with the Riemann integral, the most efficient way to evaluate an integral with respect to a product measure is as an iterated integral. A result which asserts the equality of an integral with respect to a product measure and an iterated integral is often referred to as a Fubini theorem. For example, if $f : S \times T \to \mathbf{R}$ is $\mathcal{R}$-simple and $\mu \times \nu$ integrable, it follows from the definition of the product measure that

$$\int_{S \times T} f d\mu \times \nu = \int_T \int_S f(s,t) d\mu(s) d\nu(t) = \int_S \int_T f(s,t) d\nu(t) d\mu(s),$$

i.e., Fubini's Theorem holds for such functions. We now use Mikusinski's characterization of the Lebesgue integral to extend Fubini's Theorem to $\mu \times \nu$-integrable functions.

Henceforth, we assume that the measures μ and ν are both complete.

If $f : S \times T \to \mathbf{R}^*$, for $s \in S$ ($t \in T$) we denote by $f(s,\cdot)$ ($f(\cdot,t)$) the function $f(s,\cdot) : T \to \mathbf{R}^*$ ($f(\cdot,t) : S \to \mathbf{R}^*$) defined by $f(s,\cdot)(t) = f(s,t)$ ($f(\cdot,t)(s) = f(s,t)$). We follow the usual procedure in Fubini's Theorem of agreeing that if a function φ (ψ) is defined μ-a.e. in S (ν-a.e. in T), then φ (ψ) is extended to all of $S(T)$ by setting it equal to 0 on the μ-null (ν-null) set where it is undefined. Thus, if $\{\varphi_k\}$ is a sequence of S-measurable functions which converge pointwise μ-a.e. in S, we may assume that there is an S-measurable function φ defined on S such that $\varphi_k \to \varphi$ μ-a.e. [the completeness of μ is used here]. This situation is encountered several times in the proof of Fubini's Theorem.

Theorem 3 Let $f : S \times T \to \mathbf{R}$ be $\mu \times \nu$-integrable. Then

(i) $f(s,\cdot)$ is ν-integrable for μ almost all $s \in S$,

(ii) $s \to \int_T f(s,\cdot) d\nu$ is μ-integrable, and

(iii) $\int_{S \times T} f d\mu \times \nu = \int_S \int_T f(s,\cdot) d\nu d\mu(s) = \int_S \int_T f(s,t) d\nu(t) d\mu(s)$.

Proof: By Theorem 3.8.3, there is a sequence $\{\psi_k\}$ of $\mathcal{S} \times \mathcal{T}$-simple functions such that $\sum_{k=1}^{\infty} \int_{S \times T} |\psi_k| \, d\mu \times \nu < \infty$ and $f(s,t) = \sum_{k=1}^{\infty} \psi_k(s,t)$ for all (s,t) for which the series $\sum_{k=1}^{\infty} |\psi_k(s,t)|$ converges and

$$\int_{S \times T} f \, d\mu \times \nu = \sum_{k=1}^{\infty} \int_{S \times T} \psi_k \, d\mu \times \nu. \tag{3.1}$$

By the MCT

$$\sum_{k=1}^{\infty} \int_{S \times T} |\psi_k| \, d\mu \times \nu = \sum_{k=1}^{\infty} \int_{S} \int_{T} |\psi_k(s,t)| \, d\nu(t) d\mu(s) \tag{3.2}$$

$$= \int_{S} \sum_{k=1}^{\infty} \int_{T} |\psi_k(s,t)| \, d\nu(t) d\mu(s) < \infty$$

so there is a μ-null set E such that $\sum_{k=1}^{\infty} \int_T |\psi_k(s,t)| \, d\nu(t) < \infty$ for $s \notin E$. For $s \notin E$, the MCT implies that

$$\sum_{k=1}^{\infty} \int_T |\psi_k(s,t)| \, d\nu(t) = \int_T \sum_{k=1}^{\infty} |\psi_k(s,t)| \, d\nu(t)$$

so $\sum_{k=1}^{\infty} \psi_k(s,t)$ is absolutely convergent in **R** for ν-almost all $t \in T$. [The ν-null set may depend on s.] For such s, t, $f(s,t) = \sum_{k=1}^{\infty} \psi_k(s,t)$; in particular, for $s \notin E$, $f(s,\cdot) = \sum_{k=1}^{\infty} \psi_k(s,\cdot)$ ν-a.e., and since

$$\left| \sum_{k=1}^{n} \psi_k(s,\cdot) \right| \le \sum_{k=1}^{\infty} |\psi_k(s,\cdot)|,$$

the DCT implies that $f(s,\cdot)$ is ν-integrable and $\int_T f(s,\cdot) d\nu = \sum_{k=1}^{\infty} \int_T \psi_k(s,\cdot) d\nu$. Further, for $s \notin E$,

$$\left| \sum_{k=1}^{n} \int_T \psi_k(s,\cdot) d\nu \right| \le \sum_{k=1}^{\infty} \int_T |\psi_k(s,\cdot)| \, d\nu$$

and the function on the right hand side of this inequality is μ-integrable by (2) so the DCT implies that $s \to \int_T f(s,\cdot) d\nu$ is μ-integrable with

$$\int_S \int_T f(s,\cdot) d\nu d\mu(s) = \sum_{k=1}^{\infty} \int_S \int_T \psi_k(s,t) d\nu(t) d\mu(s) = \int_{S \times T} f \, d\mu \times \nu$$

by (1).

The main difficulty in applying Fubini's Theorem is verifying the $\mu \times \nu$-integrability of the function f. A result which is often used for this is a result called Tonelli's Theorem.

3.9. PRODUCT MEASURES AND FUBINI'S THEOREM

Theorem 4 Let μ, ν be σ-finite and $f : S \times T \to \mathbf{R}^*$ non-negative and $\mu \times \nu$-measurable. Then

(i) $f(s, \cdot)$ is T-measurable for μ-almost all $s \in S$

(ii) $s \to \int_T f(s,t) d\nu(t)$ is S-measurable

(iii) $\int_{S \times T} f d\mu \times \nu = \int_S \int_T f(s,t) d\nu(t) d\mu(s)$ [the integrals might be ∞].

[Of course, there is an analogous statement with the variables s and t reversed.]

Proof: Since μ and ν are σ-finite, $\mu \times \nu$ is σ-finite so there exists an increasing sequence of $\mu \times \nu$-measurable sets $\{E_i\}$ such that $E_i \uparrow S \times T$ and $\mu \times \nu(E_i) < \infty$. Set $f_i = (f \wedge i) C_{E_i}$. Then each f_i is $\mu \times \nu$-integrable. By Theorem 3, $f_i(s, \cdot)$ is ν-integrable for μ-almost all $s \in S$ and since $f_i(s, \cdot) \uparrow f(s, \cdot)$ on T, $f(s, \cdot)$ is T-measurable for μ-almost all $s \in S$ and by the MCT,

$$\int_T f_i(s,t) d\nu(t) \uparrow \int_T f(s,t) d\nu(t) \text{ for } \mu\text{-almost all } s \in S. \quad (3.3)$$

By Theorem 3, the function $s \to \int_T f_i(s,t) d\nu(t)$ is S-measurable so by (3) $s \to \int_T f(s,t) d\nu(t)$ is S-measurable and the MCT applied to (3) implies that

$$\int_S \int_T f_i(s,t) d\nu(t) d\mu(s) \uparrow \int_S \int_T f(s,t) d\nu(t) d\mu(s). \quad (3.4)$$

Since $f_i \uparrow f$ on $S \times T$, the MCT implies

$$\int_{S \times T} f_i d\mu \times \nu \uparrow \int_{S \times T} f d\mu \times \nu. \quad (3.5)$$

By Theorem 3,

$$\int_{S \times T} f_i d\mu \times \nu = \int_S \int_T f_i(s,t) d\nu(t) d\mu(s)$$

so (4) and (5) imply

$$\int_S f d\mu \times \nu = \int_S \int_T f(s,t) d\nu(t) d\mu(s).$$

Tonelli's Theorem can be used to check the integrability of a measurable function with respect to a product measure. If μ and ν are σ-finite and $f : S \times T \to \mathbf{R}$ is $\mu \times \nu$-measurable, we can apply Tonelli's Theorem to the function $|f|$ to check its $\mu \times \nu$-integrability. If $|f|$ is $\mu \times \nu$-integrable, then Fubini's Theorem can often be applied to evaluate the product integral $\int_{S \times T} f d\mu \times \nu$.

As an application of Tonelli's Theorem we show how the product measure of a set can be computed as an integral. If $E \subset S \times T$ and $s \in S$, the s-section of E at s is defined to be $E_s = \{t \in T : (s,t) \in E\}$. Similarly, if $t \in T$, the t-section of E at t is defined to be $E^t = \{s \in S : (s,t) \in E\}$. We have the following elementary proposition.

Proposition 5 *Let $E \subset S \times T$ and $s \in S$. Then*

(i) $C_E(s,t) = C_{E_s}(t)$,

(ii) $(E^c)_s = (E_s)^c$,

(iii) $(\bigcup_a E_a)_s = \bigcup_a (E_a)_s$, where $E_a \subset S \times T$,

(iv) $(\bigcap_a E_a)_s = \bigcap_a (E_a)_s$, where $E_a \subset S \times T$.

From Proposition 5 and Fubini's theorem, we have

Theorem 6 *Let $E \subset S \times T$ be $\mu \times \nu$-measurable and have finite $\mu \times \nu$ measure. Then*

(i) *for μ-almost all $s \in S$ the section $E_s \in \mathcal{T}$,*

(ii) *the function $s \to \nu(E_s)$ is μ-integrable, and*

(iii) $\int_S \nu(E_s) d\mu(s) = \mu \times \nu(E)$.

Remark 7 If μ and ν are σ-finite, by Proposition 5 and Tonelli's Theorem conditions (i), (ii)' $s \to \nu(E_s)$ is S-measurable and (iii) also hold for any $E \subset S \times T$ which is $\mu \times \nu$-measurable.

Remark 8 In some of the developments of Fubini's Theorem for product measures, Theorem 6 is an important intermediate step in that it establishes the result for integrable simple functions. The general result is then established by approximating the general integrable function by simple functions. The proof of Theorem 6 from basic properties of product measures can be somewhat lengthy (see [Roy] 12.4.15-18).

We now give several examples which illustrate the necessity for the various hypotheses in the Fubini and Tonelli Theorems.

Example 9 Let $S = T = [0,1]$, μ = Lebesgue measure and ν counting measure restricted to the Lebesgue measurable subsets of S. Set $E = \{(x,x) : 0 \leq x \leq 1\}$. Note that E is $\mu \times \nu$-measurable. Then

$$\int_0^1 \int_0^1 C_E(x,y) dm(x) d\nu(y) = 0$$

while

$$\int_0^1 \int_0^1 C_E(x,y) d\nu(y) dm(x) = 1.$$

This example illustrates the importance of the σ-finiteness condition in Tonelli's Theorem [also in Theorem 6 and Remark 7].

3.9. PRODUCT MEASURES AND FUBINI'S THEOREM

Example 10 Let $f(0,0) = 0$ and $f(x,y) = (x^2 - y^2)/(x^2 + y^2)^2$ for $0 \leq x \leq 1$, $0 \leq y \leq 1$, otherwise. Since $\frac{\partial}{\partial y}\left(\frac{-y}{x^2+y^2}\right) = f(x,y)$ for $x \neq 0$,

$$\int_0^1 \int_0^1 f(x,y)dy\,dx = \int_0^1 \frac{1}{1+x^2}dx = \pi/4.$$

Similarly,

$$\int_0^1 \int_0^1 f(x,y)dx\,dy = -\pi/4.$$

This example shows that the non-negativity condition in Tonelli's Theorem is important.

A similar example is given in Exercise 2.

Example 11 (Sierpinski) [Sir]. This example points out the importance of the measurability assumption in Tonelli's Theorem. A result of Sierpinski gives the existence of a 1-1 map j from $[0,1]$ onto a well ordered set W such that $j(x)$ has at most countably many predecessors in W for each $x \in [0,1]$ (this result depends on the Continuum Hypothesis). Let Q be the subset of the unit square which consists of all pairs (x,y) such that $j(x)$ precedes $j(y)$ in W. Set $f = C_Q$. For each x, Q_x contains all but countably many points of $[0,1]$ so $\int_0^1 \int_0^1 f(x,y)dy\,dx = 1$, and for each y, Q^y contains at most countably many points of $[0,1]$ so $\int_0^1 \int_0^1 f(x,y)dx\,dy = 0$.

Note that each $f(x,\cdot)$ ($f(\cdot,y)$) is a Borel function but f is not measurable, i.e., f is "separately measurable" but not "jointly measurable".

Another interesting example relative to iterated integrals is given in Exercise 3.

There is another standard method of constructing product measures which we now outline. We assume that μ and ν are σ-finite. If $E \in \sigma(\mathcal{S} \times \mathcal{T})$, then for $s \in S$ ($t \in T$) every section $E_s \in \mathcal{T}$ ($E^t \in \mathcal{S}$) and the function $s \to \nu(E_s)$ ($t \to \mu(E^t)$) is $\mathcal{S}$-measurable ($\mathcal{T}$-measurable). The set function λ defined on $\sigma(\mathcal{S} \times \mathcal{T})$ by

$$\lambda(E) = \int_S \nu(E_s)d\mu(s) \left(\int_T \mu(E^t)d\nu(t)\right)$$

is a σ-finite measure which satisfies $\lambda(A \times B) = \mu(A)\nu(B)$ when $A \in \mathcal{S}$, $B \in \mathcal{T}$, and is defined to be the product of μ and ν. In general, this measure is not complete. In particular, if $\mu = \nu =$ Lebesgue measure on $\mathbf{R}$, then λ is not complete and so is not two-dimensional Lebesgue measure on $\mathbf{R}^2$ - which is somewhat annoying. Fubini's Theorem is then established by approximating integrable functions by simple functions. See, for example, the development in [Hal].

Fubini's Theorem fails miserably for finitely additive set functions; see [HY].

Exercise 1. Let $g: S \to \mathbf{R}$ ($h: T \to \mathbf{R}$) be $\mathcal{S}$-measurable ($\mathcal{T}$-measurable) and define $f: S \times T \to \mathbf{R}$ by $f(s,t) = g(s)h(t)$. Show f is $\sigma(\mathcal{S} \times \mathcal{T})$-measurable. If g and h are integrable, show f is integrable and

$$\int_{S \times T} f\,d\mu \times \nu = \int_S g\,d\mu \int_T h\,d\nu.$$

Exercise 2. Show $\int_0^1 \int_1^\infty (e^{-xy} - 2e^{-2xy}) dx dy$ and $\int_1^\infty \int_0^1 (e^{-xy} - 2e^{-2xy}) dy dx$ both exist but are not equal.

Exercise 3. Let $f(0,0) = 0$ and $f(x,y) = xy/(x^2+y^2)^2$ otherwise. Show
$$\int_{-1}^1 \int_{-1}^1 f(x,y) dx dy = \int_{-1}^1 \int_{-1}^1 f(x,y) dy dx$$
but f is not (Lebesgue) integrable over $[-1,1] \times [-1,1]$. Hint: Consider $[0,1] \times [0,1]$.

Exercise 4. Give an example of a set in the plane which is not Lebesgue measurable. Generalize to $\mathbf{R}^n$.

Exercise 5. Show that the conclusion (i) of Theorem 6 cannot be improved to: every section $E_s \in \mathcal{T}$.

Exercise 6. If $E \in \sigma(\mathcal{S} \times \mathcal{T})$, show every section $E_s \in \mathcal{T}$. [Hint: Consider all $E \subset S \times T$ such that $E_s \in \mathcal{T}$.]

Exercise 7. Let μ, ν be σ-finite. If E, F are $\mu \times \nu$-measurable and $\mu(E_s) = \nu(F_s)$ for μ-almost all $s \in S$, show $\mu \times \nu(E) = \mu \times \nu(F)$.

Exercise 8. Show $\sigma(\mathcal{B}(\mathbf{R}) \times \mathcal{B}(\mathbf{R})) = \mathcal{B}(\mathbf{R}^2)$. [Hint: Consider the projections from $\mathbf{R}^2$ into the coordinates and use Exer. 2.5.11.]

Exercise 9. Show $\mathcal{B}(\mathbf{R}^2) \ne \mathcal{M}(\mathbf{R}^2)$.

Exercise 10. If $f : S \times T \to \mathbf{R}$ is $\sigma(\mathcal{S} \times \mathcal{T})$-measurable, show $f(s, \cdot)$ is $\mathcal{T}$-measurable for every $s \in S$. Does the same hold for $\mu \times \nu$-measurable functions?

Exercise 11. Evaluate $\int_0^\infty e^{-x^2} dx = I$ by writing I^2 as a double integral in the plane and using polar coordinates.

Exercise 12. Let $f : \mathbf{R}^n \to \mathbf{R}$ be Lebesgue integrable. Show
$$\int_{\mathbf{R}^n} f(x+y) dx = \int_{\mathbf{R}^n} f(x) dx$$
for every $y \in \mathbf{R}^n$. What is the corresponding formula for $\int_{\mathbf{R}^n} f(ax) dx$, $a \in \mathbf{R}$?

3.9. PRODUCT MEASURES AND FUBINI'S THEOREM

Exercise 13. For $p > 0$, $q > 0$, the Beta Function $B(p,q)$ is defined by
$$B(p,q) = \int_0^1 x^{p-1}(1-x)^{q-1} dx.$$
Show $B(p,q) = B(q,p)$ and $\Gamma(p)\Gamma(q) = \Gamma(p+q)B(p,q)$.

Exercise 14. Evaluate $\int_0^1 \int_x^1 e^{y^2} dy dx$.

3.10 A Geometric Interpretation of the Integral

We show that Fubini's Theorem can be used to show that the integral can be interpreted to be the "area under the curve". Let $(S, \mathcal{S}, \mu)$ be a σ-finite, complete measure space. If $f : S \to \mathbf{R}$, the *ordinate set* of f is $\Omega_f = \{(x,y) : x \in S, -\infty < y < f(x)\}$.

Proposition 1 f is $\mathcal{S}$-measurable $\Leftrightarrow$ Ω_f is $\mu \times m$-measurable.

Proof: $\Leftarrow$: For $y \in \mathbf{R}$, $\Omega_f^y = \{x \in S : (x,y) \in \Omega_f\} = \{x : f(x) > y\}$. $\Omega_f^y \in \mathcal{S}$ for m-almost all $y \in \mathbf{R}$ (Remark 3.9.7). Thus, $\Omega_f^y \in \mathcal{S}$ for y in some dense subset $D \subset \mathbf{R}$. Therefore, $\{x : f(x) > r\} \in \mathcal{S}$ for $r \in D$ and f is $\mathcal{S}$-measurable by 3.1.1.

$\Rightarrow$: Suppose $f : S \to \mathbf{R}$ is $\mathcal{S}$-measurable. For $r \in \mathbf{Q}$ let

$$A_r = \{x \in S : f(x) > r\},$$
$$B_r = \{y \in \mathbf{R} : -\infty < y \leq r\},$$

and

$$E = \cup\{A_r \times B_r : r \in \mathbf{Q}\}.$$

Then E is $\mu \times m$-measurable, and we claim that $E = \Omega_f$. First, suppose that $(x,y) \in E$. Then $(x,y) \in A_r \times B_r$ for some $r \in \mathbf{Q}$ so $-\infty < y \leq r < f(x)$ and $(x,y) \in \Omega_f$. Hence, $E \subset \Omega_f$. Next, suppose $(x,y) \in \Omega_f$. Then $-\infty < y < f(x)$ so there exists $r \in \mathbf{Q}$ such that $-\infty < y \leq r < f(x)$ and $(x,y) \in A_r \times B_r \subset E$. Hence, $\Omega_f \subset E$.

Theorem 2 Let $f : S \to \mathbf{R}$ be non-negative and $\mathcal{S}$-measurable, and let

$$H = \{(x,y) : x \in S, 0 \leq y < f(x)\}.$$

Then H is $\mu \times m$-measurable and $\mu \times m(H) = \int_S f d\mu$.

Proof: H is measurable by Proposition 1 since $H = \Omega_f \setminus S \times (-\infty, 0)$. By Remark 3.9.7,

$$\mu \times m(H) = \int_S m(H^x) d\mu(x) = \int_S f(x) d\mu(x).$$

If the basic properties of the product measure are developed before the integral is defined, the conclusion in Theorem 2 can be used to define the integral for non-negative measurable functions. This approach to the integral leads to some very quick proofs of the convergence theorems, particularly the MCT. For a description of this method of defining the Lebesgue integral see [Z1].

Exercise 1. If f is $\mathcal{S}$-measurable, show the graph of f, $G = \{(x, f(x)) : x \in S\}$, has $\mu \times m$ measure 0.

Exercise 2. Show $\int_S f d\mu = \int_0^\infty \mu(\{x : f(x) > y\}) \, dy$.

3.11 Convolution Product

As another application of Fubini's Theorem we consider the convolution product of two functions. The convolution product has many important applications in analysis, from approximation theory to integral transform theory. We give two such applications in Theorem 11 and Exercise 2.

If $f, g : \mathbf{R}^n \to \mathbf{R}$, the *convolution product* of f and g, denoted by $\underline{f * g}$, is defined by
$$f * g(x) = \int_{\mathbf{R}^n} f(x-y)g(y)dy,$$
provided the integral exists. We use Mikusinski's characterization of the Lebesgue integral to show that the convolution product of two Lebesgue integrable functions exists a.e. and defines a Lebesgue integrable function. Other conditions which guarantee the existence of the convolution product are given in the exercises.

Theorem 1 *Let $f, g \in L^1(\mathbf{R}^n)$. Then*

(i) $f * g(x) = \int_{\mathbf{R}^n} f(x-y)g(y)dy$ *exists for m-almost all $x \in \mathbf{R}^n$,*

(ii) $f * g \in L^1(\mathbf{R}^n)$,

(iii) $\|f * g\|_1 \leq \|f\|_1 \|g\|_1$.

Proof: We use Fubini's Theorem to show that the function
$$H(x,y) = f(x-y)g(y)$$
is Lebesgue integrable over $\mathbf{R}^{2n}$ and then (i) and (ii) will follow from Fubini's Theorem.

Let $\{f_i\}$ ($\{g_i\}$) be Lebesgue integrable Borel functions such that
$$\sum_{i=1}^{\infty} \|f_i\|_1 < \infty \quad (\sum_{i=1}^{\infty} \|g_i\|_1 < \infty)$$
and
$$f(x) = \sum_{i=1}^{\infty} f_i(x) \quad (g(x) = \sum_{i=1}^{\infty} g_i(x))$$
for any x for which the series is absolutely convergent (Theorem 3.8.3). Set $F_i(x,y) = f_i(x-y)$, $G_i(x,y) = g_i(y)$ and note each F_i, G_i is a Borel function on $\mathbf{R}^{2n}$.

Consider the Cauchy product, $\sum_{j=1}^{\infty} \sum_{i=0}^{j} F_i G_{j-i}$, of the two series $\sum_i F_i$ and $\sum_i G_i$. Since
$$\int_{\mathbf{R}^{2n}} |F_i G_j| = \int_{\mathbf{R}^n} |f_i| \int_{\mathbf{R}^n} |g_j|,$$

$$\sum_{j=1}^{\infty} \left| \sum_{i=0}^{j} \int_{\mathbf{R}^{2n}} F_i G_{j-i} \right| \leq \sum_{j=1}^{\infty} \sum_{i=0}^{j} \int_{\mathbf{R}^n} |f_i| \int_{\mathbf{R}^n} |g_{j-i}| < \infty,$$

this last series being the Cauchy product of the series $\sum_i \|f_i\|_1$, $\sum_i \|g_i\|_1$. If for some $(x,y) \in \mathbf{R}^2$ the series

$$\sum_{j=1}^{\infty} \sum_{i=0}^{j} F_i(x,y) G_{j-i}(x,y)$$

converges absolutely, then both series $\sum_{i=1}^{\infty} F_i(x,y)$, $\sum_{i=1}^{\infty} G_i(x,y)$ must converge absolutely ([Kn] Theorem 11, p. 89) and, therefore, must converge to $F(x,y)$ and $G(x,y)$, respectively, so

$$\sum_{j=1}^{\infty} \sum_{i=0}^{j} F_i(x,y) G_{j-i}(x,y)$$

must converge to $F(x,y)G(x,y) = H(x,y)$ ([Kn]). By Exercise 3.8.1, H is Lebesgue integrable over $\mathbf{R}^{2n}$, and (i), (ii) follow from Fubini's Thoerem.

For (iii),

$$\begin{aligned}
\int_{\mathbf{R}^n} |f*g| &\leq \int_{\mathbf{R}^n} \int_{\mathbf{R}^n} |f(x-y)g(y)| \, dy\, dx \\
&= \int_{\mathbf{R}^n} \int_{\mathbf{R}^n} |f(x-y)g(y)| \, dx\, dy \\
&= \int_{\mathbf{R}^n} |g(y)| \, dy \int_{\mathbf{R}^n} |f(x)| \, dx
\end{aligned}$$

by Tonelli's Theorem.

Note that the proof of Theorem 1 shows that the function $(x,y) \to f(x-y)$ is Lebesgue measurable (see Exer. 8; for another proof of the measurability see [HS] 21.31).

Several of the algebraic properties of the convolution product are given in the exercises. We now show that the convolution product can be used in establishing approximation results.

Definition 2 *A sequence* $\{\varphi_k\} \subset L^1(\mathbf{R}^n)$ *is called an* approximate identity *or a* δ-sequence *if*

(i) $\varphi_k(x) \geq 0$ *for all* $x \in \mathbf{R}^n$,

(ii) $\|\varphi_k\|_1 = 1$ *for all* k,

(iii) *for every neighborhood of* 0, U, *in* $\mathbf{R}^n$, $\lim_k \int_{U^c} \varphi_k(x) dx = 0$.

We give examples of δ-sequences.

Example 3 In $\mathbf{R}$, take $\varphi_k(t) = \frac{k}{2} C_{[-1/k, 1/k]}$.

Example 4 $\varphi_k(t) = \sqrt{\frac{2}{\pi}} \frac{k}{1+(kt)^2}$, $t \in \mathbf{R}$.

Example 5 In $\mathbf{R}$, $\varphi_k(t) = c_k e^{-kt^2}$ where c_k is chosen such that (ii) holds.

3.11. CONVOLUTION PRODUCT

We denote by $C_c^\infty(\mathbf{R}^n)$ the vector space of all functions φ on $\mathbf{R}^n$ with compact support which have continuous partial derivatives of all orders. It is not apparent that there are any non-zero functions in $C_c^\infty(\mathbf{R}^n)$, but examples of such functions are given in the examples below.

Example 6 In $\mathbf{R}$, pick $\varphi \in C_c^\infty(\mathbf{R})$, $\varphi \geq 0$, [like $\varphi(t) = e^{-1/(1-t^2)}$ for $|t| < 1$ and $\varphi(t) = 0$ for $|t| \geq 1$]. Set $\varphi_k(t) = c_k \varphi(kt)$, where c_k is chosen such that (ii) holds. Note that if φ is chosen as above $spt(\varphi_k) \subset [-1/k, 1/k]$.

Example 7 In $\mathbf{R}^n$, pick φ_k as in Example 5 and set $\psi_k(x_1, \ldots, x_n) = \varphi_k(x_1) \cdots \varphi_k(x_n)$.

Example 8 In $\mathbf{R}^n$, pick φ_k as in Example 6 and define ψ_k as in Example 7. Note in this case, $\psi_k \in C_c^\infty(\mathbf{R}^n)$ and $spt(\psi_k) \subset \{x : |x_k| \leq 1/k\}$.

Example 9 Define φ on $\mathbf{R}^n$ by $\varphi(x) = e^{-1/(1-\|x\|^2)}$ if $\|x\| < 1$ and $\varphi(x) = 0$ otherwise. Set $\varphi_k(x) = c_k \varphi(kx)$ where c_k is chosen to satisfy (ii). Then $\varphi_k \in C_c^\infty(\mathbf{R}^n)$ and $spt(\varphi_k) \subset \{x : \|x\| \leq 1/k\}$.

We next establish an intermediary result which is interesting in its own right.

Theorem 10 Let $f : \mathbf{R}^n \to \mathbf{R}^*$ be Lebesgue integrable and for $h \in \mathbf{R}^n$ define $f_h : \mathbf{R}^n \to \mathbf{R}^*$ by $f_h(x) = f(x+h)$. Then $\lim_{h \to 0} \|f_h - f\|_1 = 0$.

Proof: Let $\epsilon > 0$. By Theorem 3.5.6 there exists $\varphi \in C_c(\mathbf{R}^n)$ such that $\|f - \varphi\|_1 < \epsilon$. There exists $a > 0$ such that $\varphi(x) = 0$ for $\|x\| > a$. φ is uniformly continuous so there exists $1 > \delta > 0$ such that $|\varphi(x) - \varphi(y)| < \epsilon$ when $\|x - y\| < \delta$. Thus, $\|h\| < \delta$ implies

$$\int_{\mathbf{R}^n} |\varphi(x+h) - \varphi(x)| \, dx < \epsilon \cdot A,$$

where $A = m\{x : \|x\| \leq a + 1\}$. Thus, if $\|h\| < \delta$,

$$\|f_h - f\|_1 \leq \|f_h - \varphi_h\|_1 + \|\varphi_h - \varphi\|_1 + \|\varphi - f\|_1 < \epsilon + \epsilon \cdot A + \epsilon.$$

Theorem 11 Let $\{\varphi_k\}$ be an approximate identity in $L^1(\mathbf{R}^n)$. Then for every $f \in L^1(\mathbf{R}^n)$, $\lim_k \|f * \varphi_k - f\|_1 = 0$.

Proof: Let $\epsilon > 0$. By Theorem 10 there exists $\delta > 0$ such that $\|h\| \leq \delta$ implies $\|f_h - f\|_1 < \epsilon$, where $f_h(x) = f(x+h)$. By (iii) there exists N such that $k \geq N$ implies $\int_{\|x\| \geq \delta} \varphi_k(x) dx < \epsilon$. Then

$$\begin{aligned}
\|f * \varphi_k - f\|_1 &= \int_{\mathbf{R}^n} |\int_{\mathbf{R}^n} (f(x-y) - f(x)) \varphi_k(y) dy| \, dx \\
&\leq \int_{\mathbf{R}^n} \varphi_k(y) \int_{\mathbf{R}^n} |f(x-y) - f(x)| \, dx dy \\
&= \int_{\mathbf{R}^n} \varphi_k(y) \|f_{-y} - f\|_1 \, dy \\
&= \int_{\|y\| \leq \delta} \varphi_k(y) \|f_{-y} - f\|_1 \, dy + \int_{\|y\| > \delta} \varphi_k(y) \|f_{-y} - f\|_1 \, dy \\
&\leq \epsilon \int_{\|y\| \leq \delta} \varphi_k(y) dy + 2 \|f\|_1 \int_{\|y\| \geq \delta} \varphi_k(y) dy < \epsilon + 2 \|f\|_1 \epsilon.
\end{aligned}$$

Corollary 12 $C_c^\infty(\mathbf{R}^n)$ *is a dense subset of* $L^1(\mathbf{R}^n)$ *(with respect to* $\|\|_1$*).*

Proof: By Theorem 3.5.6 it suffices to show that $C_c^\infty(\mathbf{R}^n)$ is dense in $C_c(\mathbf{R}^n)$. Let $\psi \in C_c(\mathbf{R}^n)$ and $K = spt(\psi)$. Choose $\{\varphi_k\} \subset C_c^\infty(\mathbf{R}^n)$ as in Example 9. If $x \in \mathbf{R}^n$ and $dist(x,K) > 1/k$, then $\varphi_k(x-y) = 0$ for $y \in K$ so

$$\varphi_k * \psi(x) = \int_K \varphi_k(x-y)\psi(y)dy$$

vanishes for such x, i.e., $spt(\varphi_k * \psi) \subset \{x : dist(x,K) \leq 1/k\}$. Hence, $\varphi_k * \psi$ has compact support and by Exercise 3 $\varphi_k * \psi$ is infinitely differentiable so $\varphi_k * \psi \in C_c^\infty(\mathbf{R}^n)$. By Theorem 11, $\|\varphi_k * \psi - \psi\|_1 \to 0$.

Proposition 13 *Let* $K \subset \mathbf{R}^n$ *be compact and* $V \supset K$ *open. Then there exists* $\psi \in C_c^\infty(\mathbf{R}^n)$ *such that* $0 \leq \psi \leq 1$, $\psi = 1$ *on* K *and* $\psi = 0$ *on* V^c.

Proof: We may assume that V is bounded. Let $\delta = dist(K, V^c) > 0$ and let K_1 be a "$\delta/3$-neigborhood of K", i.e., $K_1 = \{x : dist(x,K) < \delta/3\}$ [a sketch at this point is useful]. If k is chosen such that $1/k < \delta/3$, the function $\psi = C_{K_1} * \varphi_k$ is in $C_c^\infty(\mathbf{R}^n)$ if φ_k is chosen as in Example 9. Clearly $0 \leq \psi \leq 1$. If $x \in K$, then (Exer. 1)

$$\psi(x) = \int_{\mathbf{R}^n} C_{K_1}(y)\varphi_k(x-y)dy = \int_{\mathbf{R}^n} \varphi_k(y)C_{K_1}(x-y)dy = 1$$

since the integration is over a ball of radius $< \delta/3$ and center x. If $x \notin V$, then $\psi(x) = 0$ since the integrand in the convolution product is 0.

Exercise 1. If $f, g, h \in L^1(\mathbf{R}^n)$, show $f * g = g * f$, $f * (g * h) = (f * g) * h$, and $f * (g + h) = f * g + f * h$. Thus, $L^1(\mathbf{R}^n)$ is a commutative algebra with convolution as product. Show that it does not have an identity.

Exercise 2. If $f \in L^1(\mathbf{R})$ and $f(t) = 0$ for $t < 0$, its *Laplace transform* is defined by $\mathcal{L}\{f\}(s) = \int_0^\infty e^{-st}f(t)dt$. Show this integral exists for $s \geq 0$. If $f, g \in L^1(\mathbf{R})$ vanish for $t < 0$, show $f * g$ vanishes for $t < 0$ and $\mathcal{L}\{f * g\} = \mathcal{L}\{f\}\mathcal{L}\{g\}$.

Exercise 3. If $f \in C_c^\infty(\mathbf{R}^n)$ and $g \in L^1(\mathbf{R}^n)$, show $f * g \in L^1(\mathbf{R}^n) \cap C^\infty(\mathbf{R})$ and give a formula for calculating the partial derivatives of $f * g$.

Exercise 4. Let $\{\varphi_k\}$ be as in Example 9. If $f : \mathbf{R}^n \to \mathbf{R}$ is bounded, continuous, show $f * \varphi_k$ is uniformly continuous and $f * \varphi_k \to f$ uniformly on compact subsets of $\mathbf{R}^n$.

Exercise 5. Let $f \in L^1(\mathbf{R}^n)$. If $\int_{\mathbf{R}^n} f\varphi dm = 0$ for every $\varphi \in C_c^\infty(\mathbf{R}^n)$, show $f = 0$ a.e.

3.11. CONVOLUTION PRODUCT

Exercise 6. Let $f \in L^1(\mathbf{R}^n)$ and $g : \mathbf{R}^n \to \mathbf{R}$ be bounded, measurable. Show $f * g$ exists everywhere in $\mathbf{R}^n$ and is uniformly continuous.

Exercise 7. Let $\{\varphi_k\}$ be an approximate identity. If $g : \mathbf{R}^n \to \mathbf{R}$ is bounded, measurable and continuous at x, show $\varphi_k * g(x) \to g(x)$.

Exercise 8. Show that if $f : \mathbf{R}^n \to \mathbf{R}$ is measurable, then the function $(x, y) \to f(x - y)$ is measurable.

Exercise 9. Let $E \subset \mathbf{R}$ be Lebesgue measurable with $m(E) > 0$. Show that $E - E$ contains an interval. Hint: Assume $E \subset [a, b]$ and define

$$f(x) = \int_a^b C_E(y) C_E(x + y) \, dy.$$

Show f is continuous at 0 and $f(0) > 0$. Consider $\{x : f(x) > 0\}$.

Exercise 10. If $f \in C_c(\mathbf{R}^n)$ and $\{\varphi_k\}$ is an approximate identity, show $\{f * \varphi_k\}$ converges to f uniformly on $\mathbf{R}$.

Exercise 11. If $f, g \in C_c(\mathbf{R}^n)$, show $spt(f * g) \subset \text{closure}(spt(f) + spt(g))$ and $f * g \in C_c(\mathbf{R}^n)$.

3.12 The Radon-Nikodym Theorem

We now consider one of the most important results in the theory of measure and integration, the Radon-Nikodym Theorem. The Radon-Nikodym Theorem essentially characterizes the integral as a set function, i.e., as an indefinite integral.

Definition 1 *Let μ, ν be signed measures on the σ-algebra $\sum$. Then ν is absolutely continuous with respect to μ, denoted by $\nu \ll \mu$, if $|\mu|(E) = 0$ implies $|\nu|(E) = 0$ whenever $E \in \sum$.*

Example 2 If μ is a measure and $f : S \to \mathbf{R}^*$ has a μ-integral and $\nu(E) = \int_E f d\mu$ for $E \in \sum$, then $\nu \ll \mu$.[3.2.10]. We denote the indefinite integral ν by $\nu = f d\mu$.

The Radon-Nikodym Theorem essentially asserts that Example 2 is a canonical example in the sense that any (σ-finite) signed measure ν which is absolutely continuous with respect to a (σ-finite) measure μ must be an indefinite integral of some function with respect to μ.

We have an elementary observation which follows directly from the Jordan Decomposition.

Proposition 3 *Let μ, ν be signed measures on $\sum$. The following are equivalent:*

(i) $\nu \ll \mu$

(ii) $\nu^+ \ll \mu$ and $\nu^- \ll \mu$.

Definition 4 *Let μ, ν be signed measures on $\sum$. ν is said to be μ-continuous if*
$$\lim_{|\mu|(E) \to 0} \nu(E) = 0.$$

By Theorem 3.2.17 indefinite integrals of μ-integrable functions are μ-continuous. Concerning absolute continuity and μ-continuity, we have

Proposition 5 (i) *If ν is μ-continuous, then $\nu \ll \mu$.*

(ii) *If ν is a finite signed measure, then $\nu \ll \mu \Leftrightarrow \nu$ is μ-continuous.*

Proof: (i) is clear. (ii): $\Leftarrow$ follows from (i). $\Rightarrow$: Suppose not. Then there exists $\epsilon > 0$ and $\{E_j\} \subset \sum$ with $|\mu|(E_j) < 1/2^j$ and $|\nu|(E_j) \geq \epsilon$. Set $E = \overline{\lim} E_j = \bigcap_{k=1}^{\infty} \bigcup_{j=k}^{\infty} E_j$. Then for each k,

$$|\mu|(E) \leq \sum_{j=k}^{\infty} |\mu|(E_j) \leq \sum_{j=k}^{\infty} 1/2^j = 1/2^{k-1}$$

3.12. THE RADON-NIKODYM THEOREM

so $|\mu|(E) = 0$. But, $|\nu|$ finite implies $|\nu|(E) = \lim_k |\nu|(\bigcup_{j=k}^\infty E_j) \geq \overline{\lim}\, |\nu|(E_k) \geq \epsilon$ so ν is not absolutely continuous with respect to μ.

The finiteness condition in (ii) cannot be dropped even when μ is finite.

Example 6 Define μ, ν on $\mathcal{P}(\mathbf{N})$ by $\mu(E) = \sum_{k \in E} 1/2^k$, $\nu(E) = \sum_{k \in E} 2^k$. Then $\nu \ll \mu$ but ν is not μ-continuous.

For the proof of the Radon-Nikodym Theorem we first require a lemma.

Lemma 7 *Let μ, ν be finite measures on $\sum$ with $\nu \ll \mu$ and $\nu \neq 0$. Then there exist $\epsilon > 0$ and $A \in \sum$ such that $\mu(A) > 0$ and A is a $(\nu - \epsilon\mu)$-positive set.*

Proof: For each k let (P_k, N_k) be a Hahn Decomposition for the signed measure $\nu - \frac{1}{k}\mu$. Set $A_0 = \bigcup_{k=1}^\infty P_k$ and $B_0 = \bigcap_{k=1}^\infty N_k$. Since $B_0 \subset N_k$ for every k,

$$0 \leq \nu(B_0) \leq \frac{1}{k}\mu(B_0)$$

so $\nu(B_0) = 0$. Thus, $\nu(A_0) > 0$ and since $\nu \ll \mu$, $\mu(A_0) > 0$. Hence, $\mu(P_k) > 0$ for some k. For such a k, set $\epsilon = 1/k$ and $A = P_k$.

We are now ready for the proof of the Radon-Nikodym Theorem. We say that a signed measure ν is σ-finite if $|\nu|$ is σ-finite.

Theorem 8 (Radon-Nikodym) *Let $\mu(\nu)$ be a σ-finite (signed) measure on $\sum$ with $\nu \ll \mu$. Then there exists a $\sum$-measurable function $f : S \to \mathbf{R}^*$ such that $\nu(E) = \int_E f d\mu$ for all $E \in \sum$; the function f is unique up to μ-a.e.*

Proof: By Proposition 3 we may assume that ν is actually a measure.
We first assume that both μ and ν are finite measures. Let

$$\mathcal{F} = \left\{ g : S \to \mathbf{R}^* : g \in L^1(\mu),\ g \geq 0 \text{ with } \int_E g d\mu \leq \nu(E) \text{ for all } E \in \sum \right\}$$

and set

$$\lambda = \sup\left\{\int_S g d\mu : g \in \mathcal{F}\right\} \leq \nu(S) < \infty.$$

Choose a sequence $\{f_k\} \subset \mathcal{F}$ such that $\lim \int_S f_k d\mu = \lambda$. We construct an increasing sequence from $\{f_k\}$ by setting $g_k = f_1 \vee \ldots \vee f_k$. Then $g_k \uparrow$ and $g_k \in L^1(\mu)$. Moreover, $g_k \in \mathcal{F}$ since for any $E \in \sum$ if $E_1 = \{t \in E : f_1(t) = g_k(t)\}$ and

$$E_{i+1} = \{t \in E : f_{i+1}(t) = g_k(t)\} \setminus E_i \text{ for } 1 \leq i \leq k-1,$$

then
$$\int_E g_k d\mu = \sum_{i=1}^{k} \int_{E_i} g_k d\mu = \sum_{i=1}^{k} \int_{E_i} f_i d\mu \leq \sum_{i=1}^{k} \nu(E_i) = \nu(E).$$

Also $\lambda \geq \int_S g_k d\mu \geq \int_S f_k d\mu$ for every k implies $\lim \int_S g_k d\mu = \lambda$. By the MCT if $f = \lim g_k$, then $\int_S f d\mu = \lambda < \infty$ so f is finite μ-a.e. and we may assume $f \in L^1(\mu)$. Since $g_k \in \mathcal{F}$, $f \in \mathcal{F}$ by the MCT.

We claim that $\nu(E) = \int_E f d\mu$ for $E \in \Sigma$. Let ν_0 be the measure defined by $\nu_0(E) = \nu(E) - \int_E f d\mu$. If ν_0 is not zero, then since $\nu_0 \ll \mu$, Lemma 7 implies that there exists $\epsilon > 0$ and $A \in \Sigma$ such that $\mu(A) > 0$ and A is $(\nu_0 - \epsilon\mu)$-positive, i.e.,
$$\int_E \epsilon C_A d\mu = \epsilon\mu(E \cap A) \leq \nu_0(E \cap A) = \nu(E \cap A) - \int_{E \cap A} f d\mu \text{ for } E \in \Sigma.$$

Thus, if $h = f + \epsilon C_A$, then
$$\int_E h d\mu = \int_E f d\mu + \epsilon\mu(E \cap A) \leq \int_{E \setminus A} f d\mu + \nu(E \cap A) \leq \nu(E)$$

for every $E \in \Sigma$ so $h \in \mathcal{F}$. But,
$$\int_S h d\mu = \int_S f d\mu + \epsilon\mu(A) = \lambda + \epsilon\mu(A) > \lambda$$

which contradicts the definition of λ. Hence, $\nu = f d\mu$.

Uniqueness of f follows from 3.2.18.

Assume that μ and ν are σ-finite. Then $S = \bigcup_{j=1}^{\infty} E_j$, where $E_j \in \Sigma$, $E_j \subset E_{j+1}$ and $\mu(E_j) < \infty$, $\nu(E_j) < \infty$. By the first part, for each j there exists a non-negative, μ-integrable function g_j such that $g_j = 0$ on E_j^c and $\nu(E \cap E_j) = \int_E g_j d\mu$ for $E \in \Sigma$. From the uniqueness of g_j, it follows that $g_j = g_{j+k}$ μ-a.e. in E_j for $k \geq 1$. Set $f_j = \max\{g_1, \ldots, g_j\}$ so $\{f_j\}$ is increasing. Then
$$\nu(E \cap E_j) = \int_E f_j d\mu \text{ for } E \in \Sigma.$$

Set $f = \lim f_j$. From Proposition 2.2.5 and the MCT,
$$\nu(E) = \lim \nu(E \cap E_j) = \lim \int_E f_j d\mu = \int_E f d\mu \text{ for } E \in \Sigma.$$

Uniqueness again follows from 3.2.18.

Note that the function f is not generally μ-integrable. In fact, f is μ-integrable if and only if ν is finite.

The σ-finiteness condition on ν can be eliminated [see [Roy] for a proof]; however, the σ-finiteness condition on μ cannot be dropped as the following example shows.

Example 9 Let $S = [0,1]$ and Σ the σ-algebra of Lebesgue measurable subsets of S. If μ is counting measure on Σ, then $m \ll \mu$ but there exists no function $f \in L^1(\mu)$ such that $m = f d\mu$ [Proposition 3.2.14]. See also Exer. 11.

3.12. THE RADON-NIKODYM THEOREM

The function f in the conclusion of Theorem 8 is called the *Radon-Nikodym derivative* of ν with respect to μ and is denoted by $\frac{d\nu}{d\mu}$. Note the Radon-Nikodym derivative is only determined up to μ-a.e. Some of the basic properties of this "derivative" are given in the exercises.

The Radon-Nikodym Theorem can be extended to a more general class of measures than the σ-finite measures. One important class of such measures are called *decomposable measures*; for a description of these measures see [HS] p. 317. Necessary and sufficient conditions are known for a measure μ to admit the conclusion of the Radon-Nikodym Theorem for any signed measure ν; such measures are called *localizable* and have been characterized by Segal [see [TT] for a description].

For other methods of proving the Radon-Nikodym Theorem see [Roy] and [R2] and §3.12.1.

The Radon-Nikodym Theorem fails for finitely additive set functions; see section 3.12.1 and Exer. 3.12.1.2.

Exercise 1. Let μ, ν_1, ν_2 be signed measures on Σ with one ν_i being finite and μ a measure. If $\nu_i \ll \mu$, show $(a\nu_1 + b\nu_2) \ll \mu$. If the measures are σ-finite, compute $\frac{d(a\nu_1 + b\nu_2)}{d\mu}$.

Exercise 2. Let $S(\mathcal{T})$ be a σ-algebra. Let α, μ (β, ν) be σ-finite measures on $S(\mathcal{T})$. If $\mu \ll \alpha$, $\nu \ll \beta$, show $\mu \times \nu \ll \alpha \times \beta$. Compute the Radon-Nikodym derivative of $\mu \times \nu$ with respect to $\alpha \times \beta$.

Exercise 3. Let μ, ν be σ-finite measures on Σ with $\nu \ll \mu$. If $f \in L^1(\nu)$, show $f \frac{d\nu}{d\mu} \in L^1(\mu)$ and $\int f d\nu = \int f \frac{d\nu}{d\mu} d\mu$.

Exercise 4. Let λ, μ, ν be σ-finite measures on Σ with $\nu \ll \mu \ll \lambda$. Under appropriate conditions, show $\frac{d\nu}{d\lambda} = \frac{d\nu}{d\mu} \frac{d\mu}{d\lambda}$ [Chain Rule].

Exercise 5. Let μ, ν be finite measures on Σ with $\mu \ll \nu$, $\nu \ll \mu$. Show $\frac{d\nu}{d\mu} \neq 0$ μ-a.e. and $\frac{d\mu}{d\nu} = 1/\frac{d\nu}{d\mu}$.

Exercise 6. Let ν, ν_k, μ be σ-finite measures on Σ with $\nu = \sum_{k=1}^{\infty} \nu_k$. If $\nu_k \ll \mu$ for every k, show $\nu \ll \mu$ and $\frac{d\nu}{d\mu} = \sum_{k=1}^{\infty} \frac{d\nu_k}{d\mu}$.

Exercise 7. Let $\mathcal{A}$ be an algebra which generates Σ. Let μ, ν be finite measures on Σ. If ν is μ-continuous on $\mathcal{A}$, show $\nu \ll \mu$ on Σ.

Exercise 8. Let ν_k, μ be signed measures on Σ. Then $\{\nu_k\}$ is *uniformly μ-continuous* if $\lim_{|\mu|(E)\to 0} \nu_k(E) = 0$ uniformly in k. If $f_k \in L^1(\mu)$ and $\nu_k = f_k d\mu$, show $\{\nu_k\}$ is uniformly μ-continuous if there exists $g \in L^1(\mu)$ such that $|f_k| \leq g$ μ-a.e.

Exercise 9. Let μ be a finite measure on Σ and $f_k \in L^1(\mu)$. Show $\{f_k\}$ is uniformly μ-integrable if and only if
$$\sup \int_S |f_k|\, d\mu < \infty$$
and $\{f_k d\mu\}$ is uniformly μ-continuous.

Exercise 10. Let $\nu : \Sigma \to \mathbf{R}^*$ be finitely additive and μ be a measure on Σ. If ν is μ-continuous, show ν is countably additive.

Exercise 11. Let $S = \mathbf{R}$ and let Σ be the σ-algebra consisting of the sets which are either countable or have countable complements. Define ν on Σ by $\nu(E) = 0$ if E is countable and $\nu(E) = 1$ if E^c is countable. Let μ be counting measure. Show ν is a measure with $\nu \ll \mu$ but the Radon-Nikodym Theorem fails for ν.

Exercise 12. Let μ, ν be measures on Σ with ν finite. If $f_k \to f$ μ-measure, show $f_k \to f$ ν-measure. Can the finiteness condition be dropped?

3.12.1 The Radon-Nikodym Theorem for Finitely Additive Set Functions

Bochner has given an extension of the Radon-Nikodym Theorem to finitely additive set functions defined on an algebra ([Bo]). Bochner's original proof depended on the countably additive version and Stone space arguments. Dubins has given an elementary proof which we present ([D]). Dubins' proof depends on order properties of the space of finitely additive set functions and is independent of the countably additive version of the Radon-Nikodym Theorem.

Let $\mathcal{A}$ be an algebra of subsets of S. We let $ba(\mathcal{A})$ be the space of all real-valued, bounded, finitely additive set functions defined on the algebra $\mathcal{A}$. $ba(\mathcal{A})$ is a vector space if we set $(\mu+\nu)(A) = \mu(A)+\nu(A)$ and $(t\mu)(A) = t\mu(A)$ for $\mu, \nu \in ba(\mathcal{A})$, $t \in \mathbf{R}$, $A \in \mathcal{A}$. We write $\mu \geq \nu$ if and only if $\mu(A) \geq \nu(A)$ for all $A \in \mathcal{A}$; if $\mu \geq 0$, we say μ is non-negative. For convenience of notation, we set $\|\mu\| = |\mu|(S)$ for $\mu \in ba(\mathcal{A})$.

If $\mu, \nu \in ba(\mathcal{A})$, the meanings of $\nu \ll \mu$ and ν being μ-continuous are as in Definitions 3.12.1 and 3.12.4. If $f : S \to \mathbf{R}$ is an $\mathcal{A}$-simple function and $\mu \in ba(\mathcal{A})$ is non-negative, then the integral of f with respect to μ is meaningful and

$$\left| \int_A f d\mu \right| \leq \sup\{|f(t)| : t \in S\} \mu(A)$$

for all $A \in \mathcal{A}$ [Remark 3.2.2] and $A \to \int_A f d\mu$ defines an element of $ba(\mathcal{A})$; we denote this set function by $fd\mu$.

Bochner's Radon-Nikodym Theorem asserts that if ν is μ-continuous, then ν can be approximated by an indefinite integral $fd\mu$ where f is a simple function. We now prove Bochner's Theorem.

We require two lemmas. The first is an interesting result related to the order defined on $ba(\mathcal{A})$ and asserts that any two elements of $ba(\mathcal{A})$ have an infimum with respect to the order defined on $ba(\mathcal{A})$.

Lemma 1 *Let $\mu, \nu \in ba(\mathcal{A})$. Then*

$$\mu \wedge \nu(A) = \inf\{\mu(E) + \nu(A\backslash E) : E \in \mathcal{A}, E \subseteq A\}$$

defines an element $\mu \wedge \nu \in ba(\mathcal{A})$ which is such that $\mu, \nu \geq \mu \wedge \nu$ and if $\alpha \in ba(\mathcal{A})$ and $\alpha \leq \mu$, $\alpha \leq \nu$, then $\alpha \leq \mu \wedge \nu$.

Proof: Let $A, B \in \mathcal{A}$, $A \cap B = \emptyset$. If $E \in \mathcal{A}$, $E \subset A \cup B$, then

$$\mu(E)+\nu((A\cup B)\backslash E) = \mu(E\cap A)+\mu(E\cap B)+\nu(A\backslash E)+\nu(B\backslash E) \geq \mu\wedge\nu(A)+\mu\wedge\nu(B)$$

so

$$\mu \wedge \nu(A \cup B) \geq \mu \wedge \nu(A) + \mu \wedge \nu(B).$$

Let $\varepsilon > 0$. There exists $E_1 \subset A$, $E_2 \subset B$, $E_i \in \mathcal{A}$ such that

$$\mu \wedge \nu(A) > \mu(E_1) + \nu(A \backslash E_1) - \varepsilon/2$$

and

$$\mu \wedge \nu(B) > \mu(E_2) + \nu(B \backslash E_2) - \varepsilon/2$$

so

$$\begin{aligned}\mu \wedge \nu(A) + \mu \wedge \nu(B) &> \mu(E_1 \cup E_2) + \nu((A \backslash E_1) \cup (B \backslash E_2)) - \varepsilon \\ &= \mu(E_1 \cup E_2) + \nu((A \cup B) \backslash (E_1 \cup E_2)) - \varepsilon \\ &\geq \mu \wedge \nu(A \cup B) - \varepsilon.\end{aligned}$$

Hence,

$$\mu \wedge \nu(A \cup B) = \mu \wedge \nu(A) + \mu \wedge \nu(B)$$

and $\mu \wedge \nu \in ba(\mathcal{A})$.

Taking $E = A$ shows $\mu \wedge \nu \geq \mu$ and taking $E = \emptyset$ show $\mu \geq \nu$. If $\alpha \leq \mu$ and $\alpha \leq \nu$, then clearly $\alpha \leq \mu \wedge \nu$.

For the next lemma we adopt some unorthodox but useful notation. If μ, $\nu \in ba(\mathcal{A})$ and $\varepsilon \geq 0$, we write $\mu \leq \nu + \varepsilon$ if $\mu(A) \leq \nu(A) + \varepsilon$ for every $A \in \mathcal{A}$; note here that μ, ν are set functions and ε is a non-negative real.

Lemma 2 *Let $\varepsilon \geq 0$, $k > 0$, μ, $w \in ba(\mathcal{A})$ with μ non-negative. If*

$$-k\mu - \varepsilon < w < k\mu + \varepsilon,$$

then for every $\varepsilon' > \varepsilon$ there exists a two-valued $\mathcal{A}$-simple function f such that

$$-\frac{k}{2}\mu - \varepsilon' < w - \int f d\mu < \frac{k}{2}\mu + \varepsilon' \qquad (3.1)$$

and an $\mathcal{A}$-simple function f such that

$$-\varepsilon' < w - \int f d\mu < \varepsilon'. \qquad (3.2)$$

Proof: Choose $A \in \mathcal{A}$ such that $w(A) > w(E) - (\varepsilon' - \varepsilon)$ for all $E \in \mathcal{A}$. Let f equal $k/2$ on A and $-k/2$ on A^c. Then

$$w(E \cap A) - \int_{E \cap A} f d\mu \leq \frac{k}{2}\mu(E \cap A) + \varepsilon, \qquad (3.3)$$

and since $w(E \cap A^c) < \varepsilon' - \varepsilon$,

$$w(E \cap A^c) - \int_{E \cap A^c} f d\mu < \varepsilon' - \varepsilon + \frac{k}{2}\mu(E \cap A^c). \qquad (3.4)$$

The inequality on the right hand side of (1) now follows from (3) and (4). The other inequality in (1) is similar.

From (1) it follows by induction that for every j and $\eta > \varepsilon$ there exists a simple function f such that

$$-\frac{k}{2^j}\mu - \eta < w - \int f d\mu < \frac{k}{2^j}\mu + \eta$$

3.12. THE RADON-NIKODYM THEOREM

and this gives (2).

We are now ready to state and prove Bochner's Radon-Nikodym Theorem which asserts that if $\nu \in ba(\mathcal{A})$ is μ-continuous for some $\mu \in ba(\mathcal{A})$, then ν can be approximated by an indefinite integral with respect to μ of an $\mathcal{A}$-simple function [condition (iii) in Theorem 3 below].

Theorem 3 (Bochner) *Let $\mu, \nu \in ba(\mathcal{A})$ with μ non-negative. The following are equivalent:*

(i) *ν is μ-continuous,*

(ii) *for every $\varepsilon > 0$ there exist $w \in ba(\mathcal{A})$ and $k > 0$ such that $-k\mu \leq w \leq k\mu$ and $\|\nu - w\| \leq \varepsilon$,*

(iii) *for every $\varepsilon > 0$ there exists an $\mathcal{A}$-simple function such that $\|\nu - fd\mu\| \leq \varepsilon$.*

Proof: (i)$\Rightarrow$(ii): By decomposing ν into $\nu^+ - \nu^-$ we may assume that ν is non-negative. There exists $\delta > 0$ such that $\nu(E) < \varepsilon$ when $\mu(E) < \delta$. Set $k = \nu(S)/\delta$. By considering the cases where $\mu(E) < \delta$ and $\mu(E) \geq \delta$, it is easily checked that

$$\nu(E) - k\mu(E) < \varepsilon \text{ for all } E \in \mathcal{A}. \tag{3.5}$$

From Lemma 1,

$$\nu \wedge k\mu(S) = \inf\{\nu(E^c) + k\mu(E) : E \in \mathcal{A}\}$$

so

$$\|\nu - \nu \wedge k\mu\| = \nu(S) - \nu \wedge k\mu(S) = \sup\{\nu(E) - k\mu(E) : E \in \mathcal{A}\} \leq \varepsilon$$

by (5). Setting $w = \nu \wedge k\mu$ gives (ii).

(ii)$\Rightarrow$(iii): Choose w to satisfy (ii) and then choose f as in (2) so $\|w - fd\mu\| \leq 2\varepsilon'$ (Proposition 2.2.1.7(v)). Then

$$\|\nu - fd\mu\| \leq \|\nu - w\| + \|w - fd\mu\| \leq \varepsilon + 2\varepsilon'$$

and (iii) holds.

(iii)$\Rightarrow$(i): Given $\varepsilon > 0$ choose f as in (iii). Then there exists $\delta > 0$ such that $|\int_E fd\mu| < \varepsilon$ whenever $\mu(E) < \delta$ [Remark 3.2.2]. Thus, if $\mu(E) < \delta$, $|\nu(E)| \leq \varepsilon + |\int_E fd\mu| < 2\varepsilon$ from Proposition 2.2.1.7.

We now derive the Radon-Nikodym Theorem for countably additive measures on σ-algebras from Bochner's result. This result follows easily from the approximation property in (iii) and the completeness of $L^1(\mu)$ with respect to mean convergence.

Let Σ be a σ-algebra of subsets of S. If μ is a measure on Σ and $f : S \to \mathbf{R}^*$ has a μ-integral, we denote the set function $E \to \int_E fd\mu$ by $fd\mu$. Note that we then have

$$\|fd\mu\| = |fd\mu|(S) = \int_S |f|\, d\mu = \|f\|_1$$

[Exer. 3.2.26].

Theorem 4 (Radon-Nikodym) *Let μ be a σ-finite measure on Σ and ν a σ-finite signed measure on Σ with $\nu \ll \mu$. Then there exists a Σ-measurable function $f : S \to \mathbf{R}$ such that $\nu(E) = \int_E f d\mu$ for all $E \in \Sigma$ [i.e., $\nu = f d\mu$]; f is unique up to μ-a.e.*

Proof: By the Jordan decomposition we may assume that ν is a measure.

We first assume that both μ and ν are finite measures. Then ν is μ-continuous by Proposition 3.12.5. By Theorem 3 for each $k \in \mathbf{N}$ there exists a simple function f_k such that $\|\nu - f_k d\mu\| < 1/k$. Since $\|f_k d\mu - f_j d\mu\| = \|f_k - f_j\|_1$, $\{f_k\}$ is a Cauchy sequence in $L^1(\mu)$ and, therefore, converges to some $f \in L^1(\mu)$ by the Riesz-Fischer Theorem [3.5.1]. Then

$$\|\nu - f d\mu\| \leq \|\nu - f_k d\mu\| + \|f_k - f\|_1$$

implies that $\nu = f d\mu$.

The extension of the result to the σ-finite case is given in the proof of Theorem 3.12.8.

Note that the proof of the Radon-Nikodym Theorem above depends on Bochner's version of the theorem for finitely additive set functions and the completeness of $L^1(\mu)$ with respect to mean convergence when μ is a countably additive measure defined on a σ-algebra. This proof of Bochner's Theorem basically depends on order properties of the space $ba(\mathcal{A})$ and can be regarded as elementary.

Exercise 1. Let $\mu, \nu \in ba(\mathcal{A})$. Show

$$\mu \vee \nu(A) = \sup\{\mu(E) + \nu(A \setminus E) : E \subset A, E \in \mathcal{A}\}$$

defines an element of $ba(\mathcal{A})$ which is the supremum of μ and ν in the order of $ba(\mathcal{A})$.

Exercise 2. Let $\mathcal{A}$ be the algebra of subsets of $\mathbf{N}$ which are either finite or have finite complements. Define ν, μ on $\mathcal{A}$ by

$$\nu(E) = 0 \left[\mu(E) = \sum_{k \in E} 1/2^k\right]$$

if E is finite and

$$\nu(E) = 1 \left[\mu(E) = 1 + \sum_{k \in E} 1/2^k\right]$$

if E is infinite. Show $\nu \ll \mu$ but there is no μ-integrable f with $\nu = f d\mu$. Given $\varepsilon > 0$, find a simple function f satisfying (iii).

3.13 Lebesgue Decomposition

Intuitively, a measure ν is absolutely continuous with respect to a measure μ if ν is small on sets with small μ-measure (Proposition 3.12.5). We now consider a concept in opposition to absolute continuity, called singularity, which was briefly considered in §2.2.2.

Let Σ be a σ-algebra of subsets of S. Recall (2.2.2.6) that two measures μ and ν on Σ are said to be *singular* if there exist $A, B \in \Sigma$ such that $A \cap B = \emptyset$, $A \cup B = S$, $\mu(A) = 0 = \nu(B)$. If μ and ν are singular, we write $\mu \perp \nu$. Two signed measures μ, ν on Σ are singular, written again $\mu \perp \nu$, if $|\mu| \perp |\nu|$.

If ν is a finite signed measure, recall that ν^+ and ν^- are singular [2.2.2.7].

To illustrate that absolute continuity and singularity are opposites, we have

Proposition 1 *Let μ, ν be signed measures on Σ. If $\mu \perp \nu$ and $\nu \ll \mu$, then $\nu = 0$.*

Proof: Since $|\nu| \perp |\mu|$, there exist $A, B \in \Sigma$, $A \cap B = \emptyset$, $A \cup B = S$ with $|\mu|(A) = 0 = |\nu|(B)$. But, $|\nu| \ll |\mu|$ implies $|\nu|(A) = 0$. Hence, $|\nu|(S) = |\nu|(A) + |\nu|(B) = 0$.

We are now going to show that given a μ, any σ-finite signed measure can be decomposed uniquely into two parts, one of which is absolutely continuous with respect to μ and another which is singular with respect to μ. This is called the Lebesgue Decomposition. For this theorem, we need a preliminary result.

Proposition 2 *Let ν_k, μ be measures on Σ. If $\nu_k \perp \mu$ for every k and $\nu = \sum\limits_{k=1}^{\infty} \nu_k$, then $\nu \perp \mu$.*

Proof: Recall that ν is a measure [Exer. 2.2.13]. For each k there exist A_k, $B_k \in \Sigma$, $A_k \cap B_k = \emptyset$, $A_k \cup B_k = S$ with $\mu(A_k) = \nu_k(B_k) = 0$. Put $A = \bigcup\limits_{k=1}^{\infty} A_k$, $B = A^c = \bigcap\limits_{k=1}^{\infty} B_k$. Then

$$\mu(A) \leq \sum_{k=1}^{\infty} \mu(A_k) = 0$$

and

$$\nu(B) = \sum_{k=1}^{\infty} \nu_k(B) \leq \sum_{k=1}^{\infty} \nu_k(B_k) = 0.$$

Theorem 3 (Lebesgue Decomposition) *Let ν be a σ-finite signed measure on Σ and μ a measure on Σ. Then there exist signed measures ν_a and ν_s on Σ such that $\nu = \nu_a + \nu_s$ and $\nu_a \ll \mu$, $\nu_s \perp \mu$. The decomposition is unique.*

Proof: First assume that ν is a finite measure. Let $\mathcal{M} = \{E \in \Sigma : \mu(E) = 0\}$ and set $a = \sup\{\nu(E) : E \in \mathcal{M}\}$. Since $\nu \geq 0$ and ν is finite, $0 \leq a < \infty$. Choose $\{E_k\} \subset \mathcal{M}$ such that $\lim \nu(E_k) = a$. Set $E = \bigcup_{k=1}^{\infty} E_k$. Then $E \in \mathcal{M}$ and $\nu(E) = a$ [clearly $\nu(E) \leq a$ since $E \in \mathcal{M}$, but $\nu(E) \geq \nu(E_k)$ for all k implies $\nu(E) \geq a$].

We claim that
$$\nu(A\backslash E) = 0 \text{ for all } A \in \mathcal{M}. \tag{3.1}$$

For otherwise, $\nu(E \cup A) = \nu(A\backslash E) + \nu(E) > a$. Since $E \cup A \in \mathcal{M}$, this would contradict the definition of a.

Define $\nu_a(A) = \nu(A\backslash E)$, $\nu_s(A) = \nu(A \cap E)$ for $A \in \Sigma$. Clearly $\nu = \nu_a + \nu_s$. (1) implies that $\nu_a \ll \mu$. Now $\mu(E) = 0$ since $E \in \mathcal{M}$ and $\nu_s(S\backslash E) = \nu((S\backslash E) \cap E) = 0$ so $\mu \perp \nu_s$.

Next assume that ν is a σ-finite measure. Let $S = \bigcup_{k=1}^{\infty} E_k$, where $E_k \in \Sigma$, $\{E_k\}$ pairwise disjoint and $\nu(E_k) < \infty$. Set $\nu_k(E) = \nu(E \cap E_k)$ for $E \in \Sigma$. Then $\nu = \sum_{k=1}^{\infty} \nu_k$. From the first part, set $\nu_a = \sum_{k=1}^{\infty} (\nu_k)_a$ and $\nu_s = \sum_{k=1}^{\infty} (\nu_k)_s$. Then ν_a and ν_s are measures from Exer. 2.2.13. Clearly $\nu_a \ll \mu$ and $\nu_s \perp \mu$ by Proposition 2.

If ν is a σ-finite signed measure, then $\nu = \nu^+ - \nu^-$. In this case set $\nu_a = (\nu^+)_a - (\nu^-)_a$ and $\nu_s = (\nu^+)_s - (\nu^-)_s$ and apply Exercises 1 and 3.12.1.

For the uniqueness, suppose $\nu = \nu_a + \nu_s = \nu'_a + \nu'_s$, where $\nu'_a \ll \mu$ and $\nu'_s \perp \mu$. If $F \in \Sigma$ is such that $|\nu|(F) < \infty$, then $\nu_a - \nu'_a = \nu'_s - \nu_s$ is both singular with respect to μ and absolutely continuous with respect to μ on $\Sigma_F = \{E \cap F : E \in \Sigma\}$ so $\nu_a - \nu'_a = \nu'_s - \nu_s = 0$ on Σ_F by Proposition 1. Since ν is σ-finite, $\nu_a - \nu'_a = \nu'_s - \nu_s = 0$ on Σ.

The Lebesgue Decomposition is often derived from the Radon-Nikodym Theorem [[Roy] p. 278]. This usually requires the assumption that both ν and μ are σ-finite. Note in the proof above it is only necessary to assume that μ is non-negative, monotone, countably subadditive and vanishes at $\emptyset$, i.e., μ is an "outer measure". The proof given here is due to Brooks ([Br]).

The σ-finiteness condition in the Lebesgue Decomposition cannot be dropped.

Example 4 Let $S = [0, 1]$ and ν be counting measure restricted to the class of Lebesgue measurable subsets of S. Then there is no Lebesgue Decomposition for ν with respect to m. For if $\nu = \nu_a + \nu_s$ as in Theorem 3, there exist measurable A, B such that $A \cap B = \emptyset$, $S = A \cup B$ and $\nu_s(A) = m(B) = 0$. Take $x \in A$. Then

$$\nu(\{x\}) = 1 = \nu_a(\{x\}) + \nu_s(\{x\}) = 0$$

since $m(\{x\}) = 0$ and $x \in A$.

There is a version of the Lebesgue Decomposition Theorem for bounded finitely additive set functions; see [RR] 6.2.4 for this and further such results.

3.13. LEBESGUE DECOMPOSITION

Exercise 1. Let ν_1, ν_2 be σ-finite signed measures on Σ with one being finite. Let μ be a signed measure on Σ with $\nu_i \perp \mu$. Show $(a\nu_1 + b\nu_2) \perp \mu$.

Exercise 2. Let E be the even integers. Let ν be counting measure and define μ on $\mathcal{P}(\mathbf{N})$ by $\mu(A) = \sum_{k \in A \cap E} 1/2^k$. Describe the Lebesgue decomposition of ν with respect to μ.

Exercise 3. Let $f(t) = e^t$ for $t \leq 1$ and $f(t) = 0$ for $t > 1$, $g(t) = \frac{1}{(t+1)^2}$ for $t \geq 0$ and $g(t) = 0$ for $t < 0$. Let $\nu = fdm$, $\mu = gdm$. Describe the Lebesgue Decomposition for ν with respect to μ.

Exercise 4. Show under appropriate hypotheses that $\alpha \perp \mu$ or $\beta \perp \nu$ implies $\alpha \times \beta \perp \mu \times \nu$.

Exercise 5. If $\alpha = \alpha_a + \alpha_s$ ($\beta = \beta_a + \beta_s$) is the Lebesgue Decomposition of α (β) with respect to μ (ν), show $(\alpha \times \beta)_a = \alpha_a \times \beta_a$,

$$(\alpha \times \beta)_s = \alpha_s \times \beta_s + \alpha_s \times \beta_a + \alpha_a \times \beta_s$$

gives the Lebesgue Decomposition of $\alpha \times \beta$ with respect to $\mu \times \nu$.

3.14 The Vitali-Hahn-Saks Theorem

In this section we prove a theorem closely related to the Nikodym Convergence Theorem, the Vitali-Hahn-Saks Theorem. Our proof of this result depends on the Nikodym Convergence Theorem, but sometimes the Vitali-Hahn-Saks Theorem is established first and then the Nikodym Convergence Theorem is derived as a consequence [see [DS] for example].

Let Σ be a σ-algebra of subsets of S. Let μ be a measure on Σ and $\{\nu_i\}$ a sequence of signed measures on Σ. Then $\{\nu_i\}$ is *uniformly μ-continuous* if $\lim_{\mu(E)\to 0} \nu_i(E) = 0$ uniformly in i.

Theorem 1 *Let ν_i be a finite signed measure on Σ and μ a measure on Σ such that each ν_i is μ-continuous. If $\{\nu_i\}$ is uniformly countably additive, then $\{\nu_i\}$ is uniformly μ-continuous.*

Proof: If the conclusion fails, there exists $\epsilon > 0$ such that for every $\delta > 0$ there exist k, $E \in \Sigma$ with $|\nu_k(E)| \geq \epsilon$ and $\mu(E) < \delta$. In particular, there exist $E_1 \in \Sigma$, n_1 such that $|\nu_{n_1}(E_1)| \geq \epsilon$ and $\mu(E_1) < 1$. There exists $\delta_1 > 0$ such that $|\nu_{n_1}(E)| < \epsilon/2$ whenever $\mu(E) < \delta_1$. There exist $E_2 \in \Sigma$, $n_2 > n_1$ such that $|\nu_{n_2}(E_2)| \geq \epsilon$ and $\mu(E_2) < \delta_1/2$. Continuing this construction produces sequences $\{E_k\} \subset \Sigma$, $\delta_{k+1} < \delta_k/2$, $n_k \uparrow$ such that $|\nu_{n_k}(E_k)| \geq \epsilon$, $\mu(E_{k+1}) < \delta_k/2$ and $|\nu_{n_k}(E)| < \epsilon/2$ whenever $\mu(E) < \delta_k$. Note that

$$\mu\left(\bigcup_{j=k+1}^{\infty} E_j\right) \leq \sum_{j=k+1}^{\infty} \mu(E_j) < \delta_k/2 + \delta_{k+1}/2 + \cdots < \delta_k/2 + \delta_k/2^2 + \cdots = \delta_k$$

so that

$$\left|\nu_{n_k}\left(E_k \cap \bigcup_{j=k+1}^{\infty} E_j\right)\right| < \epsilon/2.$$

Now set $A_k = E_k \setminus \bigcup_{j=k+1}^{\infty} E_j$. The $\{A_k\}$ are pairwise disjoint and

$$|\nu_{n_k}(A_k)| \geq |\nu_{n_k}(E_k)| - \left|\nu_{n_k}\left(E_k \cap \bigcup_{j=k+1}^{\infty} E_j\right)\right| \geq \epsilon - \epsilon/2 = \epsilon/2$$

by the observation above. However, by the uniform countable additivity of $\{\nu_i\}$, we have $\lim_k \nu_i(A_k) = 0$ uniformly for $i \in \mathbb{N}$ [Lemma 2.8.4], and we have the desired contradiction.

Remark 2 Note that only the facts that μ is non-negative, increasing and countably subadditive were used.

3.14. THE VITALI-HAHN-SAKS THEOREM

From the Nikodym Convergence Theorem and Theorem 1 we can now obtain the Vitali-Hahn-Saks Theorem.

Theorem 3 *Let ν_i be a finite signed measure on $\sum$ and μ a measure on $\sum$ such that each ν_i is μ-continuous. If $\lim_i \nu_i(E) = \nu(E)$ exists for each $E \in \sum$, then*

(i) $\{\nu_i\}$ *is uniformly μ-continuous and*

(ii) ν *is μ-continuous.*

Proof: By the Nikodym Convergence Theorem $\{\nu_i\}$ is uniformly countably additive so (i) follows from Theorem 1. (ii) is immediate from (i).

Both the Vitali-Hahn-Saks and Nikodym Theorems fail for countably additive set functions defined on algebras [see Exercises 1 and 2]. There are versions of the theorems for certain finitely additive set functions [see [DU]].

Exercise 1. Let $\mathcal{A}$ be the algebra of subsets of $[0,1)$ generated by intervals of the form $[a,b)$, $0 \le a < b \le 1$. Define ν_k on $\mathcal{A}$ by $\nu_k(A) = 2^k m([1 - 1/2^k, 1) \cap A)$ so ν_k is countably additive on $\mathcal{A}$. Show that if $b < 1$, $\nu_k[a,b) = 0$ for large k and $\nu_k[c,1) = 1$ for large k so $\lim \nu_k(A) = \nu(A)$ exists and $\nu(A) = 1$ if $[1-\delta, 1) \subset A$ for some $\delta > 0$ and $\nu(A) = 0$ otherwise. Show ν is purely finitely additive (§2.6.1) so (ii) of Theorem 2.8.5 fails. Show also that (i) of Theorem 2.8.5 and (i), (ii) of Theorem 3 fail.

Exercise 2. Let $\mathcal{A}$ be the algebra of subsets of $\mathbf{N}$ which are either finite or have finite complements. Define ν_n on $\mathcal{A}$ by $\nu_n(A) = n$ if A is finite and $n \in A$, $\nu_n(A) = -n$ if A^c is finite and $n \in A^c$ and $\nu_n(A) = 0$ otherwise. Show each ν_n is bounded, finitely additive and $\lim \nu_n(A)$ exists for each $A \in \mathcal{A}$. Let $\mu(A) = \sum_{n \in A} 1/2^n$ when A is finite and $\mu(A) = 1 + \sum_{n \in A} 1/2^n$ when A^c is finite. Show μ is bounded, finitely additive, $\nu_n \ll \mu$ but (i) fails.

Chapter 4
Differentiation and Integration

4.1 Differentiating Indefinite Integrals

We consider the first half of the Fundamental Theorem of Calculus (FTC), the differentiation of indefinite integrals. We consider indefinite integrals in $\mathbf{R}^n$. Points x in $\mathbf{R}^n$ are denoted by $x = (x_1, \ldots, x_n)$ where $x_i \in \mathbf{R}$. If $x \in \mathbf{R}^n$ and $r > 0$, the open cube with center at x and sidelength $2r$ is denoted by

$$S(x,r) = \{y : |x_i - y_i| < r, i = 1, \ldots, n\}.$$

All statements concerning measurability, integrability and almost everywhere refer to Lebesgue measure on $\mathbf{R}^n$ which is denoted by m.

We begin by establishing a covering theorem which will be used in the proof of the differentiation results.

Lemma 1 *Let $\mathcal{C}$ be any collection of open cubes in $\mathbf{R}^n$ and let $U = \cup\{I : I \in \mathcal{C}\}$. If $c < m(U)$, there exist pairwise disjoint $S_1, \ldots, S_k \in \mathcal{C}$ such that $\sum_{j=1}^{k} m(S_j) > 3^{-n}c$.*

Proof: By regularity of m, there exists a compact $K \subset U$ such that $m(K) > c$ so there exist finitely many cubes $I_1, \ldots, I_j \in \mathcal{C}$ covering K. Let S_1 be the $\{I_i\}$ with the largest sidelength. Let S_2 be the $\{I_i\}$ with the largest sidelength which is disjoint from S_1 and continue this procedure until the $\{I_i\}$ are exhausted. This gives a pairwise disjoint sequence $S_1, \ldots, S_k \in \mathcal{C}$. Let S'_i be the cube with the same center as S_i but with sidelength three times that of S_i. If I_i is not one of the $S_1, \ldots, S_k$, there exists S_ℓ such that $I_i \cap S_\ell \neq \emptyset$ and the sidelength of I_i is at most that of S_ℓ. Hence, $I_i \subset S'_\ell$. Then $K \subset \bigcup_{\ell=1}^{k} S'_\ell$ so

$$c < m(K) \leq \sum_{\ell=1}^{k} m(S'_\ell) = 3^n \sum_{\ell=1}^{k} m(S_\ell).$$

A function $f : \mathbf{R}^n \to \mathbf{R}$ is said to be *locally integrable* if f is Lebesgue measurable and is m-integrable over every compact subset of $\mathbf{R}^n$. The class of all locally integrable functions on $\mathbf{R}^n$ is denoted by $L^1_{\text{loc}}(\mathbf{R}^n)$.

Let $f \in L^1_{\text{loc}}(\mathbf{R}^n)$. We consider the problem of differentiating the indefinite integral of f, $\int_E f \, dm$. That is, for what x does the limit,

$$\lim_{r \to 0} \int_{S(x,r)} f \, dm / m(S(x,r)),$$

exist and for what x does the limit equal the integrand, $f(x)$? Note that for $n = 1$, we are asking that

$$\lim_{r \to 0^+} \int_{x-r}^{x+r} f \, dm / 2r = f(x);$$

we will see that information about the usual derivative,

$$\lim_{h \to 0} \int_x^{x+h} f \, dm / h = f(x),$$

can be derived from our general results.

The principal tool used in our proof is the Hardy-Littlewood maximal function. The Hardy-Littlewood Maximal Theorem gives bounds on the average value of a function and this result is used in the differentiation theorem.

If $f \in L^1_{\text{loc}}(\mathbf{R}^n)$, $x \in \mathbf{R}^n$ and $r > 0$, we set

$$A_r f(x) = \int_{S(x,r)} f \, dm / m(S(x,r)),$$

the *average value* of f over $S(x,r)$, and $Mf(x) = \sup_{r>0} A_r |f|(x)$. Mf is called the *Hardy-Littlewood maximal function*.

Proposition 2 *Mf is measurable. [In fact, Mf is lower semi-continuous.]*

Proof: We show $E = \{x : Mf(x) \leq a\}$ is closed for every $a > 0$. Suppose $x_k \in E$ and $x_k \to x$. Let A_k be the symmetric difference $S(x_k, r) \Delta S(x, r)$ and $f_k = C_{A_k} f$. Since $\bigcap_{k=1}^{\infty} A_k = \emptyset$, $f_k \to 0$ and $|f_k| \leq |f|$ so by the DCT $\int_{\mathbf{R}^n} |f_k| \, dm \to 0$. Since

$$S(x,r) \subset A_k \cup S(x_k, r),$$
$$A_r |f|(x) \leq \int_{A_k} |f| \, dm / m(S(x,r)) + \int_{S(x_k,r)} |f| \, dm / m(S(x_k,r))$$
$$\leq \int_{\mathbf{R}^n} |f_k| \, dm / m(S(x,r)) + a.$$

Hence, $A_r |f|(x) \leq a$ and $Mf(x) \leq a$; i.e., $x \in E$.

Theorem 3 (Maximal Theorem) *There exists a constant $C > 0$, depending only on n, such that for all $f \in L^1$ and $a > 0$, $m\{x : Mf(x) > a\} \leq (C/a) \int_{\mathbf{R}^n} |f| \, dm$.*

Proof: Let $E_a = \{x : Mf(x) > a\}$. For each $x \in E_a$ choose $r_x > 0$ such that $A_{r_x} |f|(x) > a$. The cubes $\{S(x, r_x) : x \in E_a\}$ cover E_a so if $c < m(E_a)$, by Lemma 1 there exist $x_1, \ldots, x_k \in E_a$ such that the cubes $\{S_j = S(x_j, r_{x_j}) : j = 1, \ldots, k\}$ are pairwise disjoint and $\sum_{j=1}^{k} m(S_j) > 3^{-n} c$. Then

$$c < 3^n \sum_{j=1}^{k} m(S_j) \leq \frac{3^n}{a} \sum_{j=1}^{k} \int_{S_j} |f| \, dm \leq \frac{3^n}{a} \int_{\mathbf{R}^n} |f| \, dm.$$

Since $c < m(E_a)$ is arbitrary, the desired inequality follows.

4.1. DIFFERENTIATING INDEFINITE INTEGRALS

Theorem 4 *If $f \in L^1_{\text{loc}}(\mathbf{R}^n)$, then $\lim_{r \to 0} A_r f(x) = f(x)$ for m-almost all $x \in \mathbf{R}^n$.*

Proof: It suffices to show $A_r f(x) \to f(x)$ for almost all x in the cube $S(0, N)$ for arbitrary N. For x in the cube and $r < 1$, the values of $A_r f$ only depend on the values of f in $S(0, N+1)$ so we may assume that $f \in L^1(\mathbf{R}^n)$. Let $\epsilon > 0$ and pick a continuous function with compact support g on $\mathbf{R}^n$ such that $\int_{\mathbf{R}^n} |f - g|\, dm < \epsilon$ [3.5.6]. Since g is uniformly continuous, for every $x \in \mathbf{R}^n$ and $\delta > 0$ there exists $r > 0$ such that $|g(y) - g(x)| < \delta$ when $y \in S(x, r)$ so

$$|A_r g(x) - g(x)| = \left|\int_{S(x,r)} (g(y) - g(x))dy\right| / m(S(x,r)) < \delta.$$

Hence, $\lim_{r \to 0} A_r g(x) = g(x)$ and

$$\overline{\lim_{r \to 0}} |A_r f(x) - f(x)| = \overline{\lim_{r \to 0}} |A_r(f-g)(x) + (A_r g(x) - g(x)) + (g(x) - f(x))|$$

$$\leq M(f-g)(x) + |g - f|(x).$$

Let $E_a = \{x : \overline{\lim}_{r \to 0} |A_r f(x) - f(x)| > a\}$ and $F_a = \{x : |f - g|(x) > a\}$. Then $E_a \subset F_{a/2} \cup \{x : M(f-g)(x) > a/2\}$. But $\epsilon > \int_{F_a} |f-g|\,dm \geq am(F_a)$ so by the Maximal Theorem,

$$m(E_a) \leq m(F_{a/2}) + m\{x : M(f-g)(x) > a/2\} \leq \frac{2\epsilon}{a} + \frac{2C}{a}\epsilon.$$

Hence, $m(E_a) = 0$ for all $a > 0$, and the result follows.

We now show that Theorem 4 can be improved by moving the absolute value in the conclusion,

$$\lim_{r \to 0} \left|\int_{S(x,r)} f\,dm/m(S(x,r)) - f(x)\right| = \lim_{r \to 0} \left|\int_{S(x,r)} (f - f(x))dm/m(S(x,r))\right| = 0,$$

inside the integral sign. A point x is called a *Lebesgue point* of f if

$$\lim_{r \to 0} \int_{S(x,r)} |f - f(x)|\,dm/m(S(x,r)) = 0$$

and the collection of all Lebesgue points of f is called the *Lebesgue set* of f.

Theorem 5 *If $f \in L^1_{\text{loc}}(\mathbf{R}^n)$, then almost every point of $\mathbf{R}^n$ is a Lebesgue point of f.*

Proof: By Theorem 4, for every $q \in \mathbf{Q}$ there exists an m-null set Z_q such that

$$\lim_{r \to 0} \int_{S(x,r)} |f(y) - q|\,dy/m(S(x,r)) = |f(x) - q|$$

for $x \in Z_q^c$. Set $Z = \bigcup_{q \in \mathbf{Q}} Z_q$. Then $m(Z) = 0$ and if $x \notin Z$ for any $\epsilon > 0$, there exists $p \in \mathbf{Q}$ with $|f(x) - p| < \epsilon$. Since

$$|f(y) - f(x)| \leq |f(y) - p| + \epsilon,$$

$$\varlimsup_{r \to 0} \int_{S(x,r)} |f(y) - f(x)| \, dy / m(S(x,r)) \leq |f(x) - p| + \epsilon < 2\epsilon$$

so

$$\lim_{r \to 0} \int_{S(x,r)} |f(y) - f(x)| \, dy / m(S(x,r)) = 0 \text{ for } x \notin Z.$$

Next we show that the limit in Theorem 5 above exists when more general sets than cubes are considered. These results are particularly useful in **R** where they yield differentiation results.

Definition 6 *Let $x \in \mathbf{R}^n$. A family of Borel sets, $\{E_r : r > 0\}$, shrinks regularly to x if $E_r \subset S(x,r)$ for every $r > 0$ and there exists $a > 0$ (independent of r) such that $m(E_r) \geq am(S(x,r))$ [x need not belong to E_r].*

Example 7 In **R**, the families $\{[x-r, x) : r > 0\}$, $\{[x, x+r) : r > 0\}$ shrink regularly to x [with $a = 2$]. In $\mathbf{R}^n$, if $E_r = \{y : \|x - y\| < r\}$, then $\{E_r\}$ shrinks regularly to x.

For an example of a family of rectangles in $\mathbf{R}^2$ whose intersection is $\{x\}$ but which do not shrink regularly to x see Exercise 1.

Theorem 8 *Let $f \in L^1_{\text{loc}}(\mathbf{R}^n)$. If x is in the Lebesgue set of f and $\{E_r : r > 0\}$ shrinks regularly to x, then*

$$\lim_{r \to 0} \frac{1}{m(E_r)} \int_{E_r} |f(y) - f(x)| \, dy = 0.$$

Hence,

$$\lim_{r \to 0} \frac{1}{m(E_r)} \int_{E_r} f(y) \, dy = f(x).$$

In particular, this holds for almost all $x \in \mathbf{R}^n$.

Proof: Let a be as in Definition 6. Then

$$\frac{1}{m(E_r)} \int_{E_r} |f(y) - f(x)| \, dy \leq \frac{1}{m(E_r)} \int_{S(x,r)} |f(y) - f(x)| \, dy$$
$$\leq \frac{1}{am(S(x,r))} \int_{S(x,r)} |f(y) - f(x)| \, dy \to 0 \quad (4.1)$$

by Theorem 5. The last two statements follow from (1) and Theorem 5.

From Theorem 8 and Example 7 we obtain a version of the FTC in **R**. Let $f : [a, b] \to \mathbf{R}$ be integrable and let $F(x) = \int_a^x f \, dm$ be the indefinite integral of f.

Theorem 9 (FTC) *F is differentiable at every Lebesgue point x of f with $F'(x) = f(x)$. In particular, this holds a.e. in $[a, b]$.*

4.1. DIFFERENTIATING INDEFINITE INTEGRALS

Finally, we consider the differentiation of arbitrary regular Borel measures on $\mathbf{R}^n$ [The regularity assumption is redundant (2.7.6); we make this assumption for the reader who has skipped §2.7.] Let ν be a regular Borel measure on $\mathbf{R}^n$. We say that ν is differentiable at x if $\lim_{r \to 0} \nu(S(x,r))/m(S(x,r))$ exists; the value of this limit is denoted by $D\nu(x)$ and is called the *derivative* of ν at x. If ν is an indefinite integral of a locally integrable function, this agrees with our previous definition of the derivative.

Lemma 10 *If ν is a regular Borel measure on $\mathbf{R}^n$ and $\nu(A) = 0$, then $D\nu(x) = 0$ for almost all $x \in A$.*

Proof: For $\delta > 0$ let $A_\delta = \left\{ x \in A : \overline{\lim_{r \to 0}} \frac{\nu(S(x,r))}{m(S(x,r))} > \delta \right\}$. It suffices to show that $m(A_\delta) = 0$ for every such δ. Let $\epsilon > 0$. By regularity there is an open $V \supset A$ such that $\nu(V) < \epsilon$. For each $x \in A_\delta$ there is an open cube with center at x, $S_x \subset V$, such that $\nu(S_x) > m(S_x)\delta$. By Lemma 1 if $U = \bigcup_{x \in A_\delta} S_x$ and $c < m(U)$, there exist $x_1, \ldots, x_k \in A_\delta$ such that $S_{x_1}, \ldots, S_{x_k}$ are pairwise disjoint with

$$c < 3^n \sum_{i=1}^{k} m(S_{x_i}) \leq (3^n/\delta) \sum_{i=1}^{k} \nu(S_{x_i})$$

$$\leq (3^n/\delta)\nu(V) < (3^n/\delta)\epsilon.$$

Therefore, $m(U) \leq (3^n/\delta)\epsilon$ and since $A_\delta \subset U$ and $\epsilon > 0$ is arbitrary, $m(A_\delta) = 0$.

A signed Borel measure ν is said to be regular if $|\nu|$ is regular.

Theorem 11 *Let ν be a regular signed Borel measure on $\mathbf{R}^n$ and $\nu = \nu_a + \nu_s$ be its Lebesgue Decomposition with respect to m with $f = \frac{d\nu_a}{dm}$, the Radon-Nikodym derivative of ν_a with respect to m. Then ν is differentiable a.e. with $D\nu = f$ a.e. Furthermore, for almost all x, $\lim_{r \to 0} \nu(E_r)/m(E_r) = f(x)$ whenever the family $\{E_r : r > 0\}$ shrinks regularly to x.*

Proof: By the Jordan Decomposition we may assume that ν is a measure. The first part of the Theorem follows from Theorem 4 and Lemma 10. The last statement follows from Theorem 8 and Exercise 2.

Exercise 1. Let $E_r = [-r, r] \times [-r^2, r^2]$. For $r \leq 1$, show $E_r \subset S(0, r)$, $\bigcap_{r > 0} E_r = \{(0, 0)\}$, but $\{E_r : r > 0\}$ doesn't shrink regularly to $(0, 0)$.

Exercise 2. Let ν be a Borel measure on $\mathbf{R}^n$ and $\nu(A) = 0$. Show that for almost all $x \in A$, $\lim_{r \to 0} \nu(E_r)/m(E_r) = 0$ when $\{E_r : r > 0\}$ shrinks regularly to x [Lemma 10].

Exercise 3. Show $A_r f(x)$ is continuous in r.

Exercise 4. Let $f : [a, b] \to \mathbf{R}$ be integrable and $F(x) = \int_a^x f \, dm$. If f is continuous at x, show $F'(x) = f(x)$.

4.2 Differentiation of Monotone Functions

We use the results of the previous section to establish a result of Lebesgue on the almost everywhere differentiability of a monotone function. Again all measurability statements refer to Lebesgue measure.

Let $f : \mathbf{R} \to \mathbf{R}$ be monotone [assume for definiteness that $f \uparrow$]. Then f is not continuous at x if and only if $f(x^+) - f(x^-) > 0$, i.e., if and only if f has a jump discontinuity at x. If x is a discontinuity of f, we may choose a rational r_x satisfying $f(x^-) < r_x < f(x^+)$ and obtain a 1-1 mapping $x \to r_x$ from the set of discontinuities of f into $\mathbf{Q}$. This shows that the set of points of discontinuity of a monotone function is countable. Hence, a monotone function has points of continuity in every open interval.

We first consider the case of left continuous increasing functions. Such functions induce Lebesgue-Stieltjes measures and Theorem 4.1.11 can be used to prove their a.e. differentiability.

Theorem 1 *Let* $f : \mathbf{R} \to \mathbf{R}$ *be* $\uparrow$ *and left continuous. Then f is differentiable a.e. with* $f' = D\mu_f$ *a.e.*

Proof: Let μ_f be the Lebesgue-Stieltjes measure induced by f. Then $D\mu_f$ exists a.e. [4.1.11]. The families $\{[x-h, x) : h > 0\}$ and $\{[x, x+h) : h > 0\}$ shrink regularly to x [Example 4.1.7] so the limits

$$\lim_{h \to 0^+} \mu_f[x-h, x)/h = \lim_{h \to 0^+} \frac{f(x) - f(x-h)}{h} = D\mu_f(x)$$

and

$$\lim_{h \to 0^+} \mu_f[x, x+h)/h = \lim_{h \to 0^+} \frac{f(x+h) - f(x)}{h} = D\mu_f(x)$$

exist for almost all x by Theorem 4.1.11. Thus, $f'(x)$ exists for almost all x and $f' = D\mu_f$ a.e.

We now remove the left continuity assumption from Theorem 1.

Let $f : \mathbf{R} \to \mathbf{R}$ be $\uparrow$. Define $f_* : \mathbf{R} \to \mathbf{R}$ by $f_*(x) = f(x^-) = \lim_{y \to x^-} f(y)$. Then $f_* \leq f$, $f_* \uparrow$ and f_* is left continuous [for each x there exists $x_k \uparrow x$ such that f is continuous at x_k so $f(x^-) = \lim f(x_k) = f_*(x) = \lim f_*(x_k)$]. Similarly, $f_*(x^+) - f_*(x) = f(x^+) - f(x^-)$ so f and f_* have the same points of continuity.

Lemma 2 *If f_* is differentiable at x, then f is differentiable at x with $f'_*(x) = f'(x)$.*

Proof: Let $m = f'_*(x)$. Note $f(x) = f_*(x)$ since f_* is continuous at x. Let $\epsilon > 0$. Choose $\delta > 0$ such that $m - \epsilon < (f_*(y) - f_*(x))/(y - x) < m + \epsilon$ when

4.2. DIFFERENTIATION OF MONOTONE FUNCTIONS

$0 < |y - x| < \delta$. Fix y with $0 < |y - x| < \delta$ and choose a sequence $\{y_k\}$ such that $y_k \downarrow y$ with $0 < |y_k - x| < \delta$ and f continuous at each y_k. Then

$$m - \epsilon < (f_*(y) - f(x))/(y - x) \leq (f(y) - f(x))/(y - x)$$

$$\leq (f(y^+) - f(x))/(y - x) = \lim(f_*(y_k) - f_*(x))/(y_k - x) \leq m + \epsilon$$

so

$$|(f(y) - f(x))/(y - x) - m| \leq \epsilon$$

when $0 < |y - x| < \delta$. Hence, $f'(x) = m$.

Theorem 3 (Lebesgue) *If $f : \mathbf{R} \to \mathbf{R}$ is $\uparrow$, then f' exists a.e.*

Proof: Lemma 2 and Theorem 1.

There are proofs of Lebesgue's Theorem which do not use measure theory. See [RN] I.2 for a proof due to F. Riesz.

This result cannot be improved to read that f' exists except for a countable set of points. Indeed, we have

Proposition 4 *If $E \subset [a, b]$ has measure 0, there is a continuous increasing function $f : [a, b] \to \mathbf{R}$ such that $f'(x) = \infty$ for every $x \in E$.*

Proof: For each $k \in \mathbf{N}$ choose a bounded open set $G_k \supset E$ with $m(G_k) < 1/2^k$. Set $f_k(x) = m(G_k \cap [a, x])$. Then f_k is continuous, increasing and $f_k \leq 1/2^k$ on $[a, b]$. If $f = \sum_{k=1}^{\infty} f_k$, then f is continuous, non-negative and increasing. If $x \in E$ and $h > 0$ is sufficiently small, the interval $[x, x + h]$ lies in G_k so for such h, $f_k(x + h) - f_k(x) = m(x, x + h] = h$. Thus, $(f_k(x + h) - f(x))/h = 1$ for small $h > 0$. For every N if $h > 0$ is sufficiently small,

$$(f(x + h) - f(x))/h \geq \sum_{k=1}^{N}(f_k(x + h) - f_k(x))/h = N.$$

Hence, $f'(x) = \infty$.

Notation: If I is an interval in $\mathbf{R}$ and $f : I \to \mathbf{R}$ is differentiable a.e. in I, we denote by $\dot{f}$ the function defined on I by $\dot{f}(t) = f'(t)$ if $f'(t)$ exists and $\dot{f}(t) = 0$ otherwise.

Concerning the integrability of the "derivative" of a monotone function, we have

Theorem 5 *If $f : [a, b] \to \mathbf{R}$ is increasing, then $\dot{f}$ is integrable and $\int_a^b \dot{f} \leq f(b) - f(a)$.*

Proof: Extend f to $[b, \infty)$ by setting $f(t) = f(b)$ for $t \geq b$. For each k set $g_k(t) = (f(t + 1/k) - f(t))/(1/k)$ for $a \leq t \leq b$. Note $g_k \to \dot{f}$ a.e. and $g_k \geq 0$. By Fatou's Theorem,

$$\int_a^b \dot{f} \leq \varliminf \int_a^b g_k dm = \varliminf (\int_a^b kf(t+1/k)dt - \int_a^b kf(t)dt)$$

$$= \varliminf (k\int_{a+1/k}^{b+1/k} f dm - k\int_a^b f dm) = \varliminf (k\int_b^{b+1/k} f dm - k\int_a^{a+1/k} f dm)$$

$$= \varliminf (f(b) - k\int_a^{a+1/k} f dm) = f(b) - \varlimsup k\int_a^{a+1/k} f dm \leq f(b) - f(a).$$

Strict inequality in Theorem 5 can hold.

Example 6 Let K be the Cantor set in $[0,1]$ and let $[0,1]\setminus K = \bigcup_{n=0}^{\infty}\bigcup_{k=1}^{2^n} I_k^n$ as in Example 2.5.7. Define $f(t) = \frac{2k-1}{2^n}$ for $t \in I_k^n$ [make a sketch]. Then f is increasing and continuous on K^c and the range of f is dense in $[0,1]$ so f can be extended to a continuous function, f, on $[0,1]$ [$f(x) = \inf\{f(t) : t \leq x, t \in K^c\}$]. Obviously, $f'(t) = 0$ for $t \in K^c$ so $f' = 0$ a.e. But, $\int_0^1 \dot{f} = 0 < f(1) - f(0) = 1$.

The function f constructed above is called the *Cantor function*.

Remark 7 Increasing functions f with the property that $f' = 0$ a.e. are called *singular* functions. There are examples of strictly increasing singular functions; see [Fre], [RN] p. 44-49, or [HS] p. 278-282. For two entertaining articles on singular functions see [Cat] and [Za].

We address the question of when equality holds in Theorem 5 of section 4.4.

We next present an interesting result on the termwise differentiation of a series due to Fubini.

Lemma 8 *Let $f_k : [a,b] \to \mathbf{R}$ be increasing and assume $\sum_{k=1}^{\infty} f_k$ converges pointwise to f_0. Then $\sum_{k=1}^{\infty} \dot{f}_k$ converges a.e.*

Proof: Since f_0 is also increasing, f'_k exists a.e. for all $k \geq 0$. If $E = \{t : f'_k(t)$ exists for all $k \geq 0\}$, then $m(E^c) = 0$ and $0 \leq f'_k(t) < \infty$ for $t \in E$, $k \geq 0$. Let $s_n = \sum_{k=1}^n f_k$. Since $(f_0 - s_n) \uparrow$ for $n \geq 1$, $f'_0 \geq s'_n$ on E which implies that $\lim_n s'_n$ exists on E [since $s'_{n+1} \geq s'_n$]. That is, $\sum_{k=1}^{\infty} \dot{f}_k$ converges a.e.

Theorem 9 (Fubini) *Let $f_k : [a,b] \to \mathbf{R}$ be increasing and assume $\sum_{k=1}^{\infty} f_k$ converges pointwise to f_0. Then $\dot{f}_0 = \sum_{k=1}^{\infty} \dot{f}_k$ a.e.*

4.2. DIFFERENTIATION OF MONOTONE FUNCTIONS

Proof: The series $\sum_{k=1}^{\infty} \dot{f}_k$ converges a.e. by Lemma 8. By replacing f_k by $f_k - f_k(a)$ if necessary, we may assume that $f_k \geq 0$. It suffices to show that there exists a subsequence $\{s_{n_k}\}$ of $s_k = \sum_{j=1}^{k} f_j$ such that $\dot{s}_{n_k} \to \dot{f}$ a.e. since $\dot{s}_n \leq \dot{s}_{n+1}$. For each k there exists n_k with $0 \leq f_0(b) - s_{n_k}(b) < 1/2^k$ where we may assume $n_k < n_{k+1}$. Since $(f_0 - s_{n_k}) \uparrow$, for each $t \in [a, b]$,

$$0 \leq f_0(t) - s_{n_k}(t) \leq f_0(b) - s_{n_k}(b) < 1/2^k.$$

Therefore, the series $\sum_{k=1}^{\infty}(f_0 - s_{n_k})$ converges uniformly on $[a, b]$. Since $(f_0 - s_{n_k}) \uparrow$, it follows from Lemma 8 that $\sum_{k=1}^{\infty}(\dot{f}_0 - \dot{s}_{n_k})$ converges a.e. so $\dot{s}_{n_k} \to \dot{f}$ a.e.

Dini Derivates:

In the calculus we learn that a function $f : [a, b] \to \mathbf{R}$ which has a non-negative derivative in $[a, b]$ is increasing [this follows from the Mean Value Theorem]. We consider a strengthening of this result. The material which follows is used only in section 4.3 and can be skipped by the reader who does not wish to go through this section.

Let $I = [a, b]$ and $f : I \to \mathbf{R}$. The *Dini derivates* of f at $x \in I$ are defined to be

$$d^+ f(x) = \overline{\lim_{t \to x^+}}(f(t) - f(x))/(t - x),$$

$$d_+ f(x) = \underline{\lim}_{t \to x^+}(f(t) - f(x))/(t - x),$$

$$d^- f(x) = \overline{\lim_{t \to x^-}}(f(t) - f(x))/(t - x),$$

$$d_- f(x) = \underline{\lim}_{t \to x^-}(f(t) - f(x))/(t - x).$$

The upper (lower) derivative of f at x is defined to be

$$\overline{D}f(x) = \overline{\lim}_{t \to x}(f(t) - f(x))/(t - x) \qquad (\underline{D}f(x) = \underline{\lim}_{t \to x}(f(t) - f(x))/(t - x)).$$

[See Exercise 4 for the definitions.]

Thus, f is differentiable at x if and only if all four Dini derivates of f at x are equal and finite if and only if the upper and lower derivatives are equal and finite. The derivative is then the common value. Some of the elementary properties of derivates are given in Exercise 6.

Proposition 10 *Let $f : I \to \mathbf{R}$ be continuous. If $c > (f(b) - f(a))/(b - a)$, then at uncountably many points x of (a, b) we have $d^+ f(x) \leq c$. [Similarly, if $(f(b) - f(a))/(b - a) > c$, $d^- f(x) \geq c$.]*

Proof: Set $k = c(b - a)$ and consider the function

$$g(x) = f(x) - f(a) - k(x - a)/(b - a).$$

Then $g(a) = 0$ and
$$g(b) = f(b) - f(a) - k < 0.$$
Let s be such that $0 = g(a) > s > g(b)$. Consider the set $E = \{t \in [a,b] : g(t) \geq s\}$. Since g is continuous on $[a,b]$, this set is closed, and, therefore, compact. Hence, if $t_s = \sup E$, by continuity of g, $g(t_s) = s$, and since $g(b) < s$, $t_s < b$. Since $g(t_s+h) < s$ for sufficiently small $h > 0$, $d^+g(t_s) \leq 0$ so
$$d^+f(t_s) = d^+g(t_s) + k/(b-a) \leq k/(b-a) = c$$
[Exercise 6]. Different s's generate different t_s's and there are uncountable many s's between 0 and $g(b)$ so there are uncountably many points $t \in (a,b)$ such that $d^+f(t) \leq c$.

This result is a substitute for the Mean Value Theorem for non-differentiable functions.

Theorem 11 *If f is continuous and one Dini derivate is non-negative except perhaps for a countable number of points in $[a,b]$, then f is increasing.*

Proof: Suppose $d^+f(t) \geq 0$ except possibly for countably many points t in $[a,b]$ [$d_+f(t) \geq 0$ implies this and the case $d^-f(t) \geq 0$ is similar]. If f is not increasing, there exist $x < y$ such that $f(x) > f(y)$. Use Proposition 10 with
$$(f(y) - f(x))/(y-x) < c < 0$$
on $[x,y]$ to obtain that $d^+f(t) < 0$ at uncountably many points in $[x,y]$. This contradiction establishes the result.

This result obviously covers the usual calculus test for increasing functions.
More information on Dini derivates can be found in [Boa].

Exercise 1. Show that if $f : \mathbf{R} \to \mathbf{R}$ is monotone, f is measurable.

Exercise 2. Let $D = \{t_k\}$ be any countable subset of $\mathbf{R}$. Show there exists an increasing function f whose discontinuities are exactly D. [Hint: Choose $a_k > 0$ such that $\sum_{k=1}^{\infty} a_k < \infty$. Define $f_k(t) = 0$ if $t < t_k$ and $f_k(t) = a_k$ if $t \geq t_k$. Put $f = \sum_{k=1}^{\infty} f_k$ and show f is continuous in D^c and has jump a_k at t_k.]

Exercise 3. Show that a continuous, nowhere differentiable function is not monotone on any non-degenerate interval.

Exercise 4. Let $f : [a,b] = I \to \mathbf{R}$ and $x \in I$. The limit superior of f at x, $\varlimsup_{t \to x} f(t)$, is defined by
$$\varlimsup_{t \to x} f(t) = \inf_{\varepsilon > 0} \sup \{f(t) : t \in I, 0 < |t-x| < \varepsilon\}.$$

4.2. DIFFERENTIATION OF MONOTONE FUNCTIONS

Define limit inferior, $\varliminf_{t \to x} f(t)$, and establish the analogues of the statements in Exercises 1.2.1 and 1.2.10. Define one-sided such limits.

Exercise 5. Let $f(t) = t\sin(1/t)$ if $t \neq 0$ and $f(0) = 0$. Compute the four Dini derivates of f at 0. Now construct a function all of whose Dini derivates are not equal at 0.

Exercise 6. Show $\underline{d}f(x) + \underline{d}g(x) \leq \underline{d}(f+g)(x)$ $(\overline{d}f(x) + \overline{d}g(x) \geq \overline{d}(f+g)(x))$, where $\underline{d}$ $(\overline{d})$ is any lower (upper) derivate. Show that if $f'(x)$ exists, then $\overline{d}(f+g)(x) = f'(x) + \overline{d}g(x)$, $\underline{d}(f+g)(x) = f'(x) + \underline{d}g(x)$.

Exercise 7. If $f \uparrow$, show any derivate is ≥ 0.

Exercise 8. If f assumes a local maximum at x, show $d^+ f(x) \leq 0$ and $d_- f(x) \geq 0$.

Exercise 9. If $\max\{\overline{D}f(x), \underline{D}f(x)\} < \infty$, show f is continuous at x.

Exercise 10. Show Theorem 5 can be improved to

$$\int_a^b \dot{f} \leq f\left(b^-\right) - f\left(a^+\right).$$

Exercise 11. If f in Theorem 5 is not continuous, show strict inequality holds.

4.3 Integrating Derivatives

In this section we consider the other half of the Fundamental Theorem of Calculus (FTC), the integration of derivatives. We begin by showing that the most general (and most desirable!) form of the FTC does not hold for the Lebesgue integral by giving an example of a derivative which is not Lebesgue integrable. Again all measurability statements refer to Lebesgue measure.

Example 1 Let $f(t) = t^2 \cos(\pi/t^2)$ for $0 < t \leq 1$ and $f(0) = 0$. Then f is differentiable on $[0,1]$ with

$$f'(t) = 2t\cos(\pi/t^2) + (2\pi/t)\sin(\pi/t^2)$$

for $0 < t \leq 1$ and $f'(0) = 0$. For $0 < a < b < 1$, f' is bounded on $[a,b]$ and, therefore, is (Riemann) integrable with

$$\int_a^b f' dm = b^2 \cos(\pi/b^2) - a^2 \cos(\pi/a^2).$$

If we set $b_k = 1/\sqrt{2k}$, $a_k = \sqrt{2/(4k+1)}$, then $\int_{a_k}^{b_k} f' dm = 1/(2k)$. The intervals $\{[a_k, b_k]\}$ are pairwise disjoint so $\int_0^1 |f'| dm \geq \sum_{k=1}^{\infty} \int_{a_k}^{b_k} |f'| dm \geq \sum_{k=1}^{\infty} 1/(2k) = \infty$. Hence, f' is not integrable over $[0,1]$.

We establish the most general form of the FTC for the Lebesgue integral.

Lemma 2 Let $f : [a,b] \to \mathbf{R}$. Suppose all Dini derivates are non-negative a.e. in $[a,b]$ and that no Dini derivate is $-\infty$. Then f is increasing.

Proof: Let E be the set of points where some derivate is negative. By 4.2.4 there exists an increasing function $g : [a,b] \to \mathbf{R}$ such that $g'(x) = \infty$ for every $x \in E$. Let $\epsilon > 0$ and set $h = f + \epsilon g$. At all points of $[a,b] \setminus E$ all derivates of h are non-negative [Exercises 4.2.6 and 4.2.7]. The same holds at points of E since $g'(x) = \infty$ for $x \in E$ and the derivates of f at these points are not $-\infty$. Thus, by 4.2.11, $h \uparrow$. Hence, if $x < y$, $f(x) + \epsilon g(x) \leq f(y) + \epsilon g(y)$, and letting $\epsilon \to 0$ gives $f(x) \leq f(y)$.

Theorem 3 (Fundamental Theorem of Calculus) Let $f : [a,b] \to \mathbf{R}$ be differentiable on $[a,b]$. If f' is Lebesgue integrable over $[a,b]$, then

$$\int_a^b f' dm = f(b) - f(a).$$

4.3. INTEGRATING DERIVATIVES

Proof: Define $g_k(t) = f'(t)$ if $f'(t) \leq k$ and $g_k(t) = k$ if $f'(t) > k$. Then $|g_k| \leq |f'|$ and $g_k \to f'$ pointwise on $[a, b]$. Set $r_k(t) = f(t) - \int_a^t g_k dm$. We claim that $r_k \uparrow$. First, observe that r_k is differentiable a.e. with $r_k' = f' - g_k \geq 0$ a.e. [4.1.9]. Since $g_k \leq k$ on $[a, b]$, $\int_x^{x+h} g_k dm/h \leq k$ for $x \in [a, b]$, h sufficiently small. Thus,

$$(r_k(x+h) - r_k(x))/h = (f(x+h) - f(x))/h - \int_x^{x+h} g_k dm/h \geq (f(x+h) - f(x))/h - k$$

so no derivate of r_k is $-\infty$. By Lemma 2, $r_k \uparrow$. In particular, $r_k(b) \geq r_k(a)$ or $f(b) - f(a) \geq \int_a^b g_k dm$. By the DCT, $\lim \int_a^b g_k dm = \int_a^b f' dm$ so $f(b) - f(a) \geq \int_a^b f' dm$.

Replacing f by $-f$ gives the reverse inequality.

Of course, the annoying feature of Theorem 3 is the need for the hypothesis that f' is Lebesgue integrable. [Recall that the Riemann integral suffers this same defect.] Example 1 shows that this assumption is necessary. It would be desirable to have a theory of integration for which the FTC holds in full generality, i.e., a theory for which all derivatives are integrable and the formula in the FTC holds. Two such theories were developed by Perron and Denjoy [see [Pe] for a description of these integrals]. Lately, a very simple integral which is but a slight variant of the Riemann integral, called the gauge integral, has been given independently by Kurzweil and Henstock ([Ku], [He]) for which the FTC holds in full generality [see [DeS] or [Mc] for a description of the integral].

The statement in Theorem 3 cannot be improved to "f' exists a.e.". See the Cantor function in 4.2.6. However, the statement can be improved to "f' exists except for a countable number of points" [see [Co] 6.3.10, [HS], p. 299, [Wa]].

4.4 Absolutely Continuous Functions

In this section we consider a form of the FTC in which the functions are only differentiable a.e. [again all measurability statements refer to Lebesgue measure]. In particular, for increasing functions we are asking when equality holds for the inequality in Theorem 4.2.5. Recall that if $f : [a, b] \to \mathbf{R}$ is differentiable a.e., we denote by $\dot{f}$ the function $\dot{f}(t) = f'(t)$ when $f'(t)$ exists and $\dot{f}(t) = 0$ otherwise.

Suppose $f : [a, b] \to \mathbf{R}$ is increasing and $f(x) - f(a) = \int_a^x \dot{f} dm$ for $a \leq x \leq b$, i.e., equality in Theorem 4.2.5 for all $a \leq x \leq b$. Since $\lim_{m(E) \to 0} \int_E \dot{f} dm = 0$ [3.2.17], for every $\epsilon > 0$ there exists $\delta > 0$ such that whenever $\{(a_i, b_i)\}_{i=1}^n$ is a pairwise disjoint sequence of subintervals of $[a, b]$ with $\sum_{i=1}^n (b_i - a_i) < \delta$, then

$$\left| \sum_{i=1}^n \int_{a_i}^{b_i} \dot{f} dm \right| = \sum_{i=1}^n |f(b_i) - f(a_i)| < \epsilon.$$

Functions which satisfy this condition are called absolutely continuous; we give the formal definition.

Definition 1 *Let $f : [a, b] \to \mathbf{R}$. Then f is absolutely continuous if for every $\epsilon > 0$ there exists $\delta > 0$ such that whenever $\{(a_i, b_i)\}_{i=1}^n$ is a pairwise disjoint sequence of open subintervals of $[a, b]$ with $\sum_{i=1}^n (b_i - a_i) < \delta$, then*

$$\sum_{i=1}^n |f(b_i) - f(a_i)| < \epsilon.$$

Example 2 The function $f : [a, b] \to \mathbf{R}$ satisfies a *Lipschitz condition* if there exists $L > 0$ such that $|f(x) - f(y)| \leq L|x - y|$ for $x, y \in [a, b]$. Such a function is obviously absolutely continuous. [The converse is false; see Exercise 7.] For examples of functions which satisfy a Lipschitz condition, see Exercise 1.

Proposition 3 *Let $f : [a, b] \to \mathbf{R}$ be absolutely continuous. Then (i) f is uniformly continuous and (ii) f is bounded variation.*

Proof: (i) is clear. For (ii), let $\epsilon = 1$ and δ be as in Definition 1. Partition $[a, b]$ by $P = \{a = x_0 < x_1 < \ldots < x_n = b\}$, where $x_{i+1} - x_i < \delta$. Then $Var(f : [x_i, x_{i+1}]) \leq 1$ for $i = 0, \ldots, n - 1$. Hence, $Var(f : [a, b]) \leq n$.

We show below that the converses of (i) and (ii) are false [see Examples 5 and 9].

Corollary 4 *If $f : [a, b] \to \mathbf{R}$ is absolutely continuous, then f' exists a.e. and $\dot{f}$ is integrable.*

4.4. ABSOLUTELY CONTINUOUS FUNCTIONS

Proof: By Proposition 3 and Theorem 8 of Appendix A1, f is the difference of two increasing functions, so the result follows from Theorem 4.2.3.

Example 5 The function f in Example 4.3.1 is uniformly continuous but not absolutely continuous by Corollary 4.

Algebraic properties of absolutely continuous functions are given in Exercises 2-6.

We use Lebesgue-Stieltjes measures to describe the relationship between absolutely continuous set functions and absolutely continuous functions on intervals in **R**.

Let $f : [a, b] \to \mathbf{R}$ be left continuous and have bounded variation. Then $f = g - h$, where g and h are left continuous and increasing [Appendix A1]. Extend f (g and h) to **R** by setting $f(t) = f(b)$ [$g(t) = g(b)$, $h(t) = h(b)$] for $t > b$ and $f(t) = f(a)$ [$g(t) = g(a)$, $h(t) = h(a)$] for $t < a$. Then g and h are bounded, left continuous and increasing and, therefore, induce finite Lebesgue-Stieltjes measures μ_g and μ_h. Hence, $\mu_f = \mu_g - \mu_h$ is a finite, signed measure; we call μ_f the *Lebesgue-Stieltjes signed measure* induced by f. Note $\mu_f[\alpha, \beta) = f(\beta) - f(\alpha)$ for $\alpha \le \beta$.

Theorem 6 $\mu_f \ll m$ *if and only if f is absolutely continuous.*

Proof: $\Leftarrow$: Let $A \subset (a, b)$ have m-measure 0. Let $\epsilon > 0$ and let $\delta > 0$ be as in Definition 1. There exists a pairwise disjoint sequence of open intervals in (a, b), $\{(a_i, b_i)\}$, such that $\bigcup_{i=1}^{\infty}(a_i, b_i) \supset A$ and $\sum_{i=1}^{\infty}(b_i - a_i) < \delta$ [2.5.3]. For each n

$$\sum_{i=1}^{n} |f(b_i) - f(a_i)| = \sum_{i=1}^{n} |\mu_f[a_i, b_i)| < \epsilon$$

so by Exer. 2.6.6

$$\left|\sum_{i=1}^{\infty} \mu_f[a_i, b_i)\right| = \left|\mu_f(\bigcup_{i=1}^{\infty}(a_i, b_i))\right| \le \epsilon.$$

Hence, $\mu_f(A) = 0$ and $|\mu_f|(A) = 0$ by 2.2.1.7.

$\Rightarrow$: Note $\mu_f[\alpha, \beta) = \mu_f(\alpha, \beta)$ for $\alpha < \beta$. Since μ_f is finite, for every $\epsilon > 0$ there exists $\delta > 0$ such that $m(A) < \delta$ implies $|\mu_f|(A) < \epsilon$ [3.12.5]. Suppose $\{(a_i, b_i)\}_{i=1}^{n}$ is a pairwise disjoint sequence from $[a, b]$ with $\sum_{i=1}^{n}(b_i - a_i) < \delta$. Let

$$\sigma = \{i : f(b_i) - f(a_i) \ge 0\} \text{ and } \tau = \{i : f(b_i) - f(a_i) < 0\}.$$

Then

$$\left|\mu_f\left(\bigcup_{i\in\sigma}(a_i, b_i)\right)\right| = \sum_{i\in\sigma}|f(b_i) - f(a_i)| < \epsilon$$

and

$$\left|\mu_f\left(\bigcup_{i\in\tau}(a_i, b_i)\right)\right| = \sum_{i\in\tau}|f(b_i) - f(a_i)| < \epsilon$$

so $\sum_{i=1}^{n}|f(b_i) - f(a_i)| < 2\epsilon$.

Theorem 7 *Let $f : [a,b] \to \mathbf{R}$ be differentiable a.e. Then f is absolutely continuous if and only if $\dot{f}$ is integrable over $[a,b]$ and $f(x) - f(a) = \int_a^x \dot{f} dm$ for $a \leq x \leq b$.*

Proof: $\Rightarrow$: By Theorem 6 $\mu_f \ll m$, and by Theorems 4.1.11 and 4.1.9

$$D\mu_f = \frac{d\mu_f}{dm} = f'$$

a.e. Hence, $\mu_f[a,x] = f(x) - f(a) = \int_a^x \dot{f} dm$.
$\Leftarrow$: 3.2.17.

Remark 8 Theorem 7 is sometimes referred to as the FTC for the Lebesgue integral. Note that Theorem 7 gives necessary and sufficient conditions for equality to hold in Theorem 4.2.5.

Example 9 The Cantor function [4.2.6] is continuous, increasing, and hence bounded variation, but is not absolutely continuous by Theorem 7.

A function $f : [a,b] \to \mathbf{R}$ of bounded variation which is such that $f' = 0$ a.e. is said to be a *singular function*. The Cantor function supplies an example of a non-constant singular function.

Theorem 10 *Let $f \in BV[a,b]$ be left continuous. Then f is singular if and only if $\mu_f \perp m$.*

Proof: Let $\mu_f = (\mu_f)_a + (\mu_f)_s$ be the Lebesgue Decomposition of μ_f with respect to m. By Theorems 4.1.11 and 4.1.9 $f' = D\mu_f = \frac{d(\mu_f)_a}{dm}$ a.e. The result now follows.

Theorem 11 (Lebesgue Decomposition) *Let $f \in BV[a,b]$. Then there exist g, $h : [a,b] \to \mathbf{R}$ such that $f = g + h$ with g absolutely continuous and h singular. The decomposition is unique up to a constant.*

Proof: Set $g(x) = \int_a^x \dot{f} dm$ and $h = f - g$. Then g is absolutely continuous by Theorem 3.2.17 and h is singular by 4.1.9. Uniqueness follows from Exercise 8.

Change of Variable

We use the results above to establish a change of variable theorem for the Lebesgue integral.

Theorem 12 *Let $g : [a,b] \to \mathbf{R}$ be increasing and absolutely continuous and set $\alpha = g(a)$, $\beta = g(b)$. Let $f : [\alpha, \beta] \to \mathbf{R}^*$ be Lebesgue integrable over $[\alpha, \beta]$. Then $(f \circ g) \dot{g}$ is Lebesgue integrable over $[a,b]$ and*

$$\int_a^b f(g(t)) \dot{g}(t) dt = \int_\alpha^\beta f(x) dx. \tag{4.1}$$

4.4. ABSOLUTELY CONTINUOUS FUNCTIONS

Proof: First suppose f is the characteristic function of a half closed interval $[\gamma, \delta) \subset [\alpha, \beta]$. Let $c = \inf\{t : g(t) = \gamma\}$, $d = \sup\{t : g(t) = \delta\}$. Then $f \circ g = C_{[c,d)}$ so

$$\int_\alpha^\beta f\, dm = \delta - \gamma = g(d) - g(c) = \int_c^d \dot{g}\, dm = \int_a^b (f \circ g)\, \dot{g}\, dm$$

or (1) holds. It follows that (1) holds for $\mathcal{S}$-simple functions, where $\mathcal{S}$ is the semi-ring of all half-closed intervals.

Now suppose f is Lebesgue integrable over $[\alpha, \beta]$. By Mikusinski's Theorem (3.8.3) there exists a sequence of $\mathcal{S}$-simple functions $\{f_k\}$ such that $\sum_{k=1}^\infty \int_\alpha^\beta |f_k|\, dm < \infty$, $\sum_{k=1}^\infty f_k(x) = f(x)$ for any x for which $\sum_{k=1}^\infty |f_k(x)| < \infty$ and $\int_\alpha^\beta f\, dm = \sum_{k=1}^\infty \int_\alpha^\beta f_k\, dm$. We show the sequence $\{(f_k \circ g)\, \dot{g}\}$ satisfies the conditions of Exercise 3.8.1 for the function $(f \circ g)\, \dot{g}$. First,

$$\sum_{k=1}^\infty \int_a^b |f_k \circ g|\, \dot{g}\, dm = \sum_{k=1}^\infty \int_\alpha^\beta |f_k|\, dm < \infty$$

by the part above. Suppose

$$\sum_{k=1}^\infty \left| f_k(g(t))\, \dot{g}(t) \right| < \infty.$$

If $\dot{g}(t) = 0$, then

$$\sum_{k=1}^\infty f_k(g(t))\, \dot{g}(t) = f(g(t))\, \dot{g}(t)$$

while if $\dot{g}(t) > 0$,

$$\sum_{k=1}^\infty f_k(g(t))\, \dot{g}(t) = f(g(t))\, \dot{g}(t).$$

By Exercise 3.8.1 $(f \circ g)\, \dot{g}$ is Lebesgue integrable and

$$\int_a^b (f \circ g)\, \dot{g}\, dm = \sum_{k=1}^\infty \int_a^b (f_k \circ g)\, \dot{g}\, dm = \sum_{k=1}^\infty \int_\alpha^\beta f_k\, dm = \int_\alpha^\beta f\, dm.$$

There are other useful conditions under which equation (1) holds. For example, (1) holds if f is bounded and integrable and g is absolutely continuous. See [St2] for a thorough discussion of the validity of (1). It should be pointed out that (1) does not hold in general even when f is integrable and g is absolutely continuous (Exer. 13).

For change of variable in Lebesgue integrals in $\mathbf{R}^n$, see [Ru2].

Exercise 1. Show that if $f : [a, b] \to \mathbf{R}$ has a bounded derivative, then f satisfies a Lipschitz condition.

Exercise 2. Let $f, g : [a,b] \to \mathbf{R}$ be absolutely continuous. Show $|f|$, $f+g$, fg are absolutely continuous. What about $1/f$? Thus, if $AC[a,b]$ denotes the space of all absolutely continuous functions, $AC[a,b]$ is a vector subspace of $BV[a,b]$.

Exercise 3. Show the composition of absolutely continuous functions needn't be absolutely continuous.

Exercise 4. If f and g are absolutely continuous and $g \uparrow$, show $f \circ g$ is absolutely continuous.

Exercise 5. If f satisfies a Lipschitz condition and g is absolutely continuous, show $f \circ g$ is absolutely continuous.

Exercise 6. Is the uniform limit of absolutely continuous functions necessarily absolutely continuous?

Exercise 7. Show $f(t) = t^{1/3}$, $0 \le t \le 1$, is absolutely continuous but does not satisfy a Lipschitz condition.

Exercise 8. Let $f : [a,b] \to \mathbf{R}$ be absolutely continuous. If $f' = 0$ a.e., show $f =$ constant.

Exercise 9. Let f, g be absolutely continuous on $[a,b]$. Show
$$\int_a^b f\dot{g}\,dm + \int_a^b \dot{f}g\,dm = f(b)g(b) - f(a)g(a).$$

Exercise 10. If $f : [a,b] \to \mathbf{R}$ is absolutely continuous, show $Var(f : [a,b]) = \int_a^b |\dot{f}|\,dm$.

Exercise 11. Let $f : [a,b] \to \mathbf{R}$. Show f is absolutely continuous if and only if $x \to Var(f : [a,x])$ is absolutely continuous.

Exercise 12. Let $f : [a,b] \to \mathbf{R}$. Show that f satisfies a Lipschitz condition as in Example 2 if and only if f is absolutely continuous and $|f'| \le L$ a.e.

Exercise 13. Let $f(x) = 1/\sqrt{x}$ for $x > 0$ and $f(0) = 0$, $g(t) = t^2 \sin(1/t)$ for $t \ne 0$ and $g(0) = 0$. Show f is integrable over $[g(0), g(\pi/2)]$, g is absolutely continuous but (1') fails for $[a,b] = [0, \pi/2]$.

Chapter 5

Introduction to Functional Analysis

5.1 Normed Linear Spaces

In part 6 of these notes we consider some of the classic spaces of functions, most of which are associated with either measures or integrable functions. We begin by establishing an abstract framework, essentially due to S. Banach ([B1]), in which we can study these function spaces.

Let X be a vector space over the field $\mathbf{F}$ of either real or complex numbers. A topology τ on X is a *vector topology* or *linear topology* if the maps $(x,y) \to (x+y)$ from $X \times X$ into X and $(t,x) \to tx$ from $\mathbf{F} \times X$ into X are continuous [with respect to the product topologies]. If τ is a vector topology on X, the pair (X, τ) is called a *topological vector space* (TVS) or X is called a TVS if the topology is understood. A *(semi)-metric linear space* is a TVS whose topology is given by a (semi-) metric. For an example of a semi-metric linear space, we have

Example 1 Let μ be a finite measure on a σ-algebra $\sum$ of subsets of S.

Let $L^0(\mu)$ be the space of all real-valued $\sum$-measurable functions; if m is Lebesgue measure on an interval I, we set $L^0(I) = L^0(m)$. Then

$$d(f,g) = \int_S \frac{|f-g|}{1+|f-g|} d\mu$$

defines a semi-metric on $L^0(\mu)$ [3.6.8]. Since convergence in d is exactly convergence in μ-measure [3.6.9], $L^0(\mu)$ is a (complete) semi-metric linear space under d.

Most of the spaces which we consider are normed spaces, we use the semi-metric linear space in Example 1 and the one in Example 8 to illustrate some of the important properties of normed spaces.

Definition 2 *A norm on X is a function $\|\| : X \to \mathbf{R}$ satisfying*

(i) $\|x\| \geq 0$ *for all $x \in X$ and $\|x\| = 0$ if and only if $x = 0$,*

(ii) $\|tx\| = |t|\|x\|$ for all $x \in X$ and $t \in \mathbf{F}$,

(iii) $\|x + y\| \leq \|x\| + \|y\|$ for all $x, y \in X$ *[triangle inequality]*.

If a function $\|\ \| : X \to \mathbf{R}$ satisfies (ii), (iii) and

(i') $\|x\| \geq 0$ for all $x \in X$,

then $\|\ \|$ is called a semi-norm on X *[note from (ii) that $\|0\| = 0$].*

We establish an important inequality associated with the triangle inequality. Let $x, y \in X$. Then $\|x\| \leq \|x - y\| + \|y\|$ by (iii) so $\|x\| - \|y\| \leq \|x - y\|$. By symmetry $\|y\| - \|x\| \leq \|y - x\| = \|x - y\|$ by (ii). Hence,

(iv) $\|x - y\| \geq |\|x\| - \|y\||$.

If $\|\ \|$ is a (semi-) norm on X, the pair $(X, \|\ \|)$ is called a *(semi-) normed space* [semi-NLS or NLS]; if the (semi-) norm is understood, X is called a (semi-) NLS.

If X is a (semi-) NLS, then $d(x,y) = \|x - y\|$ defines a (semi-) metric on X which is translation invariant in the sense that $d(x + z, y + z) = d(x,y)$ for $x, y, z \in X$. We always assume that a (semi-) NLS is equipped with the topology induced by the (semi-) metric d. We now show that a (semi-)NLS is a TVS.

Proposition 3 *Let X be a semi-NLS. Then*

(a) *the map $(t,x) \to tx$ from $\mathbf{F} \times X \to X$ is continuous,*

(b) *the map $(x,y) \to x + y$ from $X \times X \to X$ is continuous,*

(c) *the map $x \to \|x\|$ from X to $\mathbf{R}$ is continuous.*

Proof: (a): Let $|t_k - t| \to 0$ and $\|x_k - x\| \to 0$. Then
$$\|t_k x_k - tx\| \leq |t_k - t|\|x\| + |t_k|\|x_k - x\|$$
and (a) follows.

(b) follows from the triangle inequality and (c) follows from (iv).

A semi-NLS is said to be *complete* if it is complete under the semi-metric induced by the semi-norm, i.e., if every Cauchy sequence converges. A complete NLS is called a *Banach space* or a *B-space*. We give an interesting characterization for completeness in terms of series.

Let $\sum_{k=1}^{\infty} x_k$ be a (formal) series in a semi-NLS X. The series $\sum_{k=1}^{\infty} x_k$ is said to *converge* in X if the sequence of partial sums, $s_k = \sum_{j=1}^{k} x_j$, converges in X; we write $\sum_{k=1}^{\infty} x_k = \lim s_k$ for the sum of the series. The series is said to be *absolutely convergent* in X if $\sum_{k=1}^{\infty} \|x_k\| < \infty$.

5.1. NORMED LINEAR SPACES

Theorem 4 *A semi-NLS X is complete if and only if every absolutely convergent series in X is convergent.*

Proof: $\Rightarrow$: Let $\sum_{k=1}^{\infty} \|x_k\| < \infty$ and $s_k = \sum_{i=1}^{k} x_i$. If $k > j$, then by the triangle inequality
$$\|s_k - s_j\| = \left\| \sum_{i=j+1}^{k} x_i \right\| \le \sum_{i=j+1}^{k} \|x_i\|$$
so $\{s_k\}$ is a Cauchy sequence and must converge to some point in X.

$\Leftarrow$: Let $\{x_k\}$ be a Cauchy sequence in X. Pick a subsequence $\{x_{n_k}\}$ satisfying $\|x_{n_{k+1}} - x_{n_k}\| < 1/2^k$. Since the series $\sum_{k=1}^{\infty}(x_{n_{k+1}} - x_{n_k})$ is absolutely convergent, it converges to an element $x \in X$ so
$$\sum_{k=1}^{\infty}(x_{n_{k+1}} - x_{n_k}) = \lim_k (x_{n_{k+1}} - x_{n_1}) = x$$
and the subsequence $\{x_{n_k}\}$ converges to $x + x_{n_1}$. Since $\{x_k\}$ is Cauchy, $x_k \to x + x_{n_1}$.

We now give several examples of metric linear and normed spaces. Most of the spaces which we consider are normed spaces, but we give two examples (Examples 1 and 8) of metric linear spaces in order to illustrate some of the important properties of normed spaces.

Example 5 If $S \ne \emptyset$, let $B(S)$ be the space of all bounded, real-valued functions defined on S. $B(S)$ is a vector space when addition and scalar multiplication of functions is defined pointwise. $B(S)$ also has a natural norm, called the *sup-norm*, defined by $\|f\| = \sup\{|f(t)| : t \in S\}$ [Exercise 1]. $B(S)$ is a B-space under this norm, for suppose that $\{f_k\}$ is a Cauchy sequence in $B(S)$. For $t \in S$, $|f_k(t) - f_j(t)| \le \|f_k - f_j\|$ so $\{f_k(t)\}$ is a Cauchy sequence in $\mathbf{R}$. Let $f(t) = \lim f_k(t)$. We claim that $f \in B(S)$ and $\|f_k - f\| \to 0$. Let $\epsilon > 0$. There exists N such that $k, j \ge N$ implies $\|f_k - f_j\| < \epsilon$. Then $|f_k(t) - f_j(t)| < \epsilon$ for $k, j \ge N$ and $t \in S$. Letting $j \to \infty$ gives $|f_k(t) - f(t)| \le \epsilon$ for $k \ge N$, $t \in S$ so $\|f_k - f\| \le \epsilon$ for $k \ge N$. Hence, $f_k - f \in B(S)$ so $f \in B(S)$ and $f_k \to f$ with respect to the sup-norm.

Example 6 Let S be a compact Hausdorff space and $C(S)$ the space of all continuous functions on S. Then $C(S)$ is a linear subspace of $B(S)$. Since convergence in the sup-norm is exactly uniform convergence on S, $C(S)$ is a closed subspace of $B(S)$ and is, therefore, complete.

Example 7 Let μ be a measure on the σ-algebra, Σ, of subsets of S. As in §3.5 let $L^1(\mu)$ be the vector space of all real-valued, μ-integrable functions. Then $\|f\|_1 = \int_S |f|\, d\mu$ defines a semi-norm on $L^1(\mu)$ which is complete [Riesz-Fischer Theorem].

If $S = \mathbf{N}$ and μ is counting measure on $\mathbf{N}$, we set $L^1(\mu) = \ell^1$. Thus, ℓ^1 consists of all sequences $\{t_j\}$ such that
$$\|\{t_j\}\|_1 = \sum_{j=1}^{\infty} |t_j| < \infty.$$

168   CHAPTER 5. INTRODUCTION TO FUNCTIONAL ANALYSIS

We will consider other spaces of integrable functions in §6.1.

The examples which we give now are sequence spaces; we give further examples of function spaces and study them in detail in later chapters.

Example 8 Let s be the space of all real-valued (or complex-valued) sequences. Then s is a vector space under coordinate-wise addition and scalar multiplication. We define a metric, called the *Frechet metric*, by

$$d(\{s_j\}, \{t_j\}) = \sum_{j=1}^{\infty} |s_j - t_j|/(1 + |s_j - t_j|)2^j.$$

It is easily checked that d is a translation invariant metric on s [Lemma 3.6.7 gives the triangle inequality].

We first observe that convergence in the metric d is just coordinatewise convergence. For suppose that $x^k = \{x_j^k\}_{j=1}^{\infty}$ and $x = \{x_j\}$ are sequences in s. If $d(x^k, x) \to 0$, then $\lim_k x_j^k = x_j$ for each j since

$$d(x^k, x) \geq |x_j^k - x_j|/(1 + |x_j^k - x_j|)2^j.$$

Conversely, suppose $\lim_k x_j^k = x_j$ for each j and let $\epsilon > 0$. There exists N such that $\sum_{j=N}^{\infty} 1/2^j < \epsilon/2$. There exists M such that $k \geq M$ implies

$$\sum_{j=1}^{N-1} |x_j^k - x_j| < \epsilon/2.$$

Hence, if $k \geq M$, then

$$d(x^k, x) \leq \sum_{j=1}^{N-1} |x_j^k - x_j| + \sum_{j=N}^{\infty} 1/2^j < \epsilon.$$

It follows from this observation that s is complete under d [Exercise 2].

Example 9 Let ℓ^{∞} be the linear subspace of s which consists of the bounded sequences [so $\ell^{\infty} = B(\mathbf{N})$]. Then

$$\|\{t_k\}\|_{\infty} = \sup\{|t_k| : k \in \mathbf{N}\}$$

defines a norm on ℓ^{∞} called the *sup-norm* and ℓ^{∞} is a B-space under the sup-norm (Example 5).

Example 10 c is the subspace of ℓ^{∞} which consists of the convergent sequences. We assume that c is equipped with the sup-norm. We can show that c is a B-space by showing that it is a closed subspace of the complete space ℓ^{∞}. So suppose $x^k = \{x_j^k\}_{j=1}^{\infty}$ is a sequence in c which converges to the point $x = \{x_j\} \in \ell^{\infty}$. Let

5.1. NORMED LINEAR SPACES

$\lim_{j} x_j^k = \ell_k$ for each k. Since $\left\|x^k - x\right\|_\infty \geq \left|x_j^k - x_j\right|$ for all j, $\lim_{k} x_j^k = x_j$ uniformly for $j \in \mathbf{N}$. Hence,

$$\lim_{k}\lim_{j} x_j^k = \lim_{k} \ell_k = \lim_{j}\lim_{k} x_j^k = \lim_{j} x_j$$

so $x \in c$.

Example 11 c_0 is the subspace of c consisting of all the sequences which converge to 0. We assume that c_0 is equipped with the sup-norm.

As in Example 10, c_0 is a B-space [Exercise 3].

Example 12 Let c_{00} be the subspace of c_0 which consists of all sequences $x = \{x_j\}$ which are eventually 0, i.e., there exists N (depending on x) such that $x_j = 0$ for $j > N$. We assume that c_{00} is equipped with the sup-norm. We show that c_{00} is not complete. Let $\underline{e}^k$ be the sequence in c_{00} which has a 1 in the kth coordinate and 0 in the other coordinates. If

$$x^k = \sum_{j=1}^{k} (1/j)e^j,$$

then $\{x^k\}$ is a Cauchy sequence in c_{00} which does not converge to a point in c_{00}.

c_{00} is a very useful space for giving counterexamples to results involving completeness.

We will give further examples of sequence spaces in §6.1.

Example 13 Consider $\mathbf{R}^n$ (or $\mathbf{C}^n$). $\mathbf{R}^n$ has, of course, the usual Euclidean norm

$$\|x\|_2 = \|(x_1, \ldots, x_n)\|_2 = \sqrt{\sum_{i=1}^{n} |x_i|^2},$$

but it also has other natural norms. For example,

$$\|x\|_1 = \|(x_1, \ldots, x_n)\|_1 = \sum_{i=1}^{n} |x_i|$$

or

$$\|x\|_\infty = \|(x_1, \ldots, x_n)\|_\infty = \max\{|x_i| : 1 \leq i \leq n\}.$$

[We give a further family of norms on $\mathbf{R}^n$ in §6.1.]

If $\|\|_1$ and $\|\|_2$ are two norms on a vector space X, $\|\|_1$ and $\|\|_2$ are *equivalent* if there exist $a, b > 0$ such that $a\|\|_1 \leq \|\|_2 \leq b\|\|_1$. We show that all norms on $\mathbf{R}^n$ are equivalent so, in particular, the three norms given above in Example 13 are equivalent.

Theorem 14 *Any two norms on $\mathbf{R}^n$ are equivalent.*

Theorem 14 *Any two norms on $\mathbf{R}^n$ are equivalent.*

Proof: By Exercise 5, it suffices to show that any norm, $\|\|$, on $\mathbf{R}^n$ is equivalent to the Euclidean norm, $\|\|_2$.

Let e_i be the vector in $\mathbf{R}^n$ which has a 1 in the ith coordinate and 0 in the other coordinates. If $x = (x_1, \ldots, x_n) \in \mathbf{R}^n$, then

$$\|x\| \leq \sum_{i=1}^n |x_i| \|e_i\| \leq \|x\|_2 \left(\sum_{i=1}^n \|e_i\|^2\right)^{1/2}$$

by the Cauchy-Schwarz inequality.

Next, let $S = \{x : \|x\|_2 = 1\}$ and define $f : S \to \mathbf{R}$ by $f(x) = \|x\|$. By the inequality above the identity $(\mathbf{R}^n, \|\|_2) \to (\mathbf{R}^n, \|\|)$ is continuous and $x \to \|x\|$ is continuous from $(\mathbf{R}^n, \|\|)$ to $\mathbf{R}$ [Proposition 3] so f is continuous with respect to $\|\|_2$. Since S is compact with respect to $\|\|_2$, f attains its minimum on S, say at x_0. Note that $f(x_0) = \|x_0\| = m > 0$. If $x \in \mathbf{R}^n$, $x \neq 0$, then $x/\|x\|_2 \in S$ so

$$f(x/\|x\|_2) = \|x\|/\|x\|_2 \geq m.$$

That is, $\|x\| \geq m \|x\|_2$.

For an example of two norms which are not equivalent, consider the sup-norm, $\|\|_\infty$ on c_{00} and the norm, $\|x\|_1 = \sum_{i=1}^\infty |x_i|$, where $x = \{x_i\}$ [note this is a finite sum]. If $x^i = \frac{1}{i} \sum_{j=1}^i e^j$, then $\|x^i\|_\infty = 1/i \to 0$ while $\|x^i\|_1 = 1$ [Exercise 6].

An interesting consequence of Theorem 14 is

Corollary 15 *Any finite dimensional subspace of a NLS is closed.*

Proof: By Theorem 14 any finite dimensional subspace is complete [Exercise 6].

In $\mathbf{R}^n$ it is known that sets are compact if and only if they are closed and bounded. We now show that this condition characterizes finite dimensional NLS. For this we require a lemma of F. Riesz.

Lemma 16 (Riesz) *Let X be a NLS and X_0 a proper, closed subspace of X. Then for every $0 < \theta < 1$, there exists $x_\theta \in X$ such that $\|x_\theta\| = 1$ and $\|x - x_\theta\| \geq \theta$ for every $x \in X_0$.*

Proof: Let $x_1 \in X \setminus X_0$ and set $d = \inf\{\|x - x_1\| : x \in X_0\}$. Since X_0 is closed, $d > 0$. There exists $x_0 \in X_0$ such that $\|x_1 - x_0\| \leq d/\theta$ since $d/\theta > d$. Set $x_\theta = (x_1 - x_0)/\|x_1 - x_0\|$. Then $\|x_\theta\| = 1$ and if $x \in X_0$, $\|x_0 - x_1\| x + x_0 \in X_0$ so

$$\|x - x_\theta\| = \left\| x - \frac{x_1}{\|x_1 - x_0\|} + \frac{x_0}{\|x_1 - x_0\|} \right\| = \frac{1}{\|x_1 - x_0\|} \|(\|x_1 - x_0\| x + x_0) - x_1\|$$

$$\geq d/\|x_1 - x_0\| \geq \theta.$$

5.1. NORMED LINEAR SPACES

Example 17 In general, x_θ cannot be chosen to be distance 1 from X_0 although this is the case if X_0 is finite dimensional; see Exercise 17. Consider $C[0,1]$ equipped with the sup-norm. Let X be the subspace of $C[0,1]$ consisting of those functions x satisfying $x(0) = 0$ and $X_0 = \{x \in X : \int_0^1 x(t)dt = 0\}$. Suppose there exists $x_1 \in X$ such that $\|x_1\| = 1$ and $\|x - x_1\| \geq 1$ for all $x \in X_0$. For $y \in X \setminus X_0$, let $c = \int_0^1 x_1 / \int_0^1 y$. Then $x_1 - cy \in X_0$ so $1 \leq \|x_1 - (x_1 - cy)\| = |c| \|y\|$ which implies

$$\left| \int_0^1 y \right| \leq \|y\| \left| \int_0^1 x_1 \right|.$$

Now we can make $\left|\int_0^1 y\right|$ as close to 1 as we please and still have $\|y\| = 1$ [$y_k(t) = t^{1/k}$ as $k \to \infty$ will work]. Thus, $1 \leq \left|\int_0^1 x_1\right|$. But, since $\|x_1\| = 1$ and $x_1(0) = 0$, $\left|\int_0^1 x_1\right| < 1$.

Theorem 18 *Let X be a NLS and suppose that the unit ball $B = \{x \in X : \|x\| \leq 1\}$ is compact. Then X is finite dimensional.*

Proof: Suppose X is not finite dimensional. Let $0 \neq x_1 \in B$ and set $X_1 = span\{x_1\}$. Then $X_1 \subset X$ and X_1 is closed by Corollary 15. By Lemma 17, there exists $x_2 \in B$ such that $\|x_2 - x_1\| \geq 1/2$. Let $X_2 = span\{x_1, x_2\}$. Then X_2 is a proper, closed subspace of X so by Lemma 17 there exists $x_3 \in B$ such that $\|x_3 - x_2\| \geq 1/2$, $\|x_3 - x_1\| \geq 1/2$. Inductively, there exists a sequence $\{x_k\} \subset B$ satisfying $\|x_i - x_j\| \geq 1/2$ for $i \neq j$. Hence, B is not sequentially compact.

A subset B of a semi-NLS is *bounded* if $\sup\{\|x\| : x \in B\} < \infty$; this is equivalent to B being bounded in the semi-metric induced by the norm. From Theorems 14 and 18, we obtain

Corollary 19 *Let X be a NLS. Then X is finite dimensional if and only if closed, bounded subsets of X are compact.*

The conclusion of Corollary 19 does not hold in a metric linear space [Exercise 10].

Exercise 1. Show the sup-norm defined in Example 5 is actually a norm and convergence in the sup-norm is exactly uniform convergence on S.

Exercise 2. Show s is complete under the Frechet metric.

Exercise 3. Show c_0 is a B-space under the sup norm.

Exercise 4. Give an explicit example of a series in c_{00} which is absolutely convergent but not convergent.

Exercise 5. Let $\|\|_i$ ($i = 1, 2, 3$) be norms on a vector space X. If $\|\|_1$ is equivalent to $\|\|_2$ and $\|\|_2$ is equivalent to $\|\|_3$, show $\|\|_1$ is equivalent to $\|\|_3$.

Exercise 6. If $\|\ \|_1$ and $\|\ \|_2$ are equivalent norms, show they have the same convergent (Cauchy) sequences [generate the same topologies].

Exercise 7. Show that a subset B of a semi-NLS is bounded if and only if $\{x_k\} \subset B$ and $t_k \to 0$ implies $t_k x_k \to 0$.

Exercise 8. Show $\{e^k : k \in \mathbf{N}\}$ is a closed, bounded set in ℓ^∞ which is not compact.

Exercise 9. A subset B of a metric linear space (TVS) is *bounded* if $\{x_k\} \subset B$, $t_k \to 0$ implies $t_k x_k \to 0$ [see Exercise 7]. Show a compact subset is bounded.

Exercise 10. Show that a subset B of s is bounded if and only if B is coordinatewise bounded. Show that a subset of s is compact if and only if it is closed and bounded. Hint: Use a diagonalization procedure.

Exercise 11. What is the closure of c_{oo} in ℓ^∞?

Exercise 12. Show c_{00}, c_0, c, ℓ^1 are all separable.

Exercise 13. Show ℓ^∞ is not separable. Hint: Consider all sequences of 0's and 1's.

Exercise 14. If M is a linear subspace of a NLS X which has non-empty interior, show $M = X$.

Exercise 15. If X is an infinite dimensional Banach space, show X has uncountable algebraic dimension. [Hint: If $\{x_k\} \subset X$, consider $F_k = span\{x_1, \ldots, x_k\}$ and use the Baire Category Theorem.] Give an example of a NLS with countably infinite dimension.

Exercise 16. Let $\{x_j\} \subset X$, a B-space. Show $\{x_j\}$ is bounded if and only if $\sum t_j x_j$ converges for every $\{t_j\} \in \ell^1$.

Exercise 17. Show that if X_0 is finite dimensional in Lemma 16, then θ can be taken to be equal to 1.

5.1. NORMED LINEAR SPACES

Exercise 18. Let cs be the sequence space consisting of all sequences $x = \{x_j\}$ such that $\sum_{j=1}^{\infty} x_j$ converges. Show cs is a Banach space under the norm

$$\|x\| = \sup\left\{\left|\sum_{j=1}^{n} t_j\right| : n\right\}.$$

Exercise 19. Let X, Y be NLS. Show

$$\|(x,y)\|_1 = \|x\| + \|y\|, \ \|(x,y)\|_\infty = \max\{\|x\|, \|y\|\}, \ \|(x,y)\|_2 = \left(\|x\|^2 + \|y\|^2\right)^{1/2}$$

define equivalent norms on $X \times Y$.

Exercise 20. Let X be all $f \in C[0,1]$ such that $f' \in C[0,1]$. Show X is not complete with respect to the sup-norm but is complete with respect to the norm, $\|f\| = \sup\{|f(t)| : 0 \le t \le 1\} + \sup\{|f'(t)| : 0 \le t \le 1\}$.

5.2 Linear Mappings between Normed Linear Spaces

In this section we consider the continuity of linear mappings between normed spaces. Let X and Y be semi-NLS and $T : X \to Y$ linear.

Proposition 1 *The following are equivalent:*

(i) *T is uniformly continuous,*

(ii) *T is continuous,*

(iii) *T is continuous at 0,*

(iv) *there exists $M \geq 0$ such that*

$$(*) \qquad \|Tx\| \leq M \|x\| \text{ for all } x \in X.$$

Proof: Clearly (i) $\Rightarrow$ (ii) $\Rightarrow$ (iii). (iii) $\Rightarrow$ (iv): If (*) fails, there exist $x_k \in X$ such that $\|Tx_k\| > k^2 \|x_k\|$ for each k. If $y_k = x_k/k \|x_k\|$, then $y_k \to 0$ while $\|Ty_k\| > k$ so (iii) fails.

(iv) $\Rightarrow$ (i): If $\epsilon > 0$, put $\delta = \epsilon/(M+1)$. If $\|x - y\| < \delta$, then

$$\|Tx - Ty\| = \|T(x-y)\| \leq M \|x-y\| < \epsilon.$$

The space of all continuous linear operators from X into Y is denoted by $L(X,Y)$; $L(X,Y)$ is a vector space under the operations of pointwise addition and scalar multiplication. If $X = Y$, we write $L(X) = L(X,X)$. We define a semi-norm, called the *operator norm*, on $L(X,Y)$ by $\|T\| = \sup\{\|Tx\| : \|x\| \leq 1\}$. Note that if T satisfies (*), then $\|T\| \leq M$ and $\|T\|$ is the infimum of all such numbers M satisfying (*). In particular, we have

$$\|Tx\| \leq \|T\| \|x\| \text{ for } x \in X. \tag{5.1}$$

Proposition 2 (i) *The operator norm is a semi-norm on $L(X,Y)$.*

(ii) *$\|T\| = 0$ if and only if $\|Tx\| = 0$ for all $x \in X$.*

(iii) *If Y is a NLS, then $L(X,Y)$ is a NLS under the operator norm.*

(iv) *If Z is a semi-NLS, $T \in L(X,Y)$, $S \in L(Y,Z)$, then $ST \in L(X,Z)$ and $\|ST\| \leq \|S\| \|T\|$.*

Concerning completeness, we have

Theorem 3 *If Y is a complete NLS, then $L(X,Y)$ is complete.*

5.2. LINEAR MAPPINGS BETWEEN NORMED LINEAR SPACES

Proof: Suppose $\{T_k\}$ is Cauchy with respect to the operator norm. If $x \in X$, then
$$\|T_k x - T_j x\| \leq \|T_k - T_j\| \|x\| \tag{5.2}$$
so $\{T_k x\}$ is Cauchy in Y. Let $Tx = \lim T_k x$. Then $T : X \to Y$ is linear. We show $T \in L(X,Y)$ and $T_k \to T$ in the operator norm. Let $\epsilon > 0$. There exists N such that $k, j \geq N$ implies $\|T_k - T_j\| < \epsilon$. Let $j \to \infty$ in (2) so $\|T_k x - Tx\| \leq \epsilon \|x\|$ for $k \geq N$ and $x \in X$. Hence, $T_k - T \in L(X,Y)$ with $\|T_k - T\| \leq \epsilon$ for $k \geq N$ so $T \in L(X,Y)$ and $T_k \to T$ in operator norm.

We establish the converse of Theorem 3 in 5.6.1.7.

If X is a metric linear space (TVS), a linear map from X into the scalar field is called a *linear functional*. The *dual* of X is the space of all continuous linear functionals on X and is denoted by $\underline{X}'$, i.e., $X' = L(X, \mathbf{F})$.

If x' is a linear functional on X, we often write $x'(x) = \langle x', x \rangle$ for $x \in X$.

If X is a semi-NLS, it follows from Theorem 3 that X' is a B-space under the operator norm $\|x'\| = \sup\{|\langle x', x \rangle| : \|x\| \leq 1\}$; this norm is called the *dual norm* of X'.

If X and Y are semi-NLS, a map $U : X \to Y$ is an *isometry* if $\|Ux - Uy\| = \|x - y\|$ for $x, y \in X$, i.e., if U preserves distances. If X and Y are NLS, then X and Y are *linearly isometric* if there is a linear isometry from X onto Y; if X and Y are linearly isometric, it is customary to write $X = Y$.

In later chapters we describe the duals of many classic function spaces. As an example, we now describe the dual of the sequence space c_0. F. Riesz gave descriptions of many of the duals of classic function spaces so any result which describes the dual of a specific function space is often referred to as a "Riesz Representation Theorem".

Example 4 c_0' and ℓ^1 are linearly isometric, $c_0' = \ell^1$.

Let $f \in c_0'$. Set $y_k = \langle f, e^k \rangle$ for $k \in \mathbf{N}$ and set $y = \{y_k\}$. We claim that $y \in \ell^1$. If
$$x^n = \sum_{k=1}^{n} (\operatorname{sign} y_k) e^k,$$
then
$$\langle f, x^n \rangle = \sum_{k=1}^{n} |y_k| \leq \|f\| \|x^n\|_\infty \leq \|f\|$$
so $y \in \ell^1$ and $\|y\|_1 \leq \|f\|$. For $x = \{x_k\} \in c_0$, $x = \sum_{k=1}^{\infty} x_k e^k$ so
$$\langle f, x \rangle = \sum_{k=1}^{\infty} x_k \langle f, e^k \rangle = \sum_{k=1}^{\infty} x_k y_k$$
and
$$|\langle f, x \rangle| \leq \|x\|_\infty \sum_{k=1}^{\infty} |y_k| = \|x\|_\infty \|y\|_1$$
which implies $\|f\| \leq \|y\|_1$. Hence, $\|f\| = \|y\|_1$.

Thus, the map U which sends $f \to y$ is an isometry from c_0' into ℓ^1 which is obviously linear. We show U is onto ℓ^1. If $y = \{y_k\} \in \ell^1$, $\langle f_y, x\rangle = \sum_{k=1}^{\infty} x_k y_k$ defines a continuous linear functional f_y on c_0 and $U(f_y) = y$ so U is onto.

We give a similar description of the dual of ℓ^1.

Example 5 Let $f \in (\ell^1)'$. Set $y_k = \langle f, e^k\rangle$ and $y = \{y_k\}$. Then $y \in \ell^\infty$ with $\|y\|_\infty \leq \|f\|$ since $|\langle f, e^k\rangle| \leq \|f\|$. If $x = \{x_k\} \in \ell^1$, then $x = \sum_{k=1}^{\infty} x_k e^k$ so

$$f(x) = \sum_{k=1}^{\infty} x_k f(e^k) = \sum_{k=1}^{\infty} x_k y_k$$

and $|f(x)| \leq \|\{y_k\}\|_\infty \|x\|_1$. Hence $\|f\| \leq \|\{y_k\}\|_\infty$ and $\|f\| = \|\{y_k\}\|_\infty$.

Thus, the map U which associates $f \in (\ell^1)'$ with $y \in \ell^\infty$ is an isometry from $(\ell^1)'$ into ℓ^∞ which is obviously linear. Now U is actually onto ℓ^∞ since if $y = \{y_k\} \in \ell^\infty$, $\langle f_y, x\rangle = \sum_{k=1}^{\infty} y_k x_k$ defines a continuous linear functional f_y on ℓ^1 with $U(f_y) = y$. Hence, $(\ell^1)' = \ell^\infty$.

A more general version of this result is given in Theorem 6.2.4.

Exercise 1. If X, Y are semi-NLS and $T : X \to Y$ is linear, show T is continuous if and only if T carries bounded subsets of X into bounded subsets of Y.

Exercise 2. Define $R, L : \ell^1 \to \ell^1$ (or $\ell^\infty \to \ell^\infty$) by

$$R\{t_j\} = \{0, t_1, t_2, \ldots\} \quad \text{(right shift)},$$

$$L\{t_j\} = \{t_2, t_3, \ldots\} \quad \text{(left shift)}.$$

Show R and L are continuous and compute $\|R\|$, $\|T\|$.

Exercise 3. Let X_1 be a dense linear subspace of the NLS X and let Y be a B-space. If $T : X_1 \to Y$ is linear, continuous, show that T has a unique linear extension $\hat{T} \in L(X, Y)$ with $\|T\| = \|\hat{T}\|$. In particular, $X_1' = X'$ [equality here means linearly isometric].

Exercise 4. Describe the dual of c_{00}. Hint: Exercise 3 and Example 4.

Exercise 5. Let X be a NLS. Show X and $L(\mathbf{F}, X)$ are linearly isometric.

Exercise 6. Let X be a B-space and $T \in L(X)$. Define e^T and show $e^T \in L(X)$. If $S, T \in L(X)$ commute, show $e^{T+S} = e^T e^S$.

5.2. LINEAR MAPPINGS BETWEEN NORMED LINEAR SPACES

Exercise 7. Show that c_0 and c are linearly homeomorphic.

Exercise 8. Show that c' and ℓ^1 are linearly isometric under the correspondence which associates with each $y = \{y_k\} \in \ell^1$ the linear functional $f_y \in c'$ defined by

$$\langle f_y, \{x_k\}\rangle = \sum_{k=1}^{\infty} y_k x_{k-1}$$

where $x_0 = \lim x_k$.

Exercise 9. Show the linear functional $L : c \to \mathbf{R}$ defined by $L\{x_k\} = \lim x_k$ is continuous and compute $\|L\|$.

Exercise 10. Show that s' and c_{00} are algebraically isomorphic under the map which associates with each $y = \{y_k\} \in c_{00}$ the linear functional f_y on s defined by

$$\langle f_y, \{x_k\}\rangle = \sum_{k=1}^{\infty} y_k x_k.$$

Exercise 11. Define $f : c_{00} \to \mathbf{R}$ by $f(\{x_k\}) = \sum_{k=1}^{\infty} x_k$. Show f is linear but not continuous.

Exercise 12. Let $\{t_i\} \in s$. For $\{s_i\} \in s$, set $T\{s_i\} = \{t_i s_i\}$. Find necessary and sufficient conditions on $\{t_i\}$ so that $T \in L(\ell^1, \ell^1)$ or $T \in L(\ell^\infty, \ell^\infty)$. Compute $\|T\|$.

Exercise 13. Show that any linear map from $\mathbf{R}^n$ into a semi-NLS is continuous.

Exercise 14. Describe the dual of $\mathbf{R}^n$ (including the dual norm) for each of the norms $\|\ \|_2, \|\ \|_1, \|\ \|_\infty$.

Exercise 15. let $k : [0,1]\times[0,1] \to \mathbf{R}$ be continuous. Show $Kf(s) = \int_0^1 k(s,t)f(t)\,dt$ defines a continuous linear operator from $C[0,1]$ into $C[0,1]$. Show K carries bounded subsets of $C[0,1]$ into relatively compact subsets. (Hint: Arzela-Ascoli.)

Exercise 16. Let $a_k > 0$, $a_k \downarrow 0$, $a = \{a_k\}$. Define a norm by c_0 by $\|\{t_k\}\|_a = \sup\{|t_k a_k| : k\}$. Describe the dual of c_0 under $\|\ \|_a$ and describe the dual norm of $\|\ \|_a$.

5.3 The Uniform Boundedness Principle

In this section we establish one of the earliest abstract results in functional analysis, the Uniform Boundedness Principle (UBP), and one of its most important consequences, the Banach-Steinhaus Theorem. The Uniform Boundedness Principle is one of the three basic abstract results in functional analysis, along with the Closed Graph/Open Mapping Theorems and the Hahn-Banach Theorem; these theorems will be discussed in later sections.

Let X, Y be semi-NLS.

The proof of the UBP which we give is based on a technique called a "gliding hump" or "sliding hump" argument.

Theorem 1 (UBP) *Let X be complete. If $\mathcal{F} \subset L(X,Y)$ is pointwise bounded on X [i.e., $\{Tx : T \in \mathcal{F}\}$ is bounded for each $x \in X$], then $\{\|T\| : T \in \mathcal{F}\}$ is bounded.*

Proof: If the conclusion fails, there is a sequence $\{T_i\} \subset \mathcal{F}$ satisfying $\|T_i\| > i2^{2i}$ for each i. Then for each i there exists $x_i \in X$, $\|x_i\| \leq 1$, such that $\|T_i x_i\| > i2^{2i}$. For convenience of notation, set $S_i = 2^{-i}T_i$ and $z_i = 2^{-i}x_i$. Then $\|z_i\| \leq 2^{-i}$, $\|S_i z_i\| > i$, $\lim_j S_i z_j = 0$ for each i and $\lim_i S_i z_j = 0$ for each j by the pointwise boundedness assumption. Thus, the rows and columns of the matrix $M = [S_i z_j]$ converge to 0 so there is a subsequence $\{n_i\}$ such that $\|S_{n_i} z_{n_j}\| \leq 2^{-i-j}$ for $i \neq j$ [see Lemma 2.8.1 and its proof]. Then for each i,

$$\sum_{\substack{j=1 \\ j \neq i}}^{\infty} \|S_{n_i} z_{n_j}\| \leq \sum_{j=1}^{\infty} 2^{-i-j} = 2^{-i}$$

and $\|S_{n_i} z_{n_i}\| > n_i$, i.e., for each i the sequence $\{S_{n_i} z_{n_j}\}_{j=1}^{\infty}$ has a "hump" in the ith coordinate and the sum of the norms of the other elements in the sequence is much smaller than the "hump" and as i increases the "hump" slides to the right since $\{n_j\}$ is increasing. We now "gather" or "collect" the points $\{z_{n_j}\}$ which give rise to the humps by setting $z = \sum_{j=1}^{\infty} z_{n_j}$; note that this series converges in X since the series is absolutely convergent and X is complete [5.1.4]. We now have

$$\|S_{n_i} z\| = \left\|\sum_{j=1}^{\infty} S_{n_i} z_{n_j}\right\| \geq \|S_{n_i} z_{n_i}\| - \sum_{\substack{j=1 \\ j \neq i}}^{\infty} \|S_{n_i} z_{n_j}\| \geq n_i - 2^{-i}.$$

But, since $\{T_i z\}$ is bounded, $\{S_i z\} = \{2^{-i} T_i z\}$ should converge to 0 contradicting the inequality above.

Without some condition, such as completeness, on the domain space X the UBP can fail.

5.3. THE UNIFORM BOUNDEDNESS PRINCIPLE

Example 2 Define $f_i : c_{00} \to \mathbf{R}$ by $f_i(\{t_j\}) = it_i$. Then f_i is a continuous linear functional on c_{00} with $\|f_i\| = i$, and the sequence $\{f_i\}$ is pointwise bounded on c_{00}.

There are versions of the UBP which hold without any hypothesis on the domain space. For a discussion of such results see [AS], [Sw1], or [Sw2].

We derive an important consequence of the UBP concerning the continuity of the pointwise limit of a sequence of continuous linear operators.

Theorem 3 (Banach-Steinhaus) *Let X be complete, Y a NLS and $\{T_k\} \subset L(X,Y)$. If $\lim T_k x = Tx$ exists for each $x \in X$, then $T : X \to Y$ is linear and continuous.*

Proof: T is clearly linear. The sequence $\{T_k\}$ is pointwise bounded on X so by the UBP $\sup\{\|T_k\| : k\} = M < \infty$. Thus,

$$\|Tx\| = \lim \|T_k x\| \le M \|x\| \text{ for } x \in X$$

so $T \in L(X,Y)$ with $\|T\| \le M$.

Again without some assumption on X, this result can fail.

Example 4 Define $f_i : c_{00} \to \mathbf{R}$ by $f_i(\{t_j\}) = \sum_{j=1}^{i} t_j$. Each f_i is continuous and for $\{t_j\} \in c_{00}$,

$$f_i(\{t_j\}) \to \sum_{j=1}^{\infty} t_j = f(\{t_j\}),$$

but f is not continuous (Exer. 5.2.11).

We give an application of the Banach-Steinhaus Theorem to sequence spaces [see also Exercise 3]. A further application to Fourier series is given in Chapter 6.6.

Proposition 5 *Suppose the sequence $\{t_j\}$ is such that $\sum_{j=1}^{\infty} t_j s_j$ converges for every $\{s_j\} \in c_0$. Then $\{t_j\} \in \ell^1$, i.e., $\sum_{j=1}^{\infty} |t_j| < \infty$.*

Proof: For each k define $f_k \in c_0' = \ell^1$ [5.2.4] by $f_k(\{s_j\}) = \sum_{j=1}^{k} t_j s_j$. Then

$$\lim_k f_k(\{s_j\}) = f(\{s_j\}) = \sum_{j=1}^{\infty} t_j s_j$$

for $\{s_j\} \in c_0$. By Theorem 3, $f \in (c_0)'$. By Example 5.2.4 $\{t_j\} \in \ell^1 = (c_0)'$.

For a discussion of the evolution of the UBP see [Sw1].

Exercise 1. Let $\mathcal{F} \subset L(X,Y)$. Show $\{\|T\| : T \in \mathcal{F}\}$ is bounded if and only if $\mathcal{F}$ is uniformly bounded on bounded subsets of X if and only if $\mathcal{F}$ is equicontinuous at 0 if and only if $\mathcal{F}$ is equicontinuous on X. [$\mathcal{F}$ is equicontinuous at $x_0 \in X$ if for every $\epsilon > 0$ there exists $\delta > 0$ such that $\|x - x_0\| < \delta$ implies $\|Tx - Tx_0\| < \epsilon$ for all $T \in \mathcal{F}$.]

Exercise 2. In Theorem 3, show $\|T\| \leq \underline{\lim} \|T_k\| < \infty$.

Exercise 3. Suppose the sequence $\{t_j\}$ is such that $\sum_{j=1}^{\infty} t_j s_j$ converges for every $\{s_j\} \in \ell^1$. Show $\{t_j\} \in \ell^\infty$.

Exercise 4. Let $\{T_k\} \subset L(X,Y)$ be equicontinuous and let Z be a dense linear subspace of X. Show that if Y is complete and $\lim T_k z$ exists for each $z \in Z$, then $\lim T_k x = Tx$ exists for each $x \in X$ and $T \in L(X,Y)$.

Exercise 5. Let X, Y, Z be NLS with X a B-space. Let $B : X \times Y \to Z$ be a separately continuous bilinear map. Show B is jointly continuous [use Exer. 5.1.19]. Hint: Use the UBP to show there is a constant b such that $\|B(x,y)\| \leq b \|x\| \|y\|$.

Exercise 6. Give an example showing completeness cannot be dropped in Exercise 5.

Exercise 7. Use the Baire Category Theorem to prove the UBP. Hint: Consider $F_k = \left\{ x : \sup_n \|T_n x\| \leq k \right\}$.

5.4 Quotient Spaces

In this brief section we consider the quotient of a NLS. These results are used later in dealing with some of the classical spaces of functions.

Let X be a semi-NLS and M a linear subspace of X. If $x \in X$, we denote the coset $x + M$ in X/M by $[x] = x + M$. We define a semi-norm on X/M by

$$\|[x]\|' = \inf\{\|x + m\| : m \in M\} \qquad (= \text{distance}(x, M)). \tag{5.1}$$

Proposition 1 (i) $\|\|'$ *is a semi-norm on* X/M.

(ii) X/M *is a NLS* $\Leftrightarrow M$ *is closed.*

(iii) *The quotient map* $x \to [x]$ *from* X *onto* X/M *is norm reducing [and, therefore, continuous] and open.*

(iv) X/M *is complete if* X *is complete.*

(v) *If* X *is complete and* M *is closed,* X/M *is a B-space.*

Proof: (i): $\|[tx]\|' = \inf\{\|tx + m\| : m \in M\} = \inf\{|t|\,\|x + m/t\| : m \in M\} = |t|\,\|[x]\|'$ for $t \neq 0$.

$$\begin{aligned} \|[x] + [y]\|' &= \inf\{\|x + y + m_1 + m_2\| : m_i \in M\} \\ &\leq \inf\{\|x + m_1\| : m_1 \in M\} + \inf\{\|y + m_2\| : m_2 \in M\} \\ &= \|[x]\|' + \|[y]\|'. \end{aligned}$$

(ii): $\|[x]\|' = dist(x, M) = 0$ if and only if $x \in \overline{M}$ and $[x] = 0$ if and only if $x \in M$.

(iii): Clearly $\|x\| \geq \|[x]\|'$. To show the quotient map is open we show that $\{x : \|x\| < 1\}$ is mapped onto $\{[x] : \|[x]\|' < 1\}$. Let $\|[x]\|' = 1 - \delta < 1$. There exists $m \in M$ such that $\|x + m\| < 1$ and $[x + m] = [x]$.

(iv): Let $\sum_{k=1}^{\infty}[x_k]$ be an absolutely convergent series in X/M. For each k choose $m_k \in M$ such that

$$\|[x_k]\|' \geq \|x_k + m_k\| - 1/2^k.$$

Then

$$\sum_{k=1}^{\infty}\|x_k + m_k\| \leq \sum_{k=1}^{\infty}\left(\|[x_k]\|' + 1/2^k\right) < \infty$$

so $\sum_{k=1}^{\infty}(x_k + m_k)$ is absolutely convergent in X and, therefore, convergent to some $x \in X$ [Theorem 5.1.4]. By (iii),

$$[x] = \sum_{k=1}^{\infty}[x_k + m_k] = \sum_{k=1}^{\infty}[x_k]$$

so X/M is complete by Theorem 5.1.4.

(v) follows from (ii) and (iv).

Proposition 2 *Let $K(X) = \{x \in X : \|x\| = 0\}$. Then $K(X)$ is a closed, linear subspace of X.*

Proof: If $x, y \in K(X)$, then $\|tx + sy\| \leq |t|\|x\| + |s|\|y\| = 0$ so $K(X)$ is a linear subspace. $K(X)$ is closed by Proposition 5.1.3.

Proposition 3 *The quotient map $X \to X/K(X)$ is norm preserving [i.e., an isometry].*

Proof: For $m \in K(X)$, $\|x\| = \|x\| - \|m\| \leq \|x - m\| \leq \|x\|$ so $\|x\| = \|[x]\|'$.

Corollary 4 $X/K(X)$ *is a NLS and if X is complete, $X/K(X)$ is a B-space.*

As an example, consider $L^1(\mu)$ with the L^1-norm, $\|\ \|_1$. If $K = K(L^1(\mu))$, then $f \in K$ if and only if $f = 0$ μ-a.e. Thus, the cosets of $L^1(\mu)/K$ consist of equivalence classes of functions which are equal μ-a.e.

Exercise 1. Let X, Y be semi-NLS and $T \in L(X,Y)$. Show the induced map $\hat{T} : X/\ker T \to Y$ is continuous and $\|T\| = \|\hat{T}\|$.

5.5 The Closed Graph/Open Mapping Theorems

In this chapter we discuss another of the three basic principles of functional analysis, the Closed Graph Theorem (CGT) and its companion the Open Mapping Theorem (OMT). Let X, Y be NLS and $T : X \to Y$ linear. T is said to be *closed* if its graph $\{(x, Tx) : x \in X\}$ is closed in $X \times Y$. Thus, T is closed if and only if $x_k \to x$ in X and $Tx_k \to y$ in Y implies that $y = Tx$. Any continuous linear operator $T \in L(X, Y)$ is obviously closed; however, as the following example shows not every closed operator is continuous.

Example 1 Let $X = \{f \in C[0,1] : f' \text{ exists and is continuous on } [0,1]\}$ and assume that X is equipped with the sup-norm. Let $Y = C[0,1]$ and $D : X \to Y$ be defined by $Df = f'$. Then D is obviously linear, closed [[DeS] 11.7], but is not continuous [$\|t^k\|_\infty = 1$ while $\|(t^k)'\|_\infty = k$].

Note that the domain space in this example is not complete. Indeed, the CGT asserts that if X and Y are complete, then any closed linear operator $T : X \to Y$ is continuous.

Theorem 2 (CGT) *Let X, Y be Banach spaces and $T : X \to Y$ linear. If T is closed, then T is continuous.*

Proof: Set $Z = \{x \in X : \|Tx\| < 1\}$. Since T is linear, $X = \bigcup_{k=1}^{\infty} kZ$ so from the Baire Category Theorem (A2) some $\overline{kZ} = k\overline{Z}$ contains an interior point. Hence, $\{y : \|y - x\| < r\} = S(x, r) \subset \overline{Z}$ for some $x \in \overline{Z}, r > 0$. If $\|z\| < r$, then

$$z = \frac{1}{2}(x + z) - \frac{1}{2}(x - z) \in \frac{1}{2}\overline{Z} - \frac{1}{2}\overline{Z} = \overline{Z}$$

so $S(0, r) \subset \overline{Z}$. Hence,

$$S(0, ar) \subset \overline{aZ} = a\overline{Z} \text{ for every } a > 0. \tag{5.1}$$

We now claim that if $\|z\| < r$, then $\|Tz\| \leq 2$. Since $z \in \overline{Z}$, by (1) there exists $x_1 \in Z$ such that $\|z - x_1\| < r/2$. Since $x_1 - z \in \left(\frac{1}{2}\right)\overline{Z}$, by (1), there exists $x_2 \in \frac{1}{2}Z$ such that $\|z - x_1 - x_2\| < r/2^2$. Continuing inductively produces a sequence $\{x_k\} \subset Z$ with $\|z - x_1 - x_2 - \ldots - x_k\| < r/2^k$ and $x_k \in \left(1/2^{k-1}\right)Z$. Put $s_k = \sum_{j=1}^{k} x_j$ so that $z = \lim_k s_k = \sum_{j=1}^{\infty} x_j$ and $\|Tx_k\| \leq 1/2^{k-1}$. Now $\{Ts_k\}$ is Cauchy in Y since

$$\|Ts_k - Ts_j\| \leq \sum_{i=j+1}^{k} \|Tx_i\| \leq \sum_{i=j+1}^{k} 1/2^{i-1} < 2^{1-j}$$

for $k > j$. Therefore, there exists $y \in Y$ such that $Ts_k \to y$. Since T is closed, $Tz = y$ and

$$\|Tz\| = \|y\| = \left\|\sum_{k=1}^{\infty} Tx_k\right\| \le \sum_{k=1}^{\infty} \|Tx_k\| \le \sum_{k=1}^{\infty} 1/2^{k-1} = 2.$$

If $\|x\| \le 1$, then by the above $\|T(rx/2)\| \le 2$ or $\|Tx\| \le 4/r$. Hence, T is continuous with $\|T\| \le 4/r$.

If one wants to prove that a linear map $T : X \to Y$ is continuous, one of the usual procedures is to take a sequence $\{x_k\}$ in X which converges to some $x \in X$ and show that the sequence $\{Tx_k\}$ converges to Tx. The advantage of the CGT is that we may now assume that the sequence $\{Tx_k\}$ is convergent, and we are then required to show that it converges to the proper value, namely, Tx. An example of such an application of the CGT is given in Example 6.2.9 [see also Exercise 3, Exercise 6.1.18 and the end of §6.2].

We use the CGT to derive its companion result, the Open Mapping Theorem (OMT).

Theorem 3 *Let X, Y be Banach spaces and $T \in L(X,Y)$. If T is onto, T is open.*

Proof: Let $\widehat{T}$ be the induced map, $\widehat{T} : X/\ker T \to Y$ [Exercise 5.4.1]. Then $\widehat{T}$ is 1-1, continuous and onto Y so $\widehat{T}^{-1}$ is closed [Exercise 1]. By the CGT, $\widehat{T}^{-1}$ is continuous so $\widehat{T}$ is a homeomorphism. Since the quotient map $X \to X/\ker T$ is always open [5.4.1], T is open.

Corollary 4 *If X and Y are Banach spaces and $T \in L(X,Y)$ is 1-1 and onto, then T is a homeomorphism.*

Corollary 5 *Let X be a vector space with two complete norms $\|\|_1$, $\|\|_2$ on X. If $\|\|_1$ is stronger than $\|\|_2$ [i.e., induces a stronger topology], then $\|\|_1$ and $\|\|_2$ are equivalent.*

Proof: The identity map $(X, \|\|_1) \to (X, \|\|_2)$ is continuous so Corollary 4 gives the result.

The completeness of both norms in Corollary 5 is important, see Exercise 2.

Remark 6 The CGT is often derived from the OMT; see [TL] for such a development.

Exercise 1. Let $T \in L(X,Y)$ be 1-1, onto. Show T^{-1} is closed.

Exercise 2. Show Corollary 5 is false if both norms are not complete. [Hint: Consider $C[0,1]$ with $\|\|_\infty$ and $\|f\| = \int_a^b |f|(t)dt$.]

5.5. THE CLOSED GRAPH/OPEN MAPPING THEOREMS

Exercise 3. Let X, Y be Banach spaces and $T : X \to Y$ linear. Suppose $A \subset Y'$ separates the points of Y. If $y'T$ is continuous $\forall\, y' \in A$, show T is continuous.

Exercise 4. Suppose that $\|\|$ is a complete norm on $C[0,1]$ such that $\|f_k - f\| \to 0$ implies $f_k(t) \to f(t)$ for every $t \in [0,1]$. Show $\|\|$ is equivalent to $\|\|_\infty$.

5.6 The Hahn-Banach Theorem

In §5.6.1 we will study some of the relationships between a NLS and its dual space. In order to facilitate this study, we need a very important preliminary result called the Hahn-Banach Theorem. This is one of the three basic principles of functional analysis and has applications to a wide variety of problems. Besides using this result in 5.6.1 we give additional applications in sections 5.6.2 and 5.6.3. For an interesting discussion of the history of the Hahn-Banach Theorem see [Ho].

Definition 1 *Let X be a vector space. A function $p : X \to \mathbf{R}$ is a sublinear functional if*

(i) $p(x + y) \leq p(x) + p(y) \; \forall x, y \in X$,

(ii) $p(tx) = tp(x) \; \forall t \geq 0, \; x \in X$.

A semi-norm is obviously sublinear but not conversely.

The Hahn-Banach Theorem guarantees that any linear functional defined on a subspace of a vector space which is dominated by a sublinear functional can be extended to a linear functional defined on the entire vector space and the extension is still dominated by the sublinear functional.

Theorem 2 (Hahn-Banach; real case) *Let X be a real vector space and $p : X \to \mathbf{R}$ a sublinear functional. Let M be a linear subspace of X. If $f : M \to \mathbf{R}$ is a linear functional such that $f(x) \leq p(x) \; \forall x \in M$, then $\exists$ a linear functional $F : X \to \mathbf{R}$ such that $F(x) = f(x), \; \forall x \in M$ and $F(x) \leq p(x) \; \forall x \in X$.*

Proof: Let $\mathcal{E}$ be the class of all linear extensions g of f such that $g(x) \leq p(x)$ $\forall x \in \mathcal{D}(g)$, the domain of g, with $\mathcal{D}(g) \supseteq M$. Note $\mathcal{E} \neq \emptyset$ since $f \in \mathcal{E}$. Partial order $\mathcal{E}$ by $g < h$ if and only if h is a linear extension of g. If $\mathcal{C}$ is a chain in $\mathcal{E}$, then $\bigcup_{g \in \mathcal{C}} g \in \mathcal{E}$ is clearly an upper bound for $\mathcal{C}$ so by Zorn's Lemma $\mathcal{E}$ has a maximal element F. The result follows if we can show $\mathcal{D}(F) = X$.

Suppose $\exists x_1 \in X \setminus \mathcal{D}(F)$. Let M_1 be the linear subspace spanned by $\mathcal{D}(F)$ and x_1. Thus, if $y \in M_1$, y has a unique representation in the form $y = m + tx_1$, where $m \in \mathcal{D}(F), t \in \mathbf{R}$. If $z \in \mathbf{R}$, then

$$F_1(y) = F_1(m + tx_1) = F(m) + tz$$

defines a linear functional on M_1 which extends F. If we can show that it is possible to choose z such that $F_1(y) \leq p(y) \; \forall y \in M_1$, this will show $F_1 \in \mathcal{E}$ and contradict the maximality of F.

5.6. THE HAHN-BANACH THEOREM

In order to have

$$F_1(y) = F_1(m + tx_1) = F(m) + tz \leq p(y) = p(m + tx_1), \tag{5.1}$$

we must have for $t > 0$,

$$z \leq -\frac{1}{t}F(m) + \frac{1}{t}p(y) = -F\left(\frac{m}{t}\right) + p\left(\frac{m}{t} + x_1\right),$$

or since $m/t \in \mathcal{D}(F)$, if z satisfies

$$z \leq -F(m) + p(m + x_1) \; \forall m \in \mathcal{D}(F), \tag{5.2}$$

then (1) holds for $t \geq 0$. For $t < 0$,

$$z \geq -\frac{1}{t}F(m) + \frac{1}{t}p(m + tx_1) = F(-m/t) - p(-m/t - x_1),$$

or since $-m/t \in \mathcal{D}(F)$, if z satisfies,

$$z \geq F(m) - p(m - x_1) \; \forall m \in \mathcal{D}(F), \tag{5.3}$$

then (1) holds. Thus, z must satisfy

$$F(m_1) - p(m_1 - x_1) \leq z \leq -F(m_2) + p(m_2 + x_1) \; \forall m_1, m_2 \in \mathcal{D}(F),$$

i.e., we must have

$$F(m_1) - p(m_1 - x_1) \leq -F(m_2) + p(m_2 + x_1) \; \forall m_1, m_2 \in \mathcal{D}(F). \tag{5.4}$$

But,

$$F(m_1 + m_2) = F(m_1) + F(m_2) \leq p(m_1 + m_2) \leq p(m_1 - x_1) + p(m_2 + x_1)$$

so (4) does hold.

To obtain a complex form of the Hahn-Banach Theorem, we need the following interesting observation which shows how to write a complex linear functional in terms of its real part.

Lemma 3 (Bohnenblust-Sobczyk) *Let X be a vector space over* **C**. *Suppose $F = f + ig$ is a linear functional on X. Then for $x \in X$, $F(x) = f(x) - if(ix)$ and $f : X \to$ **R** *is* **R**-*linear. Conversely, if $f : X \to$ **R** *is* **R**-*linear, then $F(x) = f(x) - if(ix)$ defines a* **C**-*linear functional on X.*

Proof: f and g are clearly **R**-linear. Now $F(ix) = iF(x)$ implies $f(ix) + ig(ix) = if(x) - g(x)$ so $f(ix) = -g(x)$ and $F(x) = f(x) - if(ix)$.
The converse is easily checked.

Theorem 4 (Hahn-Banach; complex case) *Let X be a vector space and $p : X \to \mathbf{R}$ a semi-norm. Let M be a linear subspace of X and $f : M \to \mathbf{F}$ a linear functional. If $|f(x)| \leq p(x)$ $\forall x \in M$, then f has a linear extension $F : X \to \mathbf{F}$ such that $|F(x)| \leq p(x)$ $\forall x \in X$.*

Proof: Suppose $\mathbf{F} = \mathbf{R}$. Then $f(x) \leq |f(x)| \leq p(x)$ $\forall x \in M$ so Theorem 2 implies $\exists$ a linear extension $F : X \to \mathbf{R}$ such that $F(x) \leq p(x)$ $\forall x \in X$. But then $F(-x) = -F(x) \leq p(-x) = p(x)$ so $|F(x)| \leq p(x)$.

Suppose $\mathbf{F} = \mathbf{C}$. Then $\mathcal{R}f$, the real part of f, is an $\mathbf{R}$-linear functional on X such that $|\mathcal{R}f(x)| \leq |f(x)| \leq p(x)$ $\forall x \in M$. By the first part, $\exists$ a real linear functional $f_1 : X \to \mathbf{R}$ which extends $\mathcal{R}f$ and satisfies $|f_1(x)| \leq p(x)$ $\forall x \in X$. Set $F(x) = f_1(x) - if_1(ix)$. Then F is $\mathbf{C}$-linear and extends f by Lemma 3.

For $x \in X$, write $F(x) = |F(x)| e^{i\theta}$. Then

$$|F(x)| = e^{-i\theta} F(x) = F(e^{-i\theta} x) = \mathcal{R}F(e^{-i\theta} x) = f_1(e^{-i\theta} x) \leq p(e^{-i\theta} x) = p(x).$$

Despite its very esoteric appearance we will see in the next three sections that the Hahn-Banach Theorem has a surprisingly wide variety of applications.

Exercise 1. Give an example of a sublinear functional which is not a semi-norm.

Exercise 2. Show $p(\{t_j\}) = \varlimsup_n (t_1 + \cdots + t_n)/n$ defines a sublinear functional on ℓ^∞.

5.6.1 Applications of the Hahn-Banach Theorem in NLS

We use the Hahn-Banach Theorem to derive important properties of the dual space of a NLS. We begin by establishing an important result on extending continuous linear functionals.

Theorem 1 *Let X be a semi-NLS and M a linear subspace. If $m' \in M'$, then $\exists x' \in X'$ such that x' extends m' and $\|x'\| = \|m'\|$.*

Proof: Define a semi-norm p on X by $p(x) = \|m'\| \|x\|$. Then $|\langle m', x \rangle| \leq p(x)$ for $x \in M$. By Theorem 5.6.4 there exists a linear function x' on X extending m' such that $|\langle x', x \rangle| \leq p(x) = \|m'\| \|x\|$ for all $x \in X$. Hence, $x' \in X'$ and $\|x'\| \leq \|m'\|$. Clearly $\|m'\| \leq \|x'\|$.

Theorem 1 can be used to establish several important results on the existence of continuous linear functionals on a semi-NLS.

Theorem 2 *Let M be a linear subspace of a NLS X. Suppose $x_0 \in X$ is such that distance $(x_0, M) = d > 0$. Then there exists $x'_0 \in X'$ such that $\|x'_0\| = 1$, $x'_0(M) = 0$ and $\langle x'_0, x_0 \rangle = d > 0$.*

Proof: Set $M_0 = \text{span}\{M, x_0\}$. Define a linear functional f on M_0 by $f(m + tx_0) = td$, where $m \in M$, $t \in \mathbf{F}$. Then $f(M) = 0$ and $f(x_0) = d$. Also, $f \in M'_0$ with $\|f\| \leq 1$ since if $t \neq 0$, $m \in M$, then

$$\|m + tx_0\| = |t| \|m/t + x_0\| \geq |t| d = |f(m + tx_0)|.$$

Actually, $\|f\| = 1$ since there exists $\{m_k\} \subset M$ with $\|m_k - x_0\| \downarrow d$ so

$$d = |f(m_k - x_0)| \leq \|f\| \|m_k - x_0\| \downarrow \|f\| d$$

and $\|f\| \geq 1$. Now extend f to a continuous linear functional $x' \in X'$ with $\|x'\| = \|f\| = 1$ by Theorem 1.

Remark 3 Note that Theorem 2 is applicable if M is closed and $x_0 \notin M$.

Corollary 4 *Let X be a NLS and $0 \neq x_0 \in X$. Then there exists $x'_0 \in X'$ such that $\|x'_0\| = 1$ and $\langle x'_0, x_0 \rangle = \|x_0\|$. In particular, if $x \neq y$, there exists $x'_0 \in X'$ such that $\langle x'_0, x \rangle \neq \langle x'_0, y \rangle$, i.e., the dual of X, X', separates the points of X.*

Proof: Set $M = \{0\}$ in Theorem 2.

Corollary 4 insures that the dual of a NLS is rich in continuous linear functionals. Such phenomena does not hold in general for metric linear spaces.

Example 5 Consider $X = L^0[0,1]$ (Example 5.1.1). Suppose f is a non-zero linear functional on X. Let $\varphi \in X$ be such that $\langle f, \varphi \rangle \neq 0$. Let $J_1 = [0, 1/2]$, $J_2 = (1/2, 1]$ and set $g_i = C_{J_i}\varphi$. Then $\varphi = g_1 + g_2$ so either $\langle f, g_1 \rangle \neq 0$ or $\langle f, g_2 \rangle \neq 0$. Choose one and label it φ_1. Note $m\{t : \varphi_1(t) \neq 0\} \leq 1/2$.

Continue this bisection procedure to produce a sequence $\{\varphi_j\} \subset X$ such that $\langle f, \varphi_j \rangle \neq 0$ and $m\{t : \varphi_j(t) \neq 0\} \leq 1/2^j$. Put $h_j = \varphi_j / \langle f, \varphi_j \rangle$ so $\langle f, h_j \rangle = 1$ for all j but $h_j \to 0$ in m-measure since $m\{t : h_j(t) \neq 0\} \leq 1/2^j$. Thus, f is not continuous at 0. Hence, the dual space of X is $\{0\}$.

The dual norm of an element x' in the dual of a NLS X is defined to be the supremum of the values $|\langle x', x \rangle|$ as x varies over the unit ball $\{x : \|x\| \leq 1\}$ of X. We can use Corollary 4 to establish a result dual to this; the norm of an element $x \in X$ can be found by computing the supremum of $|\langle x', x \rangle|$ as x' varies over the unit ball of X'.

Corollary 6 *For $x \in X$, a NLS, $\|x\| = \sup\{|\langle x', x \rangle| : \|x'\| \leq 1\}$.*

Proof: If $x' \in X'$, $\|x'\| \leq 1$, then $|\langle x', x \rangle| \leq \|x\|$. On the other hand, by Corollary 4 there exists $x' \in X'$ with $\|x'\| = 1$ and $\langle x', x \rangle = \|x\|$.

As another application of Corollary 4 we establish the converse of Theorem 5.2.3.

Theorem 7 *Let X, Y be NLS with $X \neq \{0\}$. If $L(X,Y)$ is complete, then Y is complete.*

Proof: Let $\{y_k\}$ be Cauchy in Y. Choose $x_0 \in X$, $\|x_0\| = 1$. By Corollary 4 there exists $x_0' \in X'$ such that $\|x_0'\| = 1$ and $\langle x_0', x_0 \rangle = 1$. Define $T_k \in L(X,Y)$ by $T_k x = \langle x_0', x \rangle y_k$. Then
$$\|(T_k - T_j)x\| \leq \|x\| \|y_k - y_j\|$$
so
$$\|T_k - T_j\| \leq \|y_k - y_j\|$$
and $\{T_k\}$ is Cauchy in $L(X,Y)$. Suppose $T_k \to T$ in $L(X,Y)$. Then
$$\|y_k - Tx_0\| = \|T_k x_0 - T x_0\| \leq \|T_k - T\| \text{ implies } y_k \to Tx_0.$$

The Canonical Imbedding and Reflexivity:

Let X be a NLS. Let X'' be the dual of X' (with the dual norm) and assume that X'' carries its dual norm from X'. X'' is called the *second dual* or *bidual* of X. Each $x \in X$ induces an element $\hat{x} \in X''$ defined by $\langle \hat{x}, x' \rangle = \langle x', x \rangle$ for $x' \in X'$. $\hat{x}$ is obviously linear and by Corollary 6,
$$\|\hat{x}\| = \sup\{|\langle x', x \rangle| : \|x'\| \leq 1\} = \|x\|$$

5.6. THE HAHN-BANACH THEOREM

so $\hat{x} \in X''$ and $\|\hat{x}\| = \|x\|$. Thus, the map $J_X : X \to X''$ defined by $J_X x = \hat{x}$ is a linear isometry which imbeds X into its bidual X''; J_X is called the *canonical map* or canonical imbedding of X into its bidual, and if X is understood, we write $J = J_X$. A NLS is called *reflexive* if $J_X X = X''$. Note from Theorem 5.2.3 any reflexive space must be a B-space. It should also be noted that for a B-space X to be reflexive, X and X'' must be linearly isometric under the canonical imbedding J_X; R.C. James has given an example of a non-reflexive B-space X which is linearly isometric to its bidual X''.

R^n is obviously reflexive; examples of reflexive B-spaces are given in subsequent chapters. An example of a non-reflexive B-space is given below (Example 9). An interesting consequence of Corollary 6 is given by

Corollary 8 *Let X be a reflexive B-space. Then every continuous linear functional $x' \in X'$ attains its maximum on the unit ball of X.*

Proof: By Corollary 6 there exists $x'' \in X''$ such that $\|x''\| = 1$ and $\langle x'', x' \rangle = \|x'\|$. But there exists $x \in X$ such that $J_X x = x''$ so $\|x\| = 1$ and $\langle x', x \rangle = \|x'\|$.

It is an interesting result of James that the converse of Corollary 8 holds [see [J]]. We can use Corollary 8 to give an example of a non-reflexive space.

Example 9 c_0 is not reflexive. Define $f : c_0 \to \mathbf{R}$ by $f(\{t_j\}) = \sum_{j=1}^{\infty} t_j/j!$. Then $f \in c_0'$ and $\|f\| = \sum_{j=1}^{\infty} 1/j!$. However, there is no $\{t_j\} \in c_0$ with $\sum_{j=1}^{\infty} t_j/j! = \|f\|$. By Corollary 8 c_0 is not reflexive.

As an application of the canonical imbedding, we have

Theorem 10 *Every NLS X is a dense subspace of a B-space $\widetilde{X}$ (i.e., every NLS has a completion).*

Proof: Set $\widetilde{X} = \overline{JX} \subset X''$ [where we are identifying X and $J_X X$ under the linear isometry J_X].

As a further application of the canonical imbedding, we use the Uniform Boundedness Principle to derive a boundedness condition for NLS.

Theorem 11 *A subset B of a NLS X is bounded if and only if $x'(B)$ is bounded for each $x' \in X'$.*

Proof: $\Rightarrow$: This follows from the inequality $|x'(x)| \le \|x'\| \|x\|$.
$\Leftarrow$: Let J be the canonical imbedding of X into its bidual. Then $\{Jb : b \in B\}$ is pointwise bounded on X', and since X' is complete, $\{\|JB\| = \|b\| : b \in B\}$ is bounded by the UBP (5.3.1).

Several additional properties of reflexive spaces are given in the exercises.

Separability:
We consider some separability results concerning a NLS and its dual.

Theorem 12 *If the dual of a NLS X is separable, then X is separable.*

Proof: Let $\{x'_k\}$ be dense in X'. For each k choose $x_k \in X$ such that $\|x_k\| \leq 1$ and $|\langle x'_k, x_k\rangle| \geq \|x'_k\|/2$. The subspace X_1 spanned by $\{x_k\}$ is separable (Exercise 8), and we claim that X_1 is dense in X. If this is not the case, by Theorem 2, there exists $x' \in X'$, $x' \neq 0$, such that $\langle x', X_1\rangle = 0$. There exists a subsequence $\{x'_{n_k}\}$ converging to x'. Then
$$\left\|x'_{n_k} - x'\right\| \geq \left|\langle x'_{n_k} - x', x_{n_k}\rangle\right| = \left|\langle x'_{n_k}, x_{n_k}\rangle\right| \geq \left\|x'_{n_k}\right\|/2.$$
Letting $k \to \infty$ gives $0 \geq \|x'\|/2$ so $x' = 0$; the desired contradiction.

The converse of Theorem 12 is false; see Exercise 6. However, for reflexive spaces, we have

Corollary 13 *Let X be a reflexive B-space. Then X is separable $\Leftrightarrow$ X' is separable.*

Proof: $\Leftarrow$: Theorem 12.
$\Rightarrow$: $J_X X = X''$ is separable so X' is separable by Theorem 12.

Exercise 1. Show that a B-space X is reflexive if and only if X' is reflexive.

Exercise 2. Show that a closed linear subspace of a reflexive space is reflexive.

Exercise 3. If X is reflexive and X' contains a countable set which separates the points of X, show X' is separable.

Exercise 4. If X is a NLS, show $J_X X$ separates the points of X'.

Exercise 5. If X is reflexive, show X' has no proper closed subspaces which separate the points of X.

Exercise 6. Show the converse of Theorem 12 is false. [Hint: Exercise 5.1.13.]

Exercise 7. Let X, Y be NLS and $T \in L(X,Y)$. Show that $T'y' = y'T$ defines a linear operator T' from Y' into X' which is continuous and $\|T\| = \|T'\|$. T' is called the *adjoint* or *transpose* of T.

Exercise 8. Let D be a countable subset of a NLS X. If X_1 is the subspace spanned by D, show X_1 and $\overline{X}_1$ are separable.

5.6. THE HAHN-BANACH THEOREM

5.6.2 Extension of Bounded, Finitely Additive Set Functions

We give an application, due to B.J. Pettis, of the Hahn-Banach Theorem to the extension of bounded, finitely additive set functions defined on algebras.

Let $\mathcal{A}$ be an algebra of subsets of S and $\sum$ the σ-algebra generated by $\mathcal{A}$. Let $\mathcal{S}(\mathcal{A})$ [$\mathcal{S}(\sum)$] be the vector space of all real-valued $\mathcal{A}$-simple [$\sum$-simple] functions; we assume that $\mathcal{S}(\mathcal{A})$ [$\mathcal{S}(\sum)$] is equipped with the sup-norm, $\|\varphi\| = \sup\{|\varphi(t)| : t \in S\}$. Let $\mu : \mathcal{A} \to \mathbf{R}$ be a bounded, finitely additive set function. We consider the possibility of extending μ to $\sum$.

Let $f : \mathcal{S}(\mathcal{A}) \to \mathbf{R}$ be the linear functional induced by integration with respect to μ, $\langle f, \varphi \rangle = \int_S \varphi d\mu$ [Remark 3.2.2]. Since $|\langle f, \varphi \rangle| \leq \|\varphi\| |\mu|(S)$ [Remark 3.2.2], f is continuous and $\|f\| \leq |\mu|(S)$. By Theorem 5.6.1.1, f has a continuous linear extension, F, to $\mathcal{S}(\sum)$ satisfying $|\langle F, \varphi \rangle| \leq \|f\| \|\varphi\|$ for $\varphi \in \mathcal{S}(\sum)$. Define $\overline{\mu} : \sum \to \mathbf{R}$ by $\overline{\mu}(E) = \langle F, C_E \rangle$ for $E \in \sum$. $\overline{\mu}$ obviously is an extension of μ and is finitely additive. Since

$$|\overline{\mu}(E)| \leq \|F\| = \|f\| \leq |\mu|(S),$$

$\overline{\mu}$ is bounded and gives a bounded, finitely additive extension of μ to the σ-algebra $\sum$ generated by $\mathcal{A}$.

In contrast to the situation in §2.4 where we considered the extension of premeasures, a bounded, finitely additive set function defined on an algebra can have an infinite number of bounded, finitely additive extensions to the generating σ-algebra [see [HY]].

Exercise 1. Show $\|f\| = |\mu|(S)$.

Exercise 2. Show the existence of a bounded, finitely additive set function defined on a σ-algebra.

5.6.3 A Translation Invariant, Finitely Additive Set Function

We show that the Hahn-Banach Theorem can be used to show that the "easy" problem of measure discussed in §1.3 has a solution. That is, we show the existence of a non-negative, translation invariant, finitely additive set function defined on the bounded subsets of **R**. We begin by showing the existence of a Banach integral and then use the integral to construct the desired translation invariant, finitely additive set function on the power set of **R**.

Let P be the space of all bounded real-valued functions defined on **R** which have period 1 equipped with the sup norm, $\|\|$. For $f \in P$ and $t_1, \ldots, t_n \in \mathbf{R}$ set

$$u(f : t_1, \ldots, t_n) = \sup\{\sum_{k=1}^{n} f(t + t_k)/n : t \in \mathbf{R}\}$$

and

$$p(f) = \inf\{u(f : t_1, \ldots, t_n) : t_1, \ldots, t_n \in \mathbf{R}\}.$$

Note p is finite since $p(f) \leq \|f\|_\infty < \infty$.

We show that p is sublinear. Clearly, p is positive homogeneous. We show that p is subadditive. Let $\epsilon > 0$ and $f_1, f_2 \in P$. Pick $s_1, \ldots, s_m$ and $t_1, \ldots, t_n$ such that

$$u(f_1 : s_1, \ldots, s_m) < p(f_1) + \epsilon, \quad u(f_2 : t_1, \ldots, t_n) < p(f_2) + \epsilon.$$

Put $r_{ij} = s_i + t_j$ for $i = 1, \ldots, m$, $j = 1, \ldots, n$. Then

$$u(f_1 + f_2 : r_{11}, r_{12}, \ldots, r_{mn}) = \frac{1}{mn} \sup\left\{\sum_{i,j}\{f_1(t + r_{ij}) + f_2(t + r_{ij})\} : t\right\}$$

$$\leq \frac{1}{mn} \sup\left\{\sum_{i,j} f_1(t + r_{ij}) : t\right\} + \frac{1}{mn} \sup\left\{\sum_{i,j} f_2(t + r_{ij}) : t\right\}$$

$$\leq \frac{1}{n} \sum_{j=1}^{n} \sup\left\{\sum_{i=1}^{m} f_1(t + t_j + s_i)/m : t\right\} + \frac{1}{m} \sum_{i=1}^{m} \sup\left\{\sum_{j=1}^{n} f_2(t + s_i + t_j)/n : t\right\}$$

$$= u(f_1 : s_1, \ldots, s_m) + u(f_2 : t_1, \ldots, t_n) \leq p(f_1) + p(f_2) + 2\epsilon \quad (5.1)$$

so $p(f_1 + f_2) \leq p(f_1) + p(f_2)$.

By the Hahn-Banach Theorem [5.6.2], there exists a linear functional $F : P \to \mathbf{R}$ such that $F(f) \leq p(f)$ for all $f \in P$ [define F on $\{0\}$ by $F(0) = 0$ and apply 5.6.2]. Note $F(-f) = -F(f) \leq p(-f)$ so

$$-p(-f) \leq F(f) \leq p(f) \text{ for } f \in P. \quad (5.2)$$

5.6. THE HAHN-BANACH THEOREM

Now F is positive [i.e., $F(f) \geq 0$ for $f \geq 0$] since $f \geq 0$ implies $p(f) \geq 0$ and $p(-f) \leq 0$ so by (2) $F(f) \geq -p(-f) \geq 0$.

Next, we claim that F is translation invariant. Fix $h \in \mathbf{R}$. If $f \in P$, set $g(t) = f(t+h) - f(t)$ for $t \in \mathbf{R}$. We need to show that $F(g) = 0$. Take $t_k = (k-1)h$ for $k = 1, \ldots, n+1$. Then

$$p(g) \leq u(g : t_1, \ldots, t_{n+1}) = \frac{1}{n+1} \sup\{f(t + (n+1)h) - f(t) : t\} \to 0$$

as $n \to \infty$ so $p(g) \leq 0$. Similarly, $p(-g) \leq 0$. From (2), $F(g) = 0$ as desired.

Also, $F(1) = 1$ from (2) since $p(1) = 1$ and $p(-1) = -1$. And, F is bounded since from (2), $F(f) \leq p(f) \leq \|f\|$ and

$$F(-f) = -F(f) \leq p(-f) \leq \|-f\| = \|f\|$$

so $|F(f)| \leq \|f\|$ and $\|F\| \leq 1$.

We now use F, the Banach integral, to induce a finitely additive set function. If $E \subset [0,1)$, let k_E be the periodic extension of C_E to $\mathbf{R}$ and set $\mu(E) = F(k_E)$. Since F is linear, μ is finitely additive on the power set, $\mathcal{P}[0,1)$, μ is bounded (by 1) since $\|F\| \leq 1$, μ is positive since F is positive and $F(1) = \mu[0,1) = 1$. Moreover, if A and B are subsets of $[0,1)$ such that $A = B + t$ for some t, then $\mu(A) = \mu(B)$ by the translation invariance of F, i.e., μ is translation invariant in this sense.

We extend μ to the bounded subsets of $\mathbf{R}$. First, consider any subset A of an interval of the form $[j, j+1)$. Then $A - j \subset [0,1)$ so we may define $\mu(A)$ to be $\mu(A - j)$. If $B \subset \mathbf{R}$ is bounded, then $B \subseteq [-n, n)$ for some n, and we may write

$$B = \bigcup_{j=-n}^{n-1} B \cap [j, j+1)$$

and define

$$\mu(B) = \sum_{j=-n}^{n-1} \mu(B \cap [j, j+1)).$$

It is easy to check that this extension of μ has values in $[0, \infty)$, is finitely additive, translation invariant, and $\mu[0,1) = 1$.

Note that the commutativity of the group of translations was used in the computation (1). This is important; Hausdorff has shown that there is no non-trivial, non-negative, finitely additive set defined on the subsets of the unit sphere in $\mathbf{R}^3$ which is invariant under the (non-abelian) group of rotations on the sphere.

Exercise 1. Show the set function $\bar{\mu}(E) = (\mu(E) + \mu(-E))/2$ has all of the properties of μ and is also invariant under reflection. [Hence, $\bar{\mu}$ is invariant under the isometries of $\mathbf{R}$ ([Na]).]

Exercise 2. A *Banach limit* is a continuous linear functional L on ℓ^∞ satisfying

(i) $L(x) \geq 0$ if $x \geq 0$ [i.e., if $x_k \geq 0$ for all k when $x = \{x_k\}$],

(ii) if $\tau x = (x_2, x_3, \ldots)$, then $L(\tau x) = L(x)$,

(iii) $L((1,1,\ldots)) = 1$.

(a) Show that if L is a Banach limit, $\underline{\lim} x_k \leq L(x) \leq \overline{\lim} x_k$ for $x \in \ell^\infty$. Hint: For $\varepsilon > 0$ choose n such that
$$\inf x_k \leq x_n < \inf x_k + \varepsilon$$
so $x_k + \varepsilon - x_n \geq 0$. Use (i) and (iii) to show $L(x) \geq \inf x_k$. Then use (ii).

(b) Show Banach limits exist. Hint: Use the functional $p(x) = \overline{\lim}(x_1 + \cdots + x_n)/n$ of Exercise 5.6.2 and the functional $Lx = \lim x_k$ for $x \in c$.

Exercise 3. For $A \subset \mathbf{N}$ let $|A|$ be the number of points in A. If L is a Banach limit, show $\mu(A) = L(|A \cap \{1,\ldots,n\}|/n)$ defines a bounded finitely additive set function on $\mathcal{P}(\mathbf{N})$.

Exercise 4. Let $\mathcal{A}$ be an algebra of subsets of S and $T: S \to S$ such that $T^{-1}A \in \mathcal{A}$ for $A \in \mathcal{A}$. Let ν be a bounded, finitely additive set function on $\mathcal{A}$. Let L be a Banach limit and define μ on $\mathcal{A}$ by $\mu(A) = L(\nu(T^{-n}A))$. Show μ is bounded, finitely additive and T-invariant (i.e., $\mu(A) = \mu(T^{-1}A)$ for $A \in \mathcal{A}$).

5.7 Ordered Linear Spaces

As well as having natural norms or metrics, many of the classical spaces of functions also have natural orders. In this section we consider basic properties of ordered spaces.

A *partial order* on a set E is a relation, $\leq$, satisfying the following properties:

(R) $x \leq x$ holds for all $x \in E$ (reflexive),

(A) if $x \leq y$ and $y \leq x$, then $x = y$ (anti-symmetry),

(T) if $x \leq y$ and $y \leq z$, then $x \leq z$ (transitive).

We write $y \geq x$ if and only if $x \leq y$.

An *ordered vector space* is a real vector space X with a partial order, $\leq$, which is compatible with the algebraic operations, i.e.,

(i) if $x \leq y$ and $z \in X$, then $x + z \leq y + z$,

(ii) if $x \leq y$ and $t \geq 0$, then $tx \leq ty$.

An element x of X is said to be *positive* if $x \geq 0$; the set of all positive elements of X is denoted by $\underline{X}^+$.

Example 1 The space s has the natural order $\{x_k\} \leq \{y_k\}$ if and only if $x_k \leq y_k$ for all $k \in \mathbf{N}$. The subspaces c_{00}, c_0, c, ℓ^1 and ℓ^∞ inherit this natural order from s.

More generally, we have

Example 2 Let $S \neq \emptyset$ and $\mathcal{F}(S)$ the vector space of all real-valued functions defined on S [the operations of addition and scalar multiplication are defined pointwise]. If $f, g \in \mathcal{F}(S)$, we define $f \leq g$ if and only if $f(t) \leq g(t)$ for all $t \in S$. Any vector subspace of $\mathcal{F}(S)$ is an ordered vector space under the order inherited from $\mathcal{F}(S)$.

If X is an ordered vector space and $x, y \in X$, then x and y have a *supremum* (*infimum*), denoted by $x \vee y$ ($x \wedge y$), if $x \leq x \vee y$, $y \leq x \vee y$ ($x \wedge y \leq x$, $x \wedge y \leq y$) and $x \leq w$, $y \leq w$ ($w \leq x$, $w \leq y$) implies $x \vee y \leq w$ ($w \leq x \wedge y$).

A *vector lattice* or a *Riesz space* is an ordered vector space X such that every two elements of X have a supremum and an infimum. We have the following elementary identities in a Riesz space.

Proposition 3 (i) $x \vee y = -[(-x) \wedge (-y)]$, $x \wedge y = -[(-x) \vee (-y)]$,

(ii) $x \vee y + z = (x + z) \vee (y + z)$, $x \wedge y + z = (x + z) \wedge (y + z)$,

(iii) $t(x \vee y) = (tx) \vee (ty)$, $t(x \wedge y) = (tx) \wedge (ty)$ *for* $t \geq 0$.

Proof: We prove (ii) and leave the remaining statements for Exercise 2.

Set $a = x \vee y + z$, $b = (x+z) \vee (y+z)$. We show $a \leq b$, $b \leq a$. First, $a - z = x \vee y$ implies $x \leq a - z$ and $y \leq a - z$ so $x + z \leq a$ and $y + z \leq a$. Hence, $(x+z) \vee (y+z) = b \leq a$. Next, $b = (x+z) \vee (y+z)$ so $b \geq x+z$ and $b \geq y+z$ or $x \leq b - z$ and $y \leq b - z$. Hence, $x \vee y \leq b - z$ or $a = x \vee y + z \leq b$.

Remark 4 From (i) it follows that if X is an ordered vector space such that every two elements of X has a supremum (infimum), then X is a Riesz space.

If X is a Riesz space and $x \in X$, set $x^+ = x \vee 0$, the *positive part* of x, and set $x^- = (-x) \vee 0 = (-x)^+$, the *negative part* of x. Then $|x| = x \vee (-x)$ is called the *absolute value* of x. We have the following properties:

Proposition 5 (i) $x = x^+ - x^-$,

(ii) $|x| = x^+ + x^-$,

(iii) $x^+ \wedge x^- = 0$,

(iv) $x \vee y = (x-y)^+ + y$,

(v) $|x+y| \leq |x| + |y|$,

(vi) $||x| - |y|| \leq |x-y|$,

(vii) $|x^+ - y^+| \leq |x-y|$, $(x+y)^+ \leq x^+ + y^+$,

(viii) $|x \vee z - y \vee z| \leq |x-y|$, $|x \wedge z - y \wedge z| \leq |x-y|$,

(ix) $|x| \leq y$ if and only if $-y \leq x \leq y$.

Proof: (i): By Proposition 3,
$$x^- + x = (-x) \vee 0 + x = 0 \vee x = x^+.$$

(ii): By Proposition 3 and (i),
$$|x| = x \vee (-x) = (2x) \vee 0 - x = 2x^+ - x = 2x^+ - (x^+ - x^-) = x^+ + x^-.$$

(iii): By Proposition 3 and (i),
$$x^+ \wedge x^- = (x^+ - x^-) \wedge 0 + x^- = x \wedge 0 + x^-$$
$$= -((-x) \vee (-0)) + x^- = -x^- + x^- = 0.$$

We leave (iv)-(vi) to Exercise 3.

(vii): Note $x^+ = (x + |x|)/2$ so using (vi),
$$|x^+ - y^+| = |(x+|x|)/2 - (y+|y|)/2| = \tfrac{1}{2}|(x-y) + (|x|-|y|)|$$
$$\leq \tfrac{1}{2}|x-y| + \tfrac{1}{2}||x|-|y|| \leq \tfrac{1}{2}|x-y| + \tfrac{1}{2}|x-y| = |x-y|.$$

5.7. ORDERED LINEAR SPACES

$$(x+y)^+ = \tfrac{1}{2}(|x+y| + (x+y))$$
$$\leq \tfrac{1}{2}(|x| + x + |y| + y)$$
$$= x^+ + y^+.$$

(viii): From Proposition 3,

$$x \vee z - y \vee z = ((x-z) \vee 0 + z) - ((y-z) \vee 0 + z) = (x-z)^+ - (y-z)^+.$$

From (vii),

$$|x \vee z - y \vee z| = \left|(x-z)^+ - (y-z)^+\right| \leq |(x-z) - (y-z)| = |x-y|.$$

From Proposition 3 and (viii),

$$|x \wedge z - y \wedge z| = |-(-x) \vee (-z) + (-y) \vee (-z)| \leq |(-y) - (-x)| = |x-y|.$$

(ix) is left to Exercise 3.

Remark 6 From (iv) [(ii)] and Remark 4, it follows that if X is an ordered vector space such that $x \vee 0 = x^+$ [$|x|$] exists for $x \in X$, then X is a Riesz space.

Example 7 $\mathcal{F}(S)$ is a Riesz space with

$$f \vee g(t) = \max\{f(t), g(t)\}, f \wedge g(t) = \min\{f(t), g(t)\}$$

for $t \in S$. In particular, s is a Riesz space. Any vector subspace of $\mathcal{F}(S)$ which is a Riesz space under the inherited order from $\mathcal{F}(S)$ is called a *function space*. For example, c_{00}, c_0, c, ℓ^1, ℓ^∞ are all function spaces. If S is a compact Hausdorff space, then $C(S)$ is a function space; if $C^k[0,1]$ is the vector space of all functions $\varphi : [0,1] \to \mathbf{R}$ which have k continuous derivatives, then $C^k[0,1]$ is an ordered vector subspace of $C[0,1]$ but is not a function space for $k \geq 1$.

Example 8 Let $BV[a,b]$ be the vector subspace of $\mathcal{F}[a,b]$ consisting of the functions of bounded variation (A.1). $BV[a,b]$ is a function space.

If X, Y are Riesz spaces and $T : X \to Y$ is linear, then T is called a *lattice homomorphism* if $T(x \vee y) = (Tx) \vee (Ty)$ for all $x, y \in X$; if T is 1-1, T is called a *lattice isomorphism*. For equivalent conditions, see Exercise 4.

A linear map $T : X \to Y$ between two ordered vector spaces is said to be *positive* if $Tx \geq 0$ for every $x \geq 0$. A lattice homomorphism between Riesz spaces is positive.

Theorem 9 Let X, Y be Riesz spaces and $T : X \to Y$ linear, 1-1, onto. If T and T^{-1} are both positive, then T is a lattice isomorphism (onto Y).

Proof: Let $x, y \in X$. Since $x \leq x \vee y$, $y \leq x \vee y$ and T is positive, $Tx \leq T(x \vee y)$, $Ty \leq T(x \vee y)$ so $(Tx) \vee (Ty) \leq T(x \vee y)$. By symmetry, since T^{-1} is positive,

$$x \vee y = (T^{-1}(Tx)) \vee (T^{-1}(Ty)) \leq T^{-1}((Tx) \vee (Ty)).$$

Hence, $T(x \vee y) \leq (Tx) \vee (Ty)$.

We next consider the order analogue of the dual of a NLS. If X is a Riesz space, a subset $B \subset X$ is *order bounded* if there exists y such that $|x| \leq y$ for all $x \in B$. If Y is another Riesz space, a linear map $T : X \to Y$ is *order bounded* if T carries order bounded sets to order bounded sets and T is *positive* if T carries positive elements to positive elements.

Let $X\tilde{\ }$ be the set of all order bounded (real-valued) linear functionals on X. Then $X\tilde{\ }$ is a vector space under pointwise addition and scalar multiplication and is ordered by the relation, $f \leq g$ if and only if $f(x) \leq g(x)$ for all $x \in X^+$. $X\tilde{\ }$ is called the *order dual* of X. It is easily checked that $X\tilde{\ }$ is an ordered vector space under this order and F. Riesz has shown that $X\tilde{\ }$ is actually a Riesz space. For this important result we first establish a lemma.

Lemma 10 (Riesz Decomposition) *Let X be a Riesz space. If $x, y, z \in X^+$ and $0 \leq z \leq x + y$, then there exist $x_1, y_1 \in X$ such that $0 \leq x_1 \leq x$, $0 \leq y_1 \leq y$ and $z = x_1 + y_1$.*

Proof: Set $x_1 = x \wedge z$, $y_1 = z - x_1 = z - x \wedge z$. Hence, $0 \leq x_1 \leq x$, $x_1 + y_1 = z$, $0 \leq y_1$. Also, from Proposition 3,

$$0 \leq y_1 = z - x \wedge z = z + (-x) \vee (-z) = (z - x) \vee 0 \leq y \vee 0 = y.$$

Ordered vector spaces which satisfy the conditions in Lemma 10 are said to have the *decomposition property*. Thus, a Riesz space has the decomposition property, but there are ordered vector spaces with the decomposition property which are not Riesz spaces [see [P] p. 14].

Theorem 11 (F. Riesz) *If X is a Riesz space, then $X\tilde{\ }$ is a Riesz space. Moreover, if $f \in X\tilde{\ }$, $x \in X^+$, then*

(i) $f^+(x) = \sup\{f(y) : 0 \leq y \leq x\}$,

(ii) $f^-(x) = \sup\{-f(y) : 0 \leq y \leq x\}$,

(iii) $|f|(x) = \sup\{f(y) : |y| \leq x\}$.

Proof: From Remark 6, it suffices to show that $f^+ = f \vee 0$ exists for every $f \in X\tilde{\ }$. Let $f \in X\tilde{\ }$ and for $x \in X^+$ set $g(x) = \sup\{f(y) : 0 \leq y \leq x\}$ [(i)]; note that since f is order bounded, $g(x)$ is finite. Clearly

$$g(x) \geq 0,\ g(x) \geq f(x) \text{ and } g(tx) = tg(x) \text{ for } t \geq 0,\ x \in X^+. \tag{5.1}$$

We claim that $g(x+y) = g(x) + g(y)$ for $x, y \in X^+$. If $0 \leq x_1 \leq x$, $0 \leq y_1 \leq y$, then $0 \leq x_1 + y_1 \leq x + y$ so $f(x_1) + f(y_1) = f(x_1 + y_1) \leq g(x+y)$ and $g(x) + g(y) \leq g(x+y)$. On the other hand, if $0 \leq z \leq x+y$, by Lemma 10 there exist $0 \leq x_1 \leq x$, $0 \leq y_1 \leq y$

5.7. ORDERED LINEAR SPACES

such that $z = x_1 + y_1$. Then $f(z) = f(x_1 + y_1) = f(x_1) + f(y_1) \leq g(x) + g(y)$ so $g(x+y) \leq g(x) + g(y)$.

For arbitrary $x \in X$, $x = x^+ - x^-$ so we can extend g to X by setting $g(x) = g(x^+) - g(x^-)$. Note that if $x = u - v$ with $u \geq 0$, $v \geq 0$, then $x^+ + v = x^- + u$ so

$$g(x^+ + v) = g(x^+) + g(v) = g(x^-) + g(u)$$

or

$$g(x^+) - g(x^-) = g(u) - g(v),$$

and the value of g does not depend upon how x is represented as the difference of two positive elements of X. From this and (1), it is easy to verify that g is a positive linear functional on X.

We claim that $g = f^+$ ($= f \vee 0$). If $h \in X^\tilde{}$ is positive with $f \leq h$, then $f(y) \leq h(y) \leq h(x)$ for $0 \leq y \leq x$ so $g(x) \leq h(x)$. From (1), $g = f^+$.

Formula (i) was established above. The remaining two formulas follow from (i) and $f^- = (-f)^+$ and $|f| = f^+ + f^-$.

Corollary 12 *If $f \in X^\tilde{}$ and $x \in X$, then $|f(x)| \leq |f|(|x|)$.*

Proof: $|f|(|x|) = \sup\{f(y) : |y| \leq |x|\} \geq (f(x)) \vee (f(-x))$.

Since $f \vee g = (f - g)^+ + g$ [Proposition 5], for $f, g \in X^\tilde{}$ and $x \in X^+$ we have

(2) $(f \vee g)(x) = \sup\{f(y) + g(x - y) : 0 \leq y \leq x\}$ and

(3) $(f \wedge g)(x) = \inf\{f(y) + g(x - y) : 0 \leq y \leq x\}$.

We have formulas "dual" to the formulas in Theorem 11.

Theorem 13 *Let X be a Riesz space, $f \in X^\tilde{}$ positive and $x \in X$. Then*

(i) $f(x^+) = \sup\{g(x) : g \in X^\tilde{}, 0 \leq g \leq f\}$,

(ii) $f(x^-) = \sup\{-g(x) : g \in X^\tilde{}, 0 \leq g \leq f\}$,

(iii) $f(|x|) = \sup\{g(x) : g \in X^\tilde{}, |g| \leq f\}$.

Proof: (i): If $0 \leq g \leq f$, then $g(x) \leq g(x^+) \leq f(x^+)$ so

$$\sup\{g(x) : g \in X^\tilde{}, 0 \leq g \leq f\} \leq f(x^+).$$

Conversely, define $p : X \to \mathbf{R}$ by $p(u) = f(u^+)$. Then p is a sublinear map on X [Proposition 5] such that $p(u) \geq 0$ for all $u \in X$. Let $Y = \{tx : t \in \mathbf{R}\}$ and define h on Y by $h(tx) = tf(x^+)$. Then h is linear on Y and $h(y) \leq p(y)$ for all $y \in Y$. By the Hahn-Banach Theorem, h has a linear extension to X, still denoted by h, such that $h(u) \leq p(u)$ for all $u \in X$. If $u \geq 0$, then $h(u) \leq p(u) = f(u)$ and

$$-h(u) = h(-u) \leq p(-u) = f((-u)^+) = f(0) = 0$$

so $0 \leq h(u) \leq f(u)$ for all $u \in X^+$. Hence, $h \in X^\sim$ [Exercise 5], $0 \leq h \leq f$, and

$$f(x^+) = h(x) \leq \sup\{g(x) : g \in X^\sim, 0 \leq g \leq f\}.$$

Formulas (ii) and (iii) follow from $f(x^-) = f((-x)^+)$ and $f(|x|) = f(x^+) + f(x^-)$.

We now consider NLS which have an order defined on them. A semi-norm (norm) on a Riesz space X is a *lattice semi-norm (norm)* if $|x| \leq |y|$ implies $\|x\| \leq \|y\|$. A *semi-normed (normed) vector lattice* is a Riesz space with a lattice semi-norm (norm); a complete normed vector lattice is called a *Banach lattice*.

Example 14 $B(S)$, $C(S)$, ℓ^∞, c, c_0 are Banach lattices.

Example 15 c_{00} is a normed vector lattice which is not a Banach lattice.

Example 16 $L^1(\mu)$ is a complete, semi-normed vector lattice.

For an example of a norm on a Riesz space which is not a lattice norm see Exercise 7.

Proposition 17 *Let X be a semi-normed vector lattice.*

(i) $\|\,|x|\,\| = \|x\|$,

(ii) $\|x^+ - y^+\| \leq \|x - y\|$ so $x \to x^+$ is uniformly continuous on X,

(iii) $\|\,|x| - |y|\,\| \leq \|x - y\|$ so $x \to |x|$ is uniformly continuous on X.

Proof: (i) is easy. By Proposition 5, $|x^+ - y^+| \leq |x - y|$ and $||x| - |y|| \leq |x - y|$ so (ii), (iii) follow.

A linear subspace Y of a vector lattice is called a *vector sublattice* if $x \wedge y$, $x \vee y \in Y$ when $x, y \in Y$. A linear subspace A of a vector lattice X is called an *ideal (order ideal)* if $|x| \leq |y|$ and $y \in A$ implies $x \in A$. Since

$$x \vee y = \frac{1}{2}(x + y + |x - y|),$$

every order ideal is a vector sublattice.

Example 18 $B(S)$ is an ideal in $\mathcal{F}(S)$. $C(S)$ is a vector sublattice of $\mathcal{F}(S)$ but is not an ideal. Similarly, c_0 and c_{00} are order ideals in ℓ^∞ but c is not.

We now compare the order dual and the norm dual of a normed vector lattice.

Theorem 19 *Let X be a normed vector lattice. Then $X' \subset X^\sim$ and, moreover, X' is an order ideal in $X^\sim$.*

5.7. ORDERED LINEAR SPACES

Proof: Let $B \subset X$ be order bounded with $|x| \leq y$ for all $x \in B$. Then $\|x\| \leq \|y\|$ for all $x \in B$ so B is norm bounded. Hence, every $x' \in X'$ carries order bounded sets to bounded subsets and $x' \in X\tilde{}$.

Assume $x\tilde{} \in X\tilde{}$, $x' \in X'$ with $|x\tilde{}| \leq |x'|$. We must show $x\tilde{} \in X'$. Let $x \in X$, $\|x\| \leq 1$. Then for every $y \in X$ with $|y| \leq |x|$, $\|y\| \leq \|x\| \leq 1$ so

$$|x\tilde{}(x)| \leq |x\tilde{}|(|x|) \leq |x'|(|x|) = \sup\{x'(y) : |y| \leq |x|\} \leq \|x'\|$$

[Corollary 12 and Theorem 11]. Thus, $x\tilde{} \in X'$ with $\|x\tilde{}\| \leq \|x'\|$.

Remark 20 It follows from Theorem 19 that if $x' \in X'$, then x' is the difference of two positive, continuous linear functionals $(x')^+$, $(x')^-$.

The argument above shows that $|x'| \leq |y'|$ in X' implies $\|x'\| \leq \|y'\|$. Hence,

Theorem 21 *The norm dual of a normed vector lattice is a Banach lattice.*

The containment in Theorem 19 can be proper.

Example 22 Define $f : c_{00} \to \mathbf{R}$ by $f(\{t_j\}) = \sum_{j=1}^{\infty} jt_j$ [finite sum]. Then f is positive and, hence, order bounded [Exercise 5]. But, $f(e^j) = j$ so f is not continuous.

However, we do have

Theorem 23 *If X is a Banach lattice, $X' = X\tilde{}$.*

Proof: Suppose there exists $f \in X\tilde{}$ such that f is not continuous. Then there exist $x_k \in X$ such that $\|x_k\| = 1$ and $|f(x_k)| > k^3$.
Set $y_n = \sum_{k=1}^{n} |x_k|/k^2$. Then

$$\|y_{n+p} - y_n\| \leq \sum_{k=n+1}^{n+p} 1/k^2$$

so $\{y_k\}$ is Cauchy in X and, therefore, converges to some $y \in X$. Clearly, $0 \leq y_n \leq y_{n+1}$, and we claim that $y_n \leq y$. First,

$$0 \leq (y_n - y)^+ \leq (y_{n+p} - y)^+ \leq |y_{n+p} - y|$$

implies $\|(y_n - y)^+\| \leq \|y_{n+p} - y\|$ for all n, p so

$$0 \leq \|(y_n - y)^+\| \leq \lim_p \|y_{n+p} - y\| = 0.$$

Hence, $(y_n - y)^+ = 0$. But, $y_n - y \leq (y_n - y)^+ = 0$ implies $0 \leq y_n \leq y$ as claimed.
Then $y \geq y_n \geq |x_n|/n^2$ implies

$$|f|(y) \geq |f|(y_n) \geq |f|(|x_n|)/n^2 \geq |f(x_n)|/n^2 > n$$

which gives the desired contradiction.

Theorem 24 *The completion of a normed vector lattice X is a Banach lattice.*

Proof: Recall the completion of X is $\overline{J_X X} \subset X''$ [Theorem 5.6.1.10] so by Exercise 8 it suffices to show that the canonical imbedding $J_X = J$ of X into X'' preserves the lattice operations. For this it suffices to show that $(Jx)^+ = J(x^+)$ for $x \in X$.

Suppose $x \in X$ and $f \in X'$, $f \geq 0$. We write $Jx = \hat{x}$. From Theorems 11, 13 and 19, we have
$$\begin{aligned}(\hat{x})^+(f) &= \sup\{\hat{x}(g) : g \in X', 0 \leq g \leq f\} \\ &= \sup\{g(x) : g \in X', 0 \leq g \leq f\} \\ &= \sup\{g(x) : g \in X\tilde{\ }, 0 \leq g \leq f\} \\ &= f(x^+) = (x^+)\tilde{\ }(f).\end{aligned}$$
That is, $(Jx)^+ = J(x^+)$ as desired.

Exercise 1. Show X^+ is a convex cone, i.e., is convex and $x \in X^+$, $t \geq 0$ implies $tx \in X^+$.

Exercise 2. Complete the proof of Proposition 3.

Exercise 3. Prove (iv)-(vi), (ix) of Proposition 5.

Exercise 4. Let X, Y be Riesz spaces and $T : X \to Y$ linear. Show that the following are equivalent:

(i) $T(x \vee y) = (Tx) \vee (Ty)$,

(ii) $T(x \wedge y) = (Tx) \wedge (Ty)$,

(iii) $(Tx) \wedge (Ty) = 0$ when $x \wedge y = 0$ in X,

(iv) $|Tx| = T(|x|)$.

Exercise 5. If $T : X \to Y$ is positive, show T is *monotone* in the sense that $x \leq y$ implies $Tx \leq Ty$. Show every positive operator is order bounded.

Exercise 6. If X is a semi-normed vector lattice show $x \to x^-$ is uniformly continuous.

Exercise 7. Let $BV_0[a, b]$ be the linear subspace of $BV[a, b]$ which consists of the functions which vanish at a. Show $BV_0[a, b]$ is a Riesz space under the pointwise order. Show the variation norm, $\|f\| = Var(f : [a, b])$ is not a lattice norm on $BV_0[a, b]$. [See Appendix 1 for the notation and definitions.]

5.7. ORDERED LINEAR SPACES

Exercise 8. Show that the closure of a vector sublattice of a normed vector lattice is a vector sublattice.

Exercise 9. If X is a normed vector lattice, show X^+ is closed.

Exercise 10. Let X be a normed vector lattice and $\{x_k\} \subset X$ satisfy $x_k \leq x_{k+1}$. If $\|x_k - x\| \to 0$, show $x = \sup\{x_k : k\}$. Hint: $(x_n - x_m)^+ = 0$ for $m > n$.

Exercise 11. Let X be a normed vector lattice. Show $x \in X$ is positive if and only if $f(x) \geq 0$ for every positive, continuous linear functional f on X. Hint: Theorem 13.(ii).

Exercise 12. Let X be a Banach lattice and $x \geq 0$, $x \in X$. Show
$$\|x\| = \sup\{f(x) : 0 \leq f \in X', \|f\| = 1\}.$$

Exercise 13. Let X, Y be normed vector lattices and $T : X \to Y$ positive. If X is a Banach lattice, show T is continuous. Hint: Theorem 23.

Exercise 14. Show that any two complete lattice norms on a Riesz space must be equivalent.

Exercise 15. Let X be an ordered vector space and Z a linear subspace of X such that whenever $x \in X$ there exists $z \in Z$ with $z \geq x$. Suppose $F : Z \to \mathbf{R}$ is a positive linear functional. Show F has a positive linear extension to X. [This is a simple version of Kantorivich's Theorem; see [Vu] X.3.1.] [Hint: Set
$$p(x) = \inf\{F(z) : z \in Z, z \geq x\}$$
and use the Hahn-Banach Theorem.]

Exercise 16. Show there exists a bounded, finitely additive extension of Lebesgue measure on $[0, 1]$. [Hint: Let $X = B[0, 1]$,
$$Z = \{f : [0, 1] \to \mathbf{R} : f \text{ is bounded and Lebesgue measurable}\}$$
and $F : Z \to \mathbf{R}$ be $F(f) = \int_0^1 f \, dm$. Use Exercise 15.]

Chapter 6
Function Spaces

6.1 L^p-Spaces

Let μ be a measure on a σ-algebra, Σ, of subsets of S and let $0 < p < \infty$. The space $L^p(\mu)$ consists of all real-valued Σ-measurable functions, f, defined on S such that $|f|^p$ is μ-integrable. If μ is Lebesgue measure on a measurable subset E of $\mathbf{R}^n$ we write $L^p(\mu) = L^p(E)$. If $f, g \in L^p(\mu)$, then

$$|f+g|^p \leq 2^p(|f|^p + |g|^p)$$

so $L^p(\mu)$ is a vector space under pointwise addition and scalar multiplication. Moreover, $0 \leq f^+ \leq |f|$, $0 \leq f^- \leq |f|$ implies $f^+, f^- \in L^p(\mu)$ when $f \in L^p(\mu)$ so $L^p(\mu)$ is also a vector lattice under the pointwise order.

For $1 \leq p < \infty$, we set $\|f\|_p = \left(\int_S |f|^p \, d\mu\right)^{1/p}$ [the case $0 < p < 1$ is covered in the exercises]. We show that $\|\cdot\|_p$ is a semi-norm on $L^p(\mu)$; note that $\|\cdot\|_p$ is a lattice semi-norm. Only the triangle inequality needs to be checked. We do this with the aid of two preliminary results.

Lemma 1 *Let $a, b \geq 0$ and $0 < t < 1$. Then $a^t b^{1-t} \leq ta + (1-t)b$.*

Proof: Define $\varphi : (0, \infty) \to \mathbf{R}$ by $\varphi(s) = ts - s^t$. Then $\varphi'(s) < 0$ for $0 < s < 1$ and $\varphi'(s) > 0$ for $s > 1$ so φ has a minimum at $s = 1$. Thus, $t - 1 \leq ts - s^t$ for $s > 0$. Put $s = a/b$ [if $b = 0$, trivial] and multiply by b to obtain the desired inequality.

Theorem 2 (Hölder's Inequality) *Let $1 < p < \infty$ and q be such that $1/p + 1/q = 1$. If $f \in L^p(\mu)$, $g \in L^q(\mu)$, then $fg \in L^1(\mu)$ and $\|fg\|_1 \leq \|f\|_p \|g\|_q$.*

Proof: If either $\|f\|_p = 0$ or $\|g\|_q = 0$, the result is trivial so assume that both are positive. Note that if we set $t = 1/p$ in Lemma 1, the conclusion reads $AB \leq A^p/p + B^q/q$ when $A, B \geq 0$. Set $A = |f(t)| / \|f\|_p$, $B = |g(t)| / \|g\|_q$ for $t \in S$ to obtain

$$\frac{|f(t)g(t)|}{\|f\|_p \|g\|_q} \leq \frac{|f(t)|^p}{p \|f\|_p^p} + \frac{|g(t)|^q}{q \|g\|_q^q}$$

so $fg \in L^1(\mu)$ and

$$\frac{\|fg\|_1}{\|f\|_p \|g\|_q} \leq \frac{\|f\|_p^p}{p\|f\|_p^p} + \frac{\|g\|_q^q}{q\|g\|_q^q} = \frac{1}{p} + \frac{1}{q} = 1.$$

Remark 3 For $p = 2$, this inequality is often referred to as the Cauchy-Schwarz Inequality.

We can now easily obtain the triangle inequality for $\|\|_p$.

Theorem 4 (Minkowski Inequality) *If $f, g \in L^p(\mu)$, $1 \leq p < \infty$, then $\|f + g\|_p \leq \|f\|_p + \|g\|_p$.*

Proof: For $p = 1$, the inequality is clear. Assume $p > 1$. Note

$$\begin{aligned}\int_S |f+g|^p \, d\mu &= \int_S |f+g| \, |f+g|^{p-1} \, d\mu \\ &\leq \int_S |f| \, |f+g|^{p-1} \, d\mu + \int_S |g| \, |f+g|^{p-1} \, d\mu.\end{aligned} \quad (6.1)$$

Observe that $(p-1)q = p$ and apply Hölder's Inequality to both terms on the right hand side of (1) to obtain

$$\begin{aligned}\int_S |f+g|^p \, d\mu &\leq \left(\int_S |f|^p \, d\mu\right)^{1/p} \left(\int_S (|f+g|^p \, d\mu\right)^{1/q} \\ &+ \left(\int_S |g|^p \, d\mu\right)^{1/p} \left(\int_S |f+g|^p \, d\mu\right)^{1/q}.\end{aligned} \quad (6.2)$$

If $\int_S |f+g|^p \, d\mu = 0$, the result is trivial; otherwise, divide (2) by

$$\left(\int_S |f+g|^p \, d\mu\right)^{1/q}$$

to obtain the desired inequality.

We now extend the Riesz-Fischer Theorem to $p > 1$.

Theorem 5 (Riesz-Fischer) *For $1 \leq p < \infty$, $L^p(\mu)$ is complete under $\|\|_p$.*

Proof: Let $\sum\limits_{k=1}^{\infty} f_k$ be an absolutely convergent series in $L^p(\mu)$. Define $g_n(t) = \sum\limits_{k=1}^{n} |f_k(t)|$ for $t \in S$. Then

$$\|g_n\|_p \leq \sum_{k=1}^{n} \|f_k\|_p \leq \sum_{k=1}^{\infty} \|f_k\|_p = M < \infty.$$

Now $g_n \uparrow$ so if we set $g(t) = \lim g_n(t)$, then g is Σ-measurable and $g_n(t)^p \uparrow g(t)^p$, $g_n \geq 0$, so by the MCT

$$\int_S g^p \, d\mu = \lim \int_S |g_n|^p \, d\mu \leq M^p.$$

6.1. L^p-SPACES

Hence, g is finite μ-a.e. and we may assume that $g \in L^p(\mu)$. Note that this means that the series $\sum_{k=1}^{\infty} |f_k(t)|$ converges in $\mathbf{R}$ for μ-almost all $t \in S$.

Define f by $f(t) = \sum_{k=1}^{\infty} f_k(t)$ when this series converges in $\mathbf{R}$ and $f(t) = 0$ otherwise. Then f is Σ-measurable and f is the μ-a.e. limit of $\sum f_k$. Let $s_n = \sum_{k=1}^{n} f_k$. Then $|s_n| \leq g_n \leq g$ so $|f| \leq g$ μ-a.e. and $f \in L^p(\mu)$. Also, $|s_n - f|^p \leq 2^p |g|^p$ and $s_n \to f$ μ-a.e. so by the DCT, $\|s_n - f\|_p \to 0$. By Theorem 5.1.4, $L^p(\mu)$ is complete.

Note that $\|f\|_p = 0$ if and only if $f = 0$ μ-a.e. so if $K_p = \{f \in L^p(\mu) : \|f\|_p = 0\}$, then $\mathcal{L}^p(\mu) = L^p(\mu)/K_p$ is a Banach space and two functions are in the same coset of $\mathcal{L}^p(\mu)$ if and only if they are equal μ-a.e. (§5.4). Instead of working with the cosets in $\mathcal{L}^p(\mu)$, it is customary to identify two functions if they are equal almost everywhere and treat the cosets as if they were functions. Under this agreement, we would say that $\mathcal{L}^p(\mu)$ is a Banach lattice.

Comparison of L^p-spaces:

In general, there are no inclusion results for L^p-spaces. For example, if $f(t) = 1/\sqrt{t}$ for $0 < t \leq 1$ and $f(t) = 0$ otherwise, then $f \in L^1(m)$ but $f \notin L^2(m)$, while if $g(t) = 1/t$ for $t \geq 1$ and $g(t) = 0$ otherwise, then $g \in L^2(m)$ but $g \notin L^1(m)$. [See also Exercises 1 and 2.]

For finite measures, we do have

Proposition 6 *Let μ be a finite measure and $1 \leq r < s < \infty$. Then $L^s(\mu) \subset L^r(\mu)$ and the inclusion map is continuous.*

Proof: Let $h \in L^s(\mu)$. Set $p = s/r$, $f = |h|^r$ and $g = 1$ in Hölder's Inequality to obtain

$$\int_S |h|^r \, d\mu \leq \left(\int_S |h|^s\right)^{r/s} (\mu(S))^{1-r/s}.$$

Apply Theorem 5.2.1 to obtain the continuity.

If μ is counting measure on S, we set $\ell^p(S) = L^p(\mu)$, and if $S = \mathbf{N}$, we write $\ell^p(\mathbf{N}) = \ell^p$. Thus, ℓ^p consists of all sequences $\{t_j\}$ such that

$$\|\{t_j\}\|_p = \left(\sum_{j=1}^{\infty} |t_j|^p\right)^{1/p} < \infty$$

[recall Example 5.1.7]. For ℓ^p-spaces, in contrast to Proposition 6, we have

Proposition 7 *If $1 \leq r < s < \infty$, $\ell^r \subset \ell^s$ and the inclusion map is continuous. Moreover, the containment is proper.*

Proof: Suppose $x = \{t_j\} \in \ell^r$ with $\|x\|_r \leq 1$. Then $|t_j| \leq 1$ for all j so $|t_j|^s \leq |t_j|^r$ which implies $\|x\|_s \leq \|x\|_r$.

If $x = \{t_j\} \in \ell^r$ with $x \neq 0$, then $x/\|x\|_r \in \ell^r$ and $\|x/\|x\|_r\|_r = 1$ so by the observation above $\|x/\|x\|_r\|_s \leq 1$ or $\|x\|_s \leq \|x\|_r$.

Take $t_j = (1/j)^{1/r}$. Then $x = \{t_j\} \in \ell^s \setminus \ell^r$.

For a thorough discussion of the possible inclusion results for L^p-spaces see Romero ([R]).

Dense subsets of L^p :

We have the analogues of Theorems 2, 3 and 6 of §3.5 for L^p-spaces [Exercise 6].

Theorem 8 *The vector space of $\sum$-simple functions in $L^p(\mu)$ is dense in $L^p(\mu)$. Moreover, given $f \in L^p(\mu)$ there exists a sequence of simple functions $\{\varphi_k\}$ in $L^p(\mu)$ such that $\varphi_k \to f$ pointwise and $\|\varphi_k - f\|_p \to 0$; if $f \geq 0$, the $\{\varphi_k\}$ can be chosen such that $\varphi_k \uparrow f$.*

Assume that μ is a premeasure on the semi-ring S of subsets of S and that $\sum$ is the σ-algebra of μ^*-measurable subsets of S (§2.4). Let μ denote the restriction of μ^* to $\sum$.

Theorem 9 *The vector space of S-simple functions in $L^p(\mu)$ is dense in $L^p(\mu)$.*

Theorem 10 *Let S be a locally compact Hausdorff space and μ a regular Borel measure on $\mathcal{B}(S)$. Then $C_c(S)$ is dense in $L^p(\mu)$.*

As an application of Theorem 10 we give a generalization of Theorem 3.11.10 to L^p-spaces.

Theorem 11 *If $f \in L^p(\mathbf{R}^n)$, then $\lim_{h \to 0} \|f_h - f\|_p = 0$.*

Proof: Let $\epsilon > 0$. By Theorem 10, there exists $\varphi \in C_c(\mathbf{R}^n)$ such that $\|f - \varphi\|_p < \epsilon$. φ is uniformly continuous and there exists a such that $\varphi(x) = 0$ for $\|x\| \geq a$ so there exists $0 < \delta < 1$ such that $|\varphi(x+h) - \varphi(x)| < \epsilon$ for $\|h\| < \delta$. Thus, if $\|h\| < \delta$, then

$$\int_{\mathbf{R}^n} |\varphi(x+h) - \varphi(x)|^p \, dx < \epsilon^p (2(a+1))^n. \tag{6.3}$$

Now if $\|h\| < \delta$,

$$\|f_h - f\|_p \leq \|f_h - \varphi_h\|_p + \|\varphi_h - \varphi\|_p + \|\varphi - f\|_p < 2\epsilon + \|\varphi_h - \varphi\|_p$$

so the result follows from (3).

Convergence in L^p :

We give a characterization of sequential convergence in L^p for $1 \leq p < \infty$. Generalizing Proposition 3.7.1, we have

Proposition 12 *If $f_k \to f$ in $L^p(\mu)$, then $f_k \to f$ μ-measure.*

We give some additional necessary conditions that must be satisfied by a sequence converging in L^p.

6.1. L^p-SPACES

Definition 13 *Let $\mathcal{F}$ be a family of signed measures on Σ. $\mathcal{F}$ is uniformly μ-continuous if $\lim_{\mu(E) \to 0} \nu(E) = 0$ uniformly for $\nu \in \mathcal{F}$.*

Proposition 14 *Let $f_k \in L^p(\mu)$ and $\nu_k(E) = \int_E |f_k|^p \, d\mu$ for $E \in \Sigma$, $k = 0, 1, \ldots$. If $f_k \to f_0$ in $L^p(\mu)$, then $\{\nu_k : k \geq 0\}$ is uniformly μ-continuous.*

Proof: Let $\epsilon > 0$. $\exists N$ such that $k \geq N$ implies $\|f_k - f_0\|_p < \epsilon$. By 3.2.17 there exists $\delta > 0$ such that $E \in \Sigma$, $\mu(E) < \delta$ implies $\nu_k(E) < \epsilon$ for $k = 0, 1, \ldots, N$. For $k \geq N$ and $\mu(E) < \delta$,

$$\nu_k(E)^{1/p} \leq \left(\int_E |f_k - f_0|^p \, d\mu\right)^{1/p} + \left(\int_E |f_0|^p \, d\mu\right)^{1/p} \leq \|f_k - f_0\|_p + \nu_0(E)^{1/p} < \epsilon + \epsilon^{1/p}.$$

Definition 15 *Let $\mathcal{F}$ be a family of signed measures on Σ. $\mathcal{F}$ is equicontinuous from above at $\emptyset$ if $E_k \in \Sigma$, $E_k \downarrow \emptyset$ implies $\lim_k \nu(E_k) = 0$ uniformly for $\nu \in \mathcal{F}$.*

Proposition 16 *Let $f_k \in L^p(\mu)$ and $\nu_k(E) = \int_E |f_k|^p \, d\mu$ for $E \in \Sigma$, $k = 0, 1, \ldots$. If $f_k \to f_0$ in $L^p(\mu)$, then $\{\nu_k : k \geq 0\}$ is equicontinuous from above at $\emptyset$.*

Proof: Let $\epsilon > 0$. There exists N such that $n \geq N$ implies $\|f_n - f_0\|_p < \epsilon$. Let $E_k \in \Sigma$, $E_k \downarrow \emptyset$. By 3.2.9 and 2.2.5, there exists K such that $\nu_n(E_k) < \epsilon$ for $k \geq K$ and $n = 0, 1, \ldots, N-1$. For $n \geq N$ and $k \geq K$,

$$\nu_n(E_k)^{1/p} \leq \left(\int_{E_k} |f_n - f_0|^p \, d\mu\right)^{1/p} + \left(\int_{E_k} |f_0|^p \, d\mu\right)^{1/p} < \epsilon + \epsilon^{1/p}.$$

Propositions 12, 14 and 16 give three necessary conditions for a sequence to converge in L^p. It is an interesting result of Vitali that these necessary conditions are also sufficient.

Theorem 17 (Vitali) *Let $f_k \in L^p(\mu)$ and $\nu_k(E) = \int_E |f_k|^p \, d\mu$ for $E \in \Sigma$, $k = 0, 1, \ldots$. Then $f_k \to f_0$ in $L^p(\mu)$ if and only if*

(i) $f_k \to f_0$ μ-measure,

(ii) $\{\nu_k : k \geq 0\}$ is uniformly μ-continuous,

(iii) $\{\nu_k : k \geq 0\}$ is equicontinuous from above at $\emptyset$.

Proof: $\Rightarrow$ follows from the above.
$\Leftarrow$: Since $\bigcup_{k=0}^{\infty} \{t : f_k(t) \neq 0\}$ is μ σ-finite [3.2.14], we may as well assume that μ is σ-finite. Let $S = \bigcup_{k=1}^{\infty} E_k$ where $E_k \in \Sigma$, $E_k \uparrow$ and $\mu(E_k) < \infty$. Set $F_k = E_k^c$.
Let $\epsilon > 0$. From (iii) $\exists k$ such that $\int_{F_k} |f_j|^p \, d\mu < (\epsilon/2)^p$ for $j \geq 0$. Hence, for $j \geq 0$,

$$\left(\int_{F_k} |f_i - f_j|^p \, d\mu\right)^{1/p} \leq \left(\int_{F_k} |f_i|^p \, d\mu\right)^{1/p} + \left(\int_{F_k} |f_j|^p \, d\mu\right)^{1/p} < \epsilon \qquad (6.4)$$

Let $G_{ij} = \{t : |f_i(t) - f_j(t)| \geq \epsilon\}$. Then

$$\begin{aligned}\int_{E_k} |f_i - f_j|^p \, d\mu &= \int_{E_k \cap G_{ij}} |f_i - f_j|^p \, d\mu + \int_{E_k \setminus G_{ij}} |f_i - f_j|^p \, d\mu \\ &\leq \int_{E_k \cap G_{ij}} |f_i - f_j|^p \, d\mu + \epsilon^p \mu(E_k).\end{aligned} \quad (6.5)$$

From (i) and (ii) the first term on the right hand side of (5) goes to 0 as $i, j \to \infty$. From (4) and (5), it follows that $\{f_i\}$ is a Cauchy sequence in L^p-norm. By the Riesz-Fischer Theorem there exists $g \in L^p(\mu)$ such that $f_i \to g$ in $L^p(\mu)$. By Proposition 12 $f_i \to g$ in μ-measure so $f_0 = g$ μ-a.e. and $f_i \to f_0$ in $L^p(\mu)$.

Dual of L^p for $p > 1$:

We now give a characterization of the dual of $L^p(\mu)$ for $1 < p < \infty$. In what follows p is fixed, $1 < p < \infty$, and $1/p + 1/q = 1$.

Proposition 18 *If $f \in L^q(\mu)$, then $F_f : L^p(\mu) \to \mathbf{R}$, defined by $F_f(g) = \int_S fg \, d\mu$, is a continuous linear functional on $L^p(\mu)$ with $\|F_f\| = \|f\|_q$.*

Proof: By Hölder's Inequality, $|F_f(g)| \leq \|f\|_q \|g\|_p$ so F_f is continuous with $\|F_f\| \leq \|f\|_q$.

Set $g = |f|^{q-1} \operatorname{sign} f$. Then g is measurable and $|g|^p = |f|^{(q-1)p} = |f|^q$ so $g \in L^p(\mu)$ and

$$\begin{aligned}F_f(g) &= \int_S |f|^q \, d\mu = \left(\int_S |f|^q \, d\mu\right)^{1/q} \left(\int_S |f|^q \, d\mu\right)^{1/p} \\ &= \|f\|_q \left(\int_S |g|^p \, d\mu\right)^{1/p} = \|f\|_q \|g\|_p.\end{aligned}$$

Hence, $\|F_f\| = \|f\|_q$.

Thus, the map $U : L^q(\mu) \to L^p(\mu)'$, $Uf = F_f$, is a linear isometry which is order preserving. We show that U is onto, i.e., every continuous linear functional on $L^p(\mu)$ has the form F_f for some $f \in L^q(\mu)$. Hence, we can identify $L^q(\mu)$ and $L^p(\mu)'$ as Banach lattices.

We first establish a useful lemma, sometimes called the Reverse Hölder Inequality.

Lemma 19 *Let μ be σ-finite and let $f : S \to \mathbf{R}$ be measurable. Suppose*

$$M = \sup\left\{\int_S |fg| \, d\mu : \|g\|_p \leq 1\right\} < \infty.$$

Then $f \in L^q(\mu)$ and $\|f\|_q = M$.

Proof: Let $S = \bigcup_{k=1}^{\infty} A_k$ with $A_k \uparrow S$ and $\mu(A_k) < \infty$ and

$$E_k = \{t \in A_k : |f(t)| \leq k\}.$$

Set $g_k = |f|^{q-1} C_{E_k} \operatorname{sign} f$. Then g_k is bounded, measurable and vanishes outside A_k so $g_k \in L^p(\mu)$. Moreover, $g_k f = |f|^q C_{E_k}$ so

$$\left|\int_S g_k f \, d\mu\right| \leq M \|g_k\|_p = M \left(\int_S |f|^q C_{E_k} \, d\mu\right)^{1/p}$$

6.1. L^P-SPACES

and
$$\left(\int_S |f|^q C_{E_k} d\mu\right)^{1-1/p=1/q} \leq M.$$

Since $|f|^q C_{E_k} \uparrow |f|^q$, the MCT implies $f \in L^q(\mu)$ with $\|f\|_q \leq M$. The reverse inequality $M \leq \|f\|_q$ now follows from Hölder's Inequality.
See also Exercise 14.

Theorem 20 (Riesz Representation Theorem) *If $F \in L^p(\mu)'$, then there exists $f \in L^q(\mu)$ such that $F(g) = \int_S fg d\mu$ for $g \in L^p(\mu)$. Moreover, $\|F\| = \|f\|_q$.*

Proof: First assume that μ is finite. Define $\nu : \Sigma \to \mathbf{R}$ by $\nu(E) = F(C_E)$ [note $C_E \in L^p(\mu)$ since μ is finite]. Then $\nu(\emptyset) = F(0) = 0$ and ν is finitely additive since F is linear. We claim that ν is actually countably additive. For if $\{E_j\} \subset \Sigma$ are pairwise disjoint and $E = \bigcup_{j=1}^{\infty} E_j$, then

$$\left\| C_E - \sum_{j=1}^n C_{E_j} \right\|_p = \left\| C_{\bigcup_{j=n+1}^{\infty} E_j} \right\|_p = \left| \sum_{j=n+1}^{\infty} \mu(E_j) \right|^{1/p} \to 0$$

so $\nu(E) - \sum_{j=1}^n \nu(E_j) \to 0$. Since

$$|\nu(E)| \leq \|F\| \|C_E\|_p = \|F\| \mu(E)^{1/p},$$

ν is a finite signed measure which is absolutely continuous with respect to μ.

By the Radon-Nikodym Theorem, there exists a μ-integrable function f such that $\nu(E) = \int_E f d\mu = F(C_E)$ for every $E \in \Sigma$. By linearity, we have $F(\varphi) = \int_S f\varphi d\mu$ for every Σ-simple function φ. We claim that $F(g) = \int_S fg d\mu$ for every $g \in L^p(\mu)$. For this we may assume $g \geq 0$. Let $A = \{t : f(t) \geq 0\}$, $B = \{t : f(t) < 0\}$. Choose a sequence of non-negative, simple functions $\{\varphi_k\}$ such that $\varphi_k \uparrow g$ and $\|\varphi_k - g\|_p \to 0$ [Theorem 8]. Then $\varphi_k C_A f = \varphi_k f^+ \uparrow gf^+$ and $\|\varphi_k C_A - gC_A\|_p \to 0$ so by the continuity of F,

$$F(\varphi_k C_A) = \int_A f\varphi_k d\mu = \int_S f^+ \varphi_k d\mu \to F(gC_A)$$

and by the MCT $gf^+ \in L^1(\mu)$ and

$$\int_A f\varphi_k d\mu = \int_S f^+ \varphi_k d\mu \uparrow \int_S f^+ g d\mu.$$

Hence, $F(gC_A) = \int_S gf^+ d\mu$. Similarly, $F(gC_B) = \int_S gf^- d\mu$ so $F(g) = \int_S fg d\mu$.
That $f \in L^q(\mu)$ and $\|F\| = \|f\|_q$ follows from Lemma 19.
Now let μ be an arbitrary measure. For $E \in \Sigma$, let

$$L^p(E) = C_E L^p(\mu) = \{f \in L^p(\mu) : f = 0 \text{ outside } E\}.$$

Set $\mathcal{E} = \{E \in \Sigma : \mu(E) < \infty\}$. By the part above, for every $E \in \mathcal{E}$ there exists $f_E \in L^q(E)$ [unique up to μ-a.e.] such that $F(C_E g) = \int_S f_E g d\mu$ for $g \in L^p(\mu)$. Moreover,
$$\|f_E\|_q = \sup\{|F(C_E g)| : \|g\|_p \leq 1\} \leq \|F\|.$$
Let $a = \sup\{\|f_E\|_q : E \in \mathcal{E}\} \leq \|F\|$. If $A, B \in \mathcal{E}$ and $A \subset B$, then $f_A = f_B$ μ-a.e. in S so $|f_A| \leq |f_B|$ μ-a.e. and $\|f_A\|_q \leq \|f_B\|_q$. Hence, there exists an increasing sequence $\{E_k\} \subset \mathcal{E}$ such that $\|f_{E_k}\|_q \uparrow a$. Let $f(t) = \overline{\lim} f_{E_k}(t)$. Since $f_{k+1} = f_k$ μ-a.e. in E_k, $\lim f_{E_k}(t)$ exists for μ-almost all $t \in S$, $f = f_k$ μ-a.e. in E_k and $f = 0$ off $E = \bigcup_{k=1}^{\infty} E_k$.

We claim that F vanishes on $L^p(E^c)$. For if F is not zero on $L^p(E^c)$, then since the simple functions are dense in $L^p(\mu)$, there exists $B \in \mathcal{E}$ such that $B \subset E^c$ and F does not vanish on $L^p(B)$. Therefore, $f_B \neq 0$ μ-a.e., and since $B \cap E_k = \emptyset$,
$$a^q \geq \|f_{B \cup E_k}\|_q^q = \|f_B\|_q^q + \|f_{E_k}\|_q^q$$
which implies $a^q \geq \|f_B\|_q^q + a^q$ and gives the desired contradiction.

Let $g \in L^p(\mu)$. Since $f \in L^q(\mu)$, $fg \in L^1(\mu)$. Since $gf_{E_k} \to gf$ μ-a.e. and $|gf_{E_k}| \leq |gf|$, the DCT implies $\int_S gf_{E_k} d\mu \to \int_S gf d\mu$. Similarly, $\|gC_{E_k} - gC_E\|_p \to 0$ so the continuity of F implies $F(gC_{E_k}) \to F(gC_E)$. Hence,
$$\lim F(gC_{E_k}) = \lim \int_S gf_{E_k} d\mu = \int_S fg d\mu = F(C_E g) = F(C_E g + C_{E^c} g) = F(g).$$

Corollary 21 *If $1 < p < \infty$, then $L^p(\mu)$ is reflexive.*

We describe the dual of certain L^1-spaces in §6.2.

We give two interesting applications of duality. The first is an extension of Minkowski's Inequality to integrals. Minkowski's Inequality asserts that the L^p-norm of a sum is less than or equal to the sum of the L^p-norms. We give a generalization where the sum is replaced by an integral.

Theorem 22 (Minkowski's Inequality for Integrals). *Let $(S, \mathcal{S}, \mu)$, $(T, \mathcal{T}, \nu)$ be σ-finite measure spaces and $f : S \times T \to \mathbf{R}$ measurable with respect to the σ-algebra of $(\mu \times \nu)^*$-measurable sets. Let $1 \leq p < \infty$ and assume that $f(\cdot, t) \in L^p(\mu)$ for $t \in T$ and $t \to \|f(\cdot, t)\|_p$ belongs to $L^1(\nu)$. Then $f(s, \cdot) \in L^1(\nu)$ for μ-almost all $s \in S$, the function $s \to \int_T f(s, \cdot) d\nu$ belongs to $L^p(\mu)$ and*
$$\left\| \int_T f(\cdot, t) d\nu(t) \right\|_p \leq \int_T \|f(\cdot, t)\|_p d\nu(t).$$

Proof: The case $p = 1$ follows from Fubini's Theorem. Let $1 < p < \infty$ and $h \in L^q(\mu)$. By Fubini's Theorem and Hölder's Inequality,

$$\begin{aligned}\int_S \left\{\int_T |f(s,t)| d\nu(t)\right\} |h(s)| d\mu(s) &= \int_T \int_S |f(s,t)| |h(s)| d\mu(s) d\nu(t) \\ &\leq \int_T \left(\int_S |f(s,t)|^p d\mu(s)\right)^{1/p} \|h\|_q d\nu(t) \\ &= \int_T \|f(\cdot,t)\|_p \|h\|_q d\nu(t).\end{aligned}$$

6.1. L^p-SPACES

By Lemma 19, $s \to \int_T f(s,t)d\nu(t)$ belongs to $L^p(\mu)$ with

$$\left\| \int_T f(\cdot, t) d\nu(t) \right\|_p \leq \int_T \|f(\cdot, t)\|_p \, d\nu(t).$$

Application to Convolution:
Recall the convolution product of two functions $f, g : \mathbf{R}^n \to \mathbf{R}$ is defined to be

$$f * g(x) = \int_{\mathbf{R}^n} f(x - y)g(y) dy$$

[§3.11].

Proposition 23 *Let $1 < p < \infty$. If $f \in L^1(\mathbf{R}^n)$, $g \in L^p(\mathbf{R}^n)$, then $f * g \in L^p(\mathbf{R}^n)$ and $\|f * g\|_p \leq \|f\|_1 \|g\|_p$.*

Proof: Let $1/p + 1/q = 1$ and $h \in L^q(\mathbf{R}^n)$. Then, using Exercise 3.11.8 and Fubini's Theorem,

$$\begin{aligned}
\int_{\mathbf{R}^n} \int_{\mathbf{R}^n} |f(x-y)g(y)h(x)| \, dy dx &= \int_{\mathbf{R}^n} |h(x)| \int_{\mathbf{R}^n} |f(x-y)g(y)| \, dy dx \\
&= \int_{\mathbf{R}^n} |h(x)| \int_{\mathbf{R}^n} |f(t)g(x-t)| \, dt dx \\
&= \int_{\mathbf{R}^n} |f(t)| \int_{\mathbf{R}^n} |h(x)g(x-t)| \, dx dt \\
&\leq \int_{\mathbf{R}^n} |f(t)| \, \|g_{-t}\|_p \, \|h\|_q \, dt \\
&= \int_{\mathbf{R}^n} |f(t)| \, \|g\|_p \, \|h\|_q \, dt \\
&= \|f\|_1 \|g\|_p \|h\|_q .
\end{aligned} \quad (6.6)$$

Since h can be taken to be non-zero everywhere $\left[e^{-\|x\|^2}\right]$, (6) shows that $f * g$ is finite a.e. The inequality in (6) shows that $f * g$ induces a continuous linear functional on $L^q(\mathbf{R}^n)$. By Theorem 20, $f * g \in L^p(\mathbf{R}^n)$ and by (6), $\|f * g\|_p \leq \|f\|_1 \|g\|_p$.

We can combine Theorem 22 and Proposition 23 to obtain an extension of Theorem 3.11.11 to the case where $1 < p < \infty$.

Theorem 24 *Let $\{\varphi_k\}$ be an approximate identity and $f \in L^p(\mathbf{R}^n)$, $1 \leq p < \infty$. Then $\|f * \varphi_k - f\|_p \to 0$.*

Proof: $f * \varphi_k \in L^p(\mathbf{R}^n)$ by Proposition 23. Denote the function $t \to f(x+t)$ by f_x. By Minkowski's Inequality for Integrals,

$$\|f * \varphi_k - f\|_p = \left\| \int_{\mathbf{R}^n} (f(x-y) - f(x))\varphi_k(y) dy \right\|_p \leq \int_{\mathbf{R}^n} \|f_{-y} - f\|_p \varphi_k(y) dy. \quad (6.7)$$

The function $g(y) = \|f_{-y} - f\|_p$ is bounded, continuous [Theorem 11] and $g(0) = 0$ so the right hand side of (7) converges to 0 as $k \to \infty$ by Exercise 3.11.7.

Exercise 1. Let $S = (0, 1/2]$, $1 \leq p < \infty$. Show $f(t) = t^{-1/p}(\ln(1/t))^{-2/p}$ is in $L^p(S)$ but not in $L^r(S)$ for $r > p$.

Exercise 2. Let $S = [0, \infty)$, $1 \leq p < \infty$. Show $f(t) = t^{-1/2}(1 + \ln|t|)^{-1}$ is in $L^2(S)$ but not in $L^p(S)$ for $p \neq 2$.

Exercise 3. If $f_k \to f$ in $L^p(\mu)$ and $g_k \to g$ in $L^q(\mu)$ for $1 < p < \infty$, $1/p + 1/q = 1$, show that $f_k g_k \to fg$ in $L^1(\mu)$.

Exercise 4. If $f \in L^r(\mu) \cap L^s(\mu)$, $1 < r < s < \infty$, show $f \in L^p(\mu)$ for $r \leq p \leq s$.

Exercise 5. Let g_n, g be measurable functions such that $|g_n| \leq M$ and $g_n \to g$ μ-a.e. show that if $f_n \to f$ in $L^p(\mu)$, then $f_n g_n \to fg$ in $L^p(\mu)$.

Exercise 6. Prove Theorems 8-10.

Exercise 7. Show the polynomials are dense in $L^p[a,b]$ for $1 \leq p < \infty$. Generalize to $\mathbf{R}^n$.

Exercise 8. Let $1/p + 1/q = 1$, $f \in L^p(\mathbf{R})$, $g \in L^q(\mathbf{R})$. Show $F(t) = \int_{\mathbf{R}} f(x+t)g(x)dx$ is uniformly continuous on $\mathbf{R}$.

Exercise 9. For $f \in L^p(\mu)$, $0 < p < 1$, set $|f|_p = \int_S |f|^p d\mu$. Show $d(f,g) = |f - g|_p$ defines a complete semi-metric on $L^p(\mu)$. Show $|\ \ |_p$ is not a semi-norm.

Exercise 10. If $f_k \to f$ in $L^p(\mu)$ for some $1 \leq p < \infty$, show there exists a subsequence of $\{f_k\}$ which converges μ-a.e. to f.

Exercise 11. Let $f_k, f, g \in L^p(\mu)$ and $f_k \to f$ μ-measure. If $|f_k| \leq g$ μ-a.e., show $f_k \to f$ in $L^p(\mu)$ [Dominated Convergence Theorem].

Exercise 12. Let $1 < p < \infty$, $1/p + 1/q = 1$. Suppose $\{f_k\} \subset L^p(\mu)$ and $\lim \int_S f_k g d\mu$ exists for every $g \in L^q(\mu)$. Show there exists $f \in L^p(\mu)$ such that $\lim \int_S f_k g d\mu = \int_S fg d\mu$.

Exercise 13. Let $f_k, f_0 \in L^2[a,b]$ and $\|f_k - f_0\|_2 \to 0$. Set $F_k(t) = \int_a^t f_k$. Show $F_k \to F_0$ uniformly on $[a,b]$. Is $p = 2$ important?

Exercise 14. Let μ be σ-finite and $1 < p < \infty$. Suppose f is measurable and $fg \in L^1(\mu)$ for every $g \in L^q(\mu)$. Show $f \in L^p(\mu)$. Show σ-finiteness cannot be dropped. [Hint: Exercise 2.2.16.]

6.1. L^p-SPACES

Exercise 15. If μ is finite, show condition (iii) of Theorem 17 can be dropped.

Exercise 16. Let $k : S \times S \to \mathbf{R}$ be in $L^2(\mu \times \mu)$. Show

(i) $y(s) = \int_S k(s,t) x(t) \, d\mu(t)$ exists for μ-almost all $s \in S$ when $x \in L^2(\mu)$ and

(ii) $y \in L^2(\mu)$ with $\|y\|_2 \leq \|k\|_2 \|x\|_2$.

The map $K : x \to y$ is a continuous linear map from $L^2(\mu)$ into itself and is called an *integral operator*. The function k is called the *kernel* of K.

Exercise 17. Show $\ell^p(S)$, $1 \leq p < \infty$, is separable if and only if S is countable.

Exercise 18. Let $A = [a_{ij}]$ be an infinite matrix and suppose A maps ℓ^r into ℓ^s in the sense that the sequence $Ax = \left\{ \sum_{j=1}^{\infty} a_{ij} x_j \right\}_i \in \ell^s$ for every $x = \{x_j\} \in \ell^r$. Use the CGT to show that A is continuous.

6.2 The Space $L^\infty(\mu)$

Let μ be a measure on a σ-algebra Σ of subsets of S. The space $L^\infty(\mu)$ consists of all Σ-measurable functions $f : S \to \mathbf{R}$ such that there exists a μ-null set N with f bounded on $S \setminus N$. $L^\infty(\mu)$ is a vector space under pointwise addition and scalar multiplication and is a vector lattice under the pointwise order. We define a semi-norm on $L^\infty(\mu)$ by

$$\|f\|_\infty = \inf\{\sup\{|f(t)| : t \in S \setminus N\} : \mu(N) = 0\}. \tag{6.1}$$

The functions in $L^\infty(\mu)$ are called μ-*essentially bounded functions* and the semi-norm in (1) is called the μ-*essential sup* of f. If $E \subset \mathbf{R}^n$ is Lebesgue measurable and m is Lebesgue measure on E, we write $L^\infty(E) = L^\infty(m)$. The following properties should be clear.

Proposition 1 (i) $\|\ \|_\infty$ *is a semi-norm on* $L^\infty(\mu)$ *and* $\|f\|_\infty = 0$ *if and only if* $f = 0$ μ-*a.e.*

(ii) $\|f\|_\infty = \inf\{M > 0 : \mu\{t : |f(t)| \geq M\} = 0\}$.

(iii) *If* $|f| \leq |g|$ μ-*a.e., then* $\|f\|_\infty \leq \|g\|_\infty$ *so* $\|\ \|_\infty$ *is a lattice norm on* $L^\infty(\mu)$.

As was the case for the other L^p-spaces, we agree to identity functions in $L^\infty(\mu)$ which are equal μ-a.e. so we regard $L^\infty(\mu)$ as a NLS. The analogue of the Riesz-Fischer Theorem holds for $L^\infty(\mu)$.

Theorem 2 $L^\infty(\mu)$ *is complete under* $\|\ \|_\infty$.

Proof: Let $\{f_k\}$ be Cauchy in $L^\infty(\mu)$. Then there exists a μ-null set N such that

$$|f_k(t) - f_j(t)| \leq \|f_k - f_j\|_\infty \text{ for } t \notin N. \tag{6.2}$$

Therefore, $\{f_k(t)\}$ is Cauchy for $t \in S \setminus N$; set $f(t) = \lim f_k(t)$ for $t \in S \setminus N$ and $f(t) = 0$ for $t \in N$. Then f is measurable. Let $\epsilon > 0$. There exists n such that $k, j \geq n$ implies $\|f_k - f_j\|_\infty < \epsilon$. Fixing $k \geq n$ and letting $j \to \infty$ in (2) implies that $\|f_k - f\|_\infty \leq \epsilon$ so $f \in L^\infty(\mu)$ and $f_k \to f$ in $L^\infty(\mu)$.

Thus, $L^\infty(\mu)$ is a Banach lattice.

We now show that the dual of $L^1(\mu)$ is $L^\infty(\mu)$ for σ-finite μ.

From Exercise 1, we have

Proposition 3 *Let* $f \in L^\infty(\mu)$. *Define* $F_f : L^1(\mu) \to \mathbf{R}$ *by* $F_f(g) = \int_S fg\,d\mu$. *Then* F_f *is a continuous linear functional on* $L^1(\mu)$ *with* $\|F_f\| \leq \|f\|_\infty$.

For σ-finite measures we show that every continuous linear functional on $L^1(\mu)$ has the form F_f for some $f \in L^\infty(\mu)$.

6.2. THE SPACE $L^\infty(\mu)$

Theorem 4 (Riesz Representation Theorem) *Let μ be σ-finite. If $F \in (L^1(\mu))'$, there exists $f \in L^\infty(\mu)$ such that $F(g) = \int_S fg d\mu$ for $g \in L^1(\mu)$ and $\|F\| = \|f\|_\infty$ [f is unique up to μ-a.e.].*

Proof: First assume that μ is finite. Then as in the proof of Theorem 6.1.20 there exists $f \in L^1(\mu)$ such that $F(g) = \int_S fg d\mu$ for $g \in L^1(\mu)$.

We claim that $f \in L^\infty(\mu)$ and $\|F\| = \|f\|_\infty$. Let $\epsilon > 0$. Set

$$A = \{t : \|F\| + \epsilon < f(t)\}.$$

Then

$$(\|F\| + \epsilon)\mu(A) \le \int_A |f|\, d\mu = \int_A f(\text{sign } f) d\mu$$
$$= F(C_A \text{sign } f) \le \|F\| \|C_A \text{sign } f\|_1 = \|F\|\mu(A).$$

Hence, $\mu(A) = 0$ and $f \in L^\infty(\mu)$ with $\|f\|_\infty \le \|F\| + \epsilon$ so $\|f\|_\infty \le \|F\|$. By Proposition 3, $\|F\| \le \|f\|_\infty$ so $\|f\|_\infty = \|F\|$.

Assume that μ is σ-finite. Let $\{E_k\}$ be an increasing sequence from $\sum$ with $E_k \uparrow S$ and $\mu(E_k) < \infty$. By the first part, there exists a sequence $\{f_k\} \subset L^\infty(\mu)$ such that f_k vanishes outside E_k, $\|f_k\|_\infty \le \|F\|$, $f_k = f_{k+1}$ in E_k and $F(g) = \int_S f_k g d\mu$ for every $g \in L^1(\mu)$ which vanishes outside E_k. Let $f = \lim f_k$ so $f = f_k$ in E_k and $\|f\|_\infty \le \|F\|$. Therefore, $F(C_{E_k}g) = \int_S f(C_{E_k}g) d\mu$ for every $g \in L^1(\mu)$. But $C_{E_k}g \to g$ pointwise and $|C_{E_k}g| \le |g|$ so by the DCT and the continuity of F,

$$F(g) = \lim F(C_{E_k}g) = \lim \int_S f(C_{E_k}g) d\mu = \int_S fg d\mu.$$

From Proposition 3, $\|F\| \le \|f\|_\infty$ so $\|F\| = \|f\|_\infty$.

Thus, when μ is σ-finite the map $U : L^\infty(\mu) \to (L^1(\mu))'$, $U(f) = F_f$ is a linear isometry from $L^\infty(\mu)$ onto $(L^1(\mu))'$ which is obviously order preserving. As before, we write $L^\infty(\mu) = (L^1(\mu))'$.

In contrast to the situation when $1 < p < \infty$, Theorem 4 is false if the σ-finiteness condition is dropped.

Example 5 Let $S = \mathbf{R}$ and let $\sum$ be the σ-algebra which consists of the sets which are countable or have countable complements. Let μ be counting measure on $\sum$. Then $L^1(\mu)$ consists of the functions g on $\mathbf{R}$ which vanish outside a countable set and satisfy $\|g\|_1 = \sum_{t \in \mathbf{R}} |g(t)| < \infty$. Define $F : L^1(\mu) \to \mathbf{R}$ by $F(g) = \sum_{t > 0} g(t)$. Then $F \in (L^1(\mu))'$ and $\|F\| = 1$. If $f : S \to \mathbf{R}$ were to satisfy

$$F(g) = \int_S fg d\mu = \sum_{t > 0} g(t) \text{ for } g \in L^1(\mu),$$

then $f = C_{(0,\infty)}$. But this function is not $\sum$-measurable. [See, however, Exercise 5].

Remark 6 There is a description of $L^1(\mu)'$ for arbitrary measures due to J. Schwartz ([Sc]). There are necessary and sufficient conditions known for the measure μ to satisfy $L^1(\mu)' = L^\infty(\mu)$; see [TT].

Recall that for $1 < p < \infty$, $L^p(\mu)$ is reflexive [6.1.21]. However, $L^1(\mu)$ is not generally reflexive, even for finite measures.

Example 7 Let $S = [0,1]$ with Lebesgue measure. Then $C[0,1]$ is a closed subspace of $L^\infty[0,1]$ (Exercise 2). Define $F : C[0,1] \to \mathbf{R}$ by $F(\varphi) = \varphi(0)$. Then F is a continuous linear functional on $C[0,1]$ with $\|F\| = 1$. By the Hahn-Banach Theorem, F can be extended to a continuous linear functional $F : L^\infty[0,1] \to \mathbf{R}$ with $\|F\| = 1$.

Now there exists no $f \in L^1[0,1]$ such that $F(g) = \int_0^1 fg \, dm$ for all $g \in L^\infty[0,1]$. For, choose continuous g_k such that $|g_k(t)| \le 1$ for $0 \le t \le 1$, $g_k(0) = 1$ and $g_k(t) \to 0$ for $0 < t \le 1$. Thus, if such an f exists, by the DCT $\lim \int_0^1 fg_k \, dm = 0$ while $F(g_k) = 1$ for each k.

See also Exercise 3 for another proof.

It follows from Exercises 3 and 5.6.1.1 that $L^\infty[0,1]$ is not reflexive. We describe the dual of L^∞ in §6.3.

We establish a result which in some sense justifies the notation for L^∞-spaces.

Proposition 8 *Let μ be a finite measure. If $f \in L^\infty(\mu)$, then $\|f\|_\infty = \lim_{p\to\infty} \|f\|_p$.*

Proof: If $\|f\|_\infty = 0$, the result is trivial so assume $\|f\|_\infty > 0$ and let $0 < k < \|f\|_\infty$. Set $H = \{t : |f(t)| \ge k\}$. Then $\mu(H) > 0$ and

$$k^p \mu(H) \le \int_H |f|^p \, d\mu \le \int_S |f|^p \, d\mu \le \|f\|_\infty^p \mu(S)$$

so

$$k\mu(H)^{1/p} \le \|f\|_p \le \|f\|_\infty \mu(S)^{1/p}.$$

Letting $p \to \infty$ gives $k \le \lim_{p\to\infty} \|f\|_p \le \|f\|_\infty$. Hence, the equality follows.

If $f \in L^\infty(\mu)$, then $fg \in L^p(\mu)$ for every $g \in L^p(\mu)$. As an application of the CGT, we establish the converse of this result.

Example 9 Let $f : S \to \mathbf{R}$ be measurable and suppose $fg \in L^p(\mu)$ for every $g \in L^p(\mu)$. Then $f \in L^\infty(\mu)$. Define a linear map $T : L^p(\mu) \to L^p(\mu)$ by $Tg = fg$. We claim that T is closed. Suppose $g_k \to g$ and $Tg_k = fg_k \to h$ in $L^p(\mu)$. Then there is a subsequence $\{g_{n_k}\}$ converging μ-a.e. to g so $fg_{n_k} \to fg = Tg$ μ-a.e. [6.1.12, 3.6.3]. Hence, $Tg = h$ and T is closed. By the CGT, T is continuous, and we may assume $\|T\| \le 1$. Let $\delta > 0$ and set $E = \{t : |f(t)| \ge 1 + \delta\}$. Since $\|T^n g\| \le \|g\|$ for $n \ge 1$,

$$\int_S |g|^p \, d\mu \ge \int_S |f^n g|^p \, d\mu \ge \int_E (1+\delta)^{np} |g|^p \, d\mu,$$

and since $(1+\delta)^{np} \to \infty$, $\int_E |g|^p \, d\mu = 0$ for every $g \in L^p(\mu)$ so $\mu(E) = 0$. Hence, $f \in L^\infty(\mu)$ with $\|f\|_\infty \le 1$.

6.2. THE SPACE $L^\infty(\mu)$

Integrating Vector-Valued Functions

We indicate how the CGT and the duality between L^1 and L^∞ can be used to define an integral, called the *Gelfand integral*, for vector-valued functions. Let X be a Banach space and μ a measure on the σ-algebra Σ of subsets of a set S. Let $f: S \to X$ be such that $x' \circ f = x'f$ is μ-integrable for every $x' \in X'$. The mapping $F: X' \to L^1(\mu)$ defined by $F(x') = x'f$ is linear and has a closed graph (Exer. 10). Hence, F is continuous by the CGT. For $E \in \Sigma$,

$$x' \to \int_E x'fd\mu = \langle C_E, F(x') \rangle$$

defines a continuous linear functional on X' since $|\langle C_E, F(x') \rangle| \leq \|F\| \|x'\|$. This continuous linear functional is called the Gelfand integral of f over E and is denoted by $\int_E fd\mu$. Thus $\int_E fd\mu \in X''$ and $\langle \int_E fd\mu, x' \rangle = \int_E x'fd\mu$. If $\int_E fd\mu \in X$ ($= J_X X$) for every $E \in \Sigma$, then f is said to be Pettis integrable and $\int_E fd\mu$ is called the *Pettis integral* of F over Σ. Of course, if X is reflexive, every Gelfand integrable function is Pettis integrable, but, in general, this is not the case.

Example 10 Let μ be counting measure on $\mathbf{N}$ and define $f: \mathbf{N} \to c_0$ by $f(k) = e^k$. Let $x' = \{t_k\} \in \ell^1 = (c_0)'$. Then $x'f(k) = t_k$ so $x'f$ is μ-integrable and

$$\int_E x'fd\mu = \sum_{k \in E} t_k = \langle C_E, x' \rangle.$$

Hence, $\int_E fd\mu = C_E$ and when E is infinite $\int_E fd\mu = C_E \in \ell^\infty \backslash c_0$ so f is not Pettis integrable.

For more information on the Gelfand and Pettis integrals, see [Du]. There is also a close analogue of the Lebesgue integral for vector-valued functions, called the Bochner integral; this integral is also discussed in [Du].

Exercise 1 (Hölder's Inequality for $p = 1$). If $f \in L^\infty(\mu), g \in L^1(\mu)$, show $fg \in L^1(\mu)$ and $\|fg\|_1 \leq \|f\|_\infty \|g\|_1$.

Exercise 2. If $\varphi \in C[0,1]$, show $\sup\{|f(t)| : 0 \leq t \leq 1\}$ equals the m-essential sup of φ. Is this equality valid for an arbitrary measure on $[0,1]$?

Exercise 3. Show $L^1[0,1]$ is separable and $L^\infty[0,1]$ is not separable. Use Corollary 5.6.1.13 to show that $L^1[0,1]$ is not reflexive.

Exercise 4. Repeat Exercise 3 for ℓ^1 and ℓ^∞.

Exercise 5. Let $S \neq \emptyset$. Let $\ell^1(S) = L^1(\mu)$, where μ is counting measure on S, and $\ell^\infty(S) = L^\infty(\mu)$ [so $\ell^\infty(S)$ consists of all bounded functions on S]. Show $\ell^1(S)' = \ell^\infty(S)$.

Exercise 6. Let $1 \leq p < \infty$ and $1/p + 1/q = 1$. Show that if $f \in L^p(\mathbf{R})$, $g \in L^q(\mathbf{R})$, then $f * g$ is defined everywhere, is uniformly continuous, and $\|f * g\|_\infty \leq \|f\|_p \|g\|_q$.

Exercise 7. Let μ be σ-finite. Show $g \in L^\infty(\mu)$ satisfies $0 \leq g \leq 1$ μ-a.e. if and only if $0 \leq \int_S fg d\mu \leq \int_S f d\mu$ for every $f \in L^1(\mu)$ with $f \geq 0$.

Exercise 8. μ is *non-atomic* if every $E \in \Sigma$ with $\mu(E) > 0$ contains a subset $A \in \Sigma$ with $0 < \mu(A) < \mu(E)$ [i.e., Lebesgue measure]. If μ is finite and non-atomic, show $L^\infty(\mu)$ is infinite dimensional. Can the non-atomic assumption be dropped?

Exercise 9. If $f \in L^\infty(\mu) \cap L^1(\mu)$, show $f \in L^p(\mu)$, $1 < p < \infty$.

Exercise 10. Show the operator $F : X' \to L^1(\mu)$ defined in the last paragraph has a closed graph.

Exercise 11. Let μ be counting measure on $\mathbf{N}$ and $f : \mathbf{N} \to c_0$ be defined by $f(k) = e^k/k$. Find the Gelfand integral of f with respect to μ. Is f Pettis integrable?

Exercise 12. Let $f \in L^\infty[a,b]$. Show there exists a sequence of continuous functions $\{f_n\}$ such that $\|f_n\|_\infty \leq \|f\|_\infty$ and $\lim_n \int_a^b f_n g \, dm = \int_a^b f g \, dm$ for every $g \in L^1[a,b]$.

6.3 The Space of Finitely Additive Set Functions

In this section we describe the dual of L^∞ as a space of bounded, finitely additive set functions.

Let $\mathcal{A}$ be an algebra of subsets of a set S. We denote by $ba(\mathcal{A})$ the space of all bounded, finitely additive set functions on $\mathcal{A}$. If we define addition and scalar multiplication pointwise on $\mathcal{A}$, $ba(\mathcal{A})$ is a vector space, and if we define $\nu \leq \mu$ to mean $\nu(A) \leq \mu(A)$ for all $A \in \mathcal{A}$, then $ba(\mathcal{A})$ becomes an ordered vector space; it follows from Lemma 3.12.1.1 that $ba(\mathcal{A})$ is a vector lattice under the pointwise order. We define a norm on $ba(\mathcal{A})$ by

$$\|\nu\| = |\nu|(S), \tag{6.1}$$

where $|\nu|$ is the variation of ν [§2.2.1; that (1) defines a norm on $ba(\mathcal{A})$ follows from Exercise 2.2.1.1]. From Proposition 2.2.1.7 (iv) it follows that (1) defines a lattice norm so $ba(\mathcal{A})$ is a normed lattice. We show that $ba(\mathcal{A})$ is a Banach lattice.

Theorem 1 $ba(\mathcal{A})$ *is complete.*

Proof: Let $\{\nu_k\}$ be a Cauchy sequence in $ba(\mathcal{A})$. Let $\epsilon > 0$. There exists n such that $k, j \geq n$ implies

$$\|\nu_k - \nu_j\| < \epsilon. \tag{6.2}$$

Then $|\nu_k(A) - \nu_j(A)| < \epsilon$ for $k, j \geq n$ and $A \in \mathcal{A}$ [Proposition 2.2.1.7 (i)] so $\{\nu_k(A)\}$ converges, to say $\nu(A)$. Fixing $k \geq n$ in (2) and letting $j \to \infty$ gives $|\nu_k(A) - \nu(A)| \leq \epsilon$ for $A \in \mathcal{A}$. By Proposition 2.2.1.7 (v) $\|\nu_k - \nu\| \leq \epsilon$ for $k \geq n$ so $\nu \in ba(\mathcal{A})$ and $\nu_k \to \nu$.

The Dual of $L^\infty(\mu)$:

Let μ be a measure on a σ-algebra, Σ, of subsets of S. Let $ba(\Sigma : \mu)$ be the subspace of $ba(\Sigma)$ which consists of those $\nu \in ba(\Sigma)$ which are absolutely continuous with respect to μ in the sense that $\mu(E) = 0$ implies $|\nu|(E) = 0$. Then $ba(\Sigma : \mu)$ is a Banach space under the norm in (1) since $ba(\Sigma : \mu)$ is a closed subspace of $ba(\Sigma)$ [Exercise 1].

We show that each $\nu \in ba(\Sigma : \mu)$ induces a continuous linear functional on $L^\infty(\mu)$. We write $\|f\|_\infty$ for the μ-essential sup of f and $\|f\|_u = \sup\{|f(t)| : t \in S\}$ for the sup-norm. If $\nu \in ba(\Sigma)$ and $h : S \to \mathbf{R}$ is bounded and Σ-simple, the integral $\int_S h\,d\nu$ is well-defined and

$$\left|\int_S h\,d\nu\right| \leq \int_S |h|\,d|\nu| \leq \|h\|_u |\nu|(S) \tag{6.3}$$

[Remark 3.2.2]. If $h : S \to \mathbf{R}$ is bounded and Σ-measurable, there exists a sequence of Σ-simple functions $\{\varphi_k\}$ such that $\varphi_k \to h$ uniformly on S and $|\varphi_k| \leq |h|$ (3.1.1.3). From (3), $\{\int_S \varphi_k\,d\nu\}$ is a Cauchy sequence so we may define $\int_S h\,d\nu = \lim \int_S \varphi_k\,d\nu$; it is easily checked that the definition of $\int_S h\,d\nu$ does not depend on the sequence $\{\varphi_k\}$ and (3) still holds [Exercise 2].

Lemma 2 *Let $\nu \in ba(\Sigma : \mu)$ and $h : S \to \mathbf{R}$ a bounded, Σ-measurable function such that $A = \{t \in S : h(t) \neq 0\}$ is μ-null. Then $\int_S h d\nu = 0$.*

Proof: $\left|\int_S h d\nu\right| \leq \int_S |h| \, d|\nu| = \int_A |h| \, d|\nu| + \int_{S \setminus A} |h| \, d|\nu| \leq \|h\|_u \, |\nu|(A) = 0$.

If $\nu \in ba(\Sigma : \mu)$ and $h \in L^\infty(\mu)$, we may define the integral of h with respect to ν as follows: Pick $f : S \to \mathbf{R}$ bounded and Σ-measurable such that $\|f - h\|_\infty = 0$ and $\|h\|_\infty = \|f\|_u$ [Exercise 3]. Now set $\int_S h d\nu = \int_S f d\nu$; by Lemma 2 this definition does not depend on the choice of f and, moreover,

$$\left|\int_S h d\nu\right| \leq \|h\|_\infty \, |\nu|(S).$$

Hence, if $\nu \in ba(\Sigma : \mu)$, then $F_\nu(h) = \int_S h d\nu$ defines a continuous linear functional F_ν on $L^\infty(\mu)$ with $\|F_\nu\| \leq |\nu|(S) = \|\nu\|$. It is actually the case that $\|F_\nu\| = \|\nu\|$. For suppose $\epsilon > 0$ is given. Choose a partition $\{E_j : 1 \leq j \leq n\}$ of S such that

$$\sum_{j=1}^n |\nu(E_j)| > \|\nu\| - \epsilon$$

[Proposition 2.2.1.7], and for each j let $t_j = \operatorname{sign} \nu(E_j)$ and set $h = \sum_{j=1}^n t_j C_{E_j}$. Then $h \in L^\infty(\mu)$, $\|h\|_\infty \leq 1$, and

$$\|F_\nu\| \geq |F_\nu(h)| = \left|\int_S h d\nu\right| = \left|\sum_{j=1}^n t_j \nu(E_j)\right| = \sum_{j=1}^n |\nu(E_j)| \geq \|\nu\| - \epsilon$$

so $\|F_\nu\| \geq \|\nu\|$ and equality holds.

Thus, the map $U : ba(\Sigma : \mu) \to L^\infty(\mu)'$, $U(\nu) = F_\nu$, is a linear isometry which obviously preserves order. We show that U is onto so we have the Riesz Representation Theorem for $L^\infty(\mu)'$.

Theorem 3 (Riesz Representation Theorem) $L^\infty(\mu)' = ba(\Sigma : \mu)$ *[as Banach lattices]*.

Proof: Let $F \in L^\infty(\mu)'$. For $E \in \Sigma$ set $\nu(E) = F(C_E)$. Then ν is obviously finitely additive, and since $|\nu(E)| \leq \|F\| \|C_E\|_\infty$, ν is bounded and absolutely continuous with respect to μ, i.e., $\nu \in ba(\Sigma : \mu)$.

Since $F(\varphi) = \int_S \varphi d\nu$ for every Σ-simple function φ, the argument above shows that $F = F_\nu = U(\nu)$.

The Dual of ℓ^∞:

If $\Sigma = \mathcal{P}(\mathbf{N})$, we write $\underline{ba} = ba(\mathcal{P}(\mathbf{N}))$. If μ is counting measure on $\mathcal{P}(\mathbf{N})$, then $\ell^\infty = L^\infty(\mu)$ and $ba = ba(\mathcal{P}(\mathbf{N}) : \mu)$ so $(\ell^\infty)'$ and ba are isometrically isomorphic as Banach lattices.

6.3. THE SPACE OF FINITELY ADDITIVE SET FUNCTIONS

Exercise 1. Show $ba(\Sigma : \mu)$ is a closed subspace of $ba(\Sigma)$.

Exercise 2. Show $\int_S h \, d\nu$ is well-defined for h bounded and measurable and $\nu \in ba(\Sigma)$ with $|\int_S h \, d\nu| \leq \|h\|_u |\nu|(S)$.

Exercise 3. (a) Show $h \in L^\infty(\mu)$ if and only if $\exists$ a bounded, Σ-measurable function f such that $\{t \in S : h(t) \neq f(t)\}$ is μ-null. (b) Show that if $h \in L^\infty(\mu)$, then

$$\|h\|_\infty = \inf\{\|f\|_u : f \text{ as in (a)}\}$$

and this inf is attained. [Hint: Show $\|f\|_u \geq \|h\|_\infty$ for such f; set $Z = \{t \in S : |h(t)| \leq \|h\|_\infty\}$ and show $\|hC_Z\|_u = \|h\|_\infty$.]

Exercise 4. Let $B(S, \Sigma)$ be the space of all bounded, Σ-measurable functions. Show $B(S, \Sigma)$ is a Banach space under the sup-norm and its dual is $ba(\Sigma)$.

Exercise 5. Show $\int_S h \, d(\mu + \nu) = \int_S h \, d\mu + \int_S h \, d\nu$ when $\mu, \nu \in ba(\Sigma)$ and $h \in B(S, \Sigma)$.

6.4 The Space of Countably Additive Set Functions

Let Σ be a σ-algebra of subsets of S and let $ca(\Sigma)$ be the space of all finite signed measures on Σ. Since any finite signed measure on a σ-algebra is bounded [2.2.1.5], $ca(\Sigma)$ is a linear subspace of $ba(\Sigma)$ and $\|\nu\| = |\nu|(S)$ is a norm on $ca(\Sigma)$. We show $ca(\Sigma)$ is complete under this norm.

Theorem 1 $ca(\Sigma)$ *is a Banach space.*

Proof: We show that $ca(\Sigma)$ is a closed subspace of $ba(\Sigma)$ [§6.3]. Let $\{\mu_k\} \subset ca(\Sigma)$ be such that $\{\mu_k\}$ converges to $\mu \in ba(\Sigma)$. Then

$$|\mu_k(E) - \mu(E)| \le \|\mu_k - \mu\|$$

for each $E \in \Sigma$ [2.2.1.7] so $\lim \mu_k(E) = \mu(E)$ exists. Let $\{E_j\}$ be pairwise disjoint from Σ and set $E = \bigcup_{j=1}^{\infty} E_j$. Let $\epsilon > 0$ and choose n such that $\|\mu_n - \mu\| < \epsilon$. Then

$$\left|\mu(E) - \sum_{i=1}^{p} \mu(E_i)\right| \le |\mu(E) - \mu_n(E)| + \left|\sum_{i=1}^{p}(\mu_n(E_i) - \mu(E_i))\right|$$

$$+ \left|\sum_{i=p+1}^{\infty} \mu_n(E_i)\right| < 2\epsilon + \left|\sum_{i=p+1}^{\infty} \mu_n(E_i)\right|. \tag{6.1}$$

The last term on the right hand side of (1) can be made $< \epsilon$ for p large so μ is countably additive.

$ca(\Sigma)$ is an ordered vector space under the pointwise order which it inherits from $ba(\Sigma)$, and we show that it is actually a vector lattice [Banach lattice].

Theorem 2 $ca(\Sigma)$ *is a Banach lattice. Moreover, if μ, $\nu \in ca(\Sigma)$, then*

$$\mu \wedge \nu(A) = \inf\left\{\mu(E) + \nu(A\backslash E) : E \in \Sigma, E \subset A\right\}$$

and

$$\mu \vee \nu(A) = \sup\left\{\mu(E) + \nu(A\backslash E) : E \in \Sigma, E \subset A\right\}$$

for $A \in \Sigma$.

Proof: By Lemma 3.12.1.1, it suffices to show that $\mu \wedge \nu$ ($\mu \vee \nu$) is countably additive. Let $\{A_i\} \subset \Sigma$ be pairwise disjoint and set $A = \bigcup_{i=1}^{\infty} A_i$. If $E \in \Sigma$, $E \subset A$, then

$$\mu(E) + \nu(A\backslash E) = \sum_{i=1}^{\infty}\{\mu(A_i \cap E) + \nu(A_i\backslash E)\} \ge \sum_{i=1}^{\infty} \mu \wedge \nu(A_i)$$

6.4. THE SPACE OF COUNTABLY ADDITIVE SET FUNCTIONS

so
$$\mu \wedge \nu(A) \geq \sum_{i=1}^{\infty} \mu \wedge \nu(A_i).$$

Let $\epsilon > 0$. For each i there exists $E_i \in \Sigma$ such that $E_i \subset A_i$ and
$$\mu(E_i) + \nu(A_i \backslash E_i) < \mu \wedge \nu(A_i) + \epsilon/2^i.$$

Now $\{E_i\}$ are pairwise disjoint and if $E = \bigcup_{i=1}^{\infty} E_i$, then $E \subset A$ and $\bigcup_{i=1}^{\infty}(A_i \backslash E_i) = A \backslash E$ so
$$\mu \wedge \nu(A) \leq \mu(E) + \nu(A \backslash E) = \sum_{i=1}^{\infty}\{\mu(E_i) + \nu(A_i \backslash E_i)\} < \sum_{i=1}^{\infty} \mu \wedge \nu(A_i) + \epsilon.$$

Hence, $\mu \wedge \nu(A) = \sum_{i=1}^{\infty} \mu \wedge \nu(A_i)$.

The other statement is similar. ∎

There is an interesting order theoretic characterization of singularity for measures.

Proposition 3 *Let $\mu, \nu \in ca(\Sigma)$ be measures. Then $\mu \perp \nu$ if and only if $\mu \wedge \nu = 0$.*

Proof: $\Rightarrow$: Let $A, B \in \Sigma$, $A \cap B = \emptyset$, $S = A \cup B$ with $\mu(A) = \nu(B) = 0$. Then
$$0 \leq \mu \wedge \nu(S) = \mu \wedge \nu(A) + \mu \wedge \nu(B) \leq \mu(A) + \nu(B) = 0$$
so $\mu \wedge \nu = 0$.

$\Leftarrow$: $\mu \wedge \nu(S) = 0$ implies that for each i there exists $E_i \in \Sigma$ such that $\mu(E_i) + \nu(E_i^c) < 1/2^i$. Let $A_n = \bigcup_{i=n}^{\infty} E_i$, $A = \bigcap_{n=1}^{\infty} A_n = \overline{\lim} E_i$, and $B = A^c$. Then
$$\mu(A) \leq \mu(A_n) \leq \sum_{i=n}^{\infty} \mu(E_i) \leq \sum_{i=n}^{\infty} 1/2^i = 1/2^{n-1}$$
implies $\mu(A) = 0$. Also
$$\nu(A_n^c) = \nu\left(\bigcap_{i=n}^{\infty} E_i^c\right) \leq \nu(E_i^c) < 1/2^i$$
for $i \geq n$ implies $\nu(A_n^c) = 0$ so
$$\nu(B) = \nu\left(\bigcup_{n=1}^{\infty} A_n^c\right) \leq \sum_{n=1}^{\infty} \nu(A_n^c) = 0.$$

Hence, $\mu \perp \nu$. ∎

Exercise 1. Let $\lambda \in ca(\Sigma)$ be positive and let $ca(\Sigma : \lambda)$ be the subspace of $ca(\Sigma)$ consisting of the elements which are absolutely continuous with respect to λ. Show $ca(\Sigma : \lambda)$ is isometrically isomorphic to $L^1(\lambda)$.

Exercise 2. Show $\nu \in ba(\Sigma)$ is purely finitely additive if and only if $\nu \perp ca(\Sigma)$.

6.5 The Space of Continuous Functions

Let S be a locally compact Hausdorff space and $\mathcal{B} = \mathcal{B}(S)$, the Borel sets of S. Recall that $C_c(S)$ is the space of all real-valued continuous functions on S which have compact support. We equip $C_c(S)$ with the sup-norm, $\|\ \|_\infty$; $C_c(S)$ is not generally complete [see Exercise 3]. Under the pointwise order, $C_c(S)$ is a normed vector lattice. We begin by considering positive linear functionals on $C_c(S)$.

If μ is a Borel measure on $\mathcal{B}$ [so μ is finite on compact subsets], then μ induces a positive linear functional, F_μ, on $C_c(S)$ defined by $< F_\mu, \varphi > = \int_S \varphi d\mu$. If $F : C_c(S) \to \mathbf{R}$ is a positive linear functional on $C_c(S)$ and $F = F_\mu$ for some Borel measure μ, we say that μ is a *representing measure* for F. We first observe that regular representing measures are unique.

Proposition 1 *Let μ, ν be regular Borel measures on $\mathcal{B}$ such that $\int_S \varphi d\mu = \int_S \varphi d\nu$ for every $\varphi \in C_c(S)$. Then $\mu = \nu$.*

Proof: Notation: If V is open and $f \in C_c(S)$, then $\underline{f \prec V}$ means $0 \leq f \leq 1$ and $spt(f) \subset V$. If K is compact and $f \in C_c(S)$, $\underline{K \prec f}$ means $0 \leq f \leq 1$ and $f = 1$ on K.

By regularity, it suffices to show that $\mu(K) = \nu(K)$ for every compact K. Let $\epsilon > 0$. Pick V open such that $K \subset V$ and $\mu(V) < \mu(K) + \epsilon$. There exists $f \in C_c(S)$ such that $K \prec f \prec V$ [3.5.5]. Since $C_K \leq f \leq C_V$,

$$\nu(K) = \int_S C_K d\nu \leq \int_S f d\nu = \int_S f d\mu \leq \int_S C_V d\mu = \mu(V) < \mu(K) + \epsilon$$

so $\nu(K) \leq \mu(K)$. By symmetry, $\nu(K) = \mu(K)$.

Without the regularity assumption, Proposition 1 can fail; see Exercise 4.

We now show that every positive linear functional on $C_c(S)$ has a (unique) regular representing measure. For the proof we require a preliminary lemma.

Lemma 2 *Let $K \subset S$ be compact and $V_1, \ldots, V_n$ open with $K \subset \bigcup_{i=1}^{n} V_i$. Then there exist continuous functions $f_1, \ldots, f_n \in C_c(S)$ such that*

(i) $0 \leq f_i \leq 1$, $i = 1, \ldots, n$,

(ii) $spt(f_i) \subset V_i$,

(iii) $\sum_{i=1}^{n} f_i(t) = 1$ *for all $t \in K$.*

$\{f_i\}$ *is called a* partition of unity.

6.5. THE SPACE OF CONTINUOUS FUNCTIONS

Proof: Let $x \in K$. Then $x \in V_i$ for some i and there exists an open neighborhood U_x of x such that $\overline{U}_x$ is compact and $x \in U_x \subset \overline{U}_x \subset V_i$. Let $x_1, \ldots, x_m \in K$ be such that $K \subset \bigcup_{i=1}^{m} U_{x_i}$ and let H_i be the union of those U_{x_j} such that $\overline{U}_{x_j} \subset V_i$. By 3.5.5 for each i there exists $g_i \in C_c(S)$ such that $\overline{H}_i \prec g_i \prec V_i$. Put

$$f_1 = g_1, f_2 = (1-g_1)g_2, \ldots, f_n = (1-g_1)(1-g_2)\cdots(1-g_{n-1})g_n.$$

Then $f_i \prec V_i$ and

$$f_1 + \ldots + f_n = 1 - (1-g_1)(1-g_2)\cdots(1-g_n). \tag{6.1}$$

Since $K \subset \bigcup_{i=1}^{m} H_i$, at least one $g_i(x) = 1$ for each $x \in K$ so (iii) holds by (1).

To motivate the construction of a representing measure for a positive linear functional, we give the following result.

Proposition 3 *If μ is a regular Borel measure and V is open, then*

$$\mu(V) = \sup\{\int_S f d\mu : f \in C_c(S), f \prec V\}.$$

Proof: Clearly $\mu(V)$ is greater than the sup above. Let $r < \mu(V)$. Choose $K \subset V$ compact such that $\mu(K) > r$. Choose $f \in C_c(S)$ such that $K \prec f \prec V$ [3.5.5]. Then $\int_S f d\mu \geq \mu(K) > r$ so the equality follows.

Theorem 4 (Riesz Representation Theorem) *Let F be a positive linear functional on $C_c(S)$. Then there exists a unique regular Borel measure μ representing F, i.e., $F = F_\mu$.*

Proof: For V open, define

$$\mu(V) = \sup\{F(f) : f \in C_c(S), f \prec V\}$$

[cf. Proposition 3]. Certainly, $0 \leq \mu(V) \leq \infty$ and $\mu(U) \leq \mu(V)$ if U is open and $U \subset V$. Extend μ to all subsets of S by setting $\mu^*(A) = \inf\{\mu(V) : V \text{ open}, V \supset A\}$; note $\mu^*(V) = \mu(V)$ for all open V.

We claim that μ^* is an outer measure on S. Clearly $0 \leq \mu^*(A) \leq \infty$ for $A \subset S$, $\mu^*(\emptyset) = 0$ and μ^* is increasing on subsets of S. It remains to show that μ^* is countably subadditive. Let $\{A_i\} \subset S$ and $A = \bigcup_{i=1}^{\infty} A_i$. If $\sum_{i=1}^{\infty} \mu^*(A_i) = \infty$, there is nothing to prove so assume $\sum_{i=1}^{\infty} \mu^*(A_i) < \infty$. Let $\epsilon > 0$. For each i choose V_i open such that $A_i \subset V_i$ and $\mu(V_i) < \mu^*(A_i) + \epsilon/2^i$. Set $V = \bigcup_{i=1}^{\infty} V_i$. If $f \prec V$, then $K = spt(f) \subset \bigcup_{i=1}^{\infty} V_i$

so there exists n such that $K \subset \bigcup_{i=1}^{n} V_i$. By Lemma 2 there exist $f_1, \ldots, f_n \in C_c(S)$ such that $f_i \prec V_i$ and $\sum_{i=1}^{n} f_i = 1$ on K. Then $f \leq \sum_{i=1}^{n} f_i$ so

$$F(f) \leq \sum_{i=1}^{n} F(f_i) \leq \sum_{i=1}^{n} \mu(V_i) \leq \sum_{i=1}^{\infty} \mu(V_i) \leq \sum_{i=1}^{\infty} \mu^*(A_i) + \epsilon$$

and

$$\mu(V) \leq \sum_{i=1}^{\infty} \mu^*(A_i) + \epsilon.$$

Hence,

$$\mu^*(A) \leq \sum_{i=1}^{\infty} \mu^*(A_i).$$

Next, we claim that $\mu^*(K) < \infty$ for each compact $K \subset S$. Choose V open, with compact closure such that $K \subset V$. There exists $g \in C_c(S)$ with $\overline{V} \prec g$ [3.5.5]. If $f \prec V$, then $f \leq g$ so $F(f) \leq F(g)$ and $\mu^*(K) \leq \mu(V) \leq F(g) < \infty$.

We now claim that μ^* is finitely additive over the family of compact sets. Let K_1, K_2 be compact with $K_1 \cap K_2 = \emptyset$. Since μ^* is countably subadditive, it suffices to show that $\mu^*(K_1) + \mu^*(K_2) \leq \mu^*(K_1 \cup K_2)$. Since $K_1 \subset K_2^c$, there exists an open V_1 with compact closure such that $K_1 \subset V_1 \subset \overline{V_1} \subset K_2^c$. Hence, $K_2 \subset \left(\overline{V_1}\right)^c$. Set $V_2 = \left(\overline{V_1}\right)^c$ and note $V_1 \cap V_2 = \emptyset$. Choose an open $V \supset K_1 \cup K_2$ with $\mu(V) \leq \mu^*(K_1 \cup K_2) + \epsilon$. Then $K_1 \subset V \cap V_1$, $K_2 \subset V \cap V_2$. Pick $f_1, f_2 \in C_c(S)$ such that $f_1 \prec V \cap V_1$, $f_2 \prec V \cap V_2$, $\mu(V \cap V_1) < F(f_1) + \epsilon$ and $\mu(V \cap V_2) < F(f_2) + \epsilon$. Note $f_1 + f_2 \prec V$ since $V_1 \cap V_2 = \emptyset$. Then

$$\mu^*(K_1) + \mu^*(K_2) \leq \mu(V \cap V_1) + \mu(V \cap V_2)$$
$$< F(f_1) + F(f_2) + 2\epsilon = F(f_1 + f_2) + 2\epsilon$$
$$\leq \mu(V) + 2\epsilon < \mu^*(K_1 \cup K_2) + 3\epsilon$$

so $\mu^*(K_1) + \mu^*(K_2) \leq \mu^*(K_1 \cup K_2)$.

We now claim that for each open V, $\mu(V) = \sup\{\mu^*(K) : V \supset K \text{ compact}\}$ [inner regularity]. Let $r < \mu(V)$. Choose $f \prec V$ with $r < F(f)$. Let $K = \mathrm{spt}(f)$. If W is open, $K \subset W$, then $f \prec W$ so $r < F(f) \leq \mu(W)$ which implies $\mu(V) \geq \mu^*(K) \geq F(f) > r$ from which the desired equality follows.

We next assert that $\mathcal{B} \subset \mathcal{M}(\mu^*)$, the class of μ^*-measurable sets. First note that $\mu^*(K) + \mu(V) = \mu^*(K \cup V)$ when K is compact, V is open and $K \cap V = \emptyset$ [if $K_1 \subset V$ is compact, then by the finite additivity of μ^* over compact sets and the countable subadditivity of μ^*,

$$\mu^*(K) + \mu^*(K_1) = \mu^*(K \cup K_1) \leq \mu^*(K \cup V) \leq \mu^*(K) + \mu^*(V)$$

and the identity follows from the inner regularity of the preceding paragraph].

It suffices to show that if V is open, then V is μ^*-measurable. Let $A \subset S$ with $\mu^*(A) < \infty$ be a test set. First assume that A is open. Let K be compact with

6.5. THE SPACE OF CONTINUOUS FUNCTIONS

$K \subset A \cap V$. Then the open set $W = A \backslash K$ satisfies $K \cap W = \emptyset$ and $A \cap V^c \subset W \subset A$ so by the observation in the paragraph above,

$$\mu^*(K) + \mu^*(A \backslash V) \leq \mu^*(K) + \mu(W) = \mu^*(K \cup W) \leq \mu^*(A).$$

The inner regularity of open sets implies that $\mu^*(A \cap V) + \mu^*(A \backslash V) \leq \mu^*(A)$ for every open A.

Now let $A \subset S$ be arbitrary. If $W \supset A$ is open, then by the above

$$\mu^*(A \cap V) + \mu^*(A \backslash V) \leq \mu(W \cap V) + \mu^*(W \backslash V) \leq \mu(W).$$

By the definition of μ^*,

$$\mu^*(A \cap V) + \mu^*(A \backslash V) \leq \mu^*(A)$$

and V is μ^*-measurable.

It follows that μ^* restricted to $\mathcal{B}$, which we denote by μ, is a measure on $\mathcal{B}$. By the properties established above μ is a Borel measure and every open set is inner regular. Every Borel set is outer regular by the definition of μ^* so μ is a regular Borel measure.

Finally, we claim that $F(f) = \int_S f d\mu$ for $f \in C_c(S)$. Let $f \in C_c(S)$, $\epsilon > 0$ and fix V open with $K = \mathrm{spt}(f) \subset V$ and $\mu(V) < \infty$. Choose $c > 0$ such that $|f(t)| < c$ for $t \in S$. Partition the interval $[-c, c]$ in the range of f by $y_i = -c + i(2c/n)$, $i = 0, 1, \ldots, n$, where n is chosen such that $2c/n < \epsilon$. Hence, $y_i - y_{i-1} = 2c/n < \epsilon$. Let $A_i = \{t \in K : y_{i-1} < f(t) \leq y_i\}$ $(i = 1, \ldots, n)$ be the partition of K induced by the $\{y_i\}$. Note the open set $W_i = \{t \in V : y_{i-1} - \epsilon < f(t) < y_i + \epsilon\}$ contains A_i. By regularity for each i there exists open V_i with $A_i \subset V_i \subset W_i$ and $\mu(V_i \backslash A_i) < \epsilon/n$. Then $K \subset \bigcup_{i=1}^n V_i \subset V$. By Lemma 2 there exists $g_1, \ldots, g_n \in C_c(S)$ such that $g_i \prec V_i$ and $\sum_{i=1}^n g_i(t) = 1$ for $t \in K$. Note that $fg_i \leq (y_i + \epsilon)g_i$ and $f = \sum_{i=1}^n fg_i$. Thus,

$$\begin{aligned}
F(f) - \int_S f d\mu &= \sum_{i=1}^n F(gf_i) - \sum_{i=1}^n \int_{A_i} f d\mu \\
&\leq \sum_{i=1}^n (y_i + \epsilon) F(g_i) - \sum_{i=1}^n (y_i - \epsilon) \mu(A_i) \\
&\leq \sum_{i=1}^n (y_i + \epsilon) \mu(V_i) - \sum_{i=1}^n (y_i - \epsilon) \mu(A_i) \\
&= \sum_{i=1}^n (y_i + \epsilon)(\mu(V_i) - \mu(A_i)) + 2\epsilon \sum_{i=1}^n \mu(A_i) \\
&\leq \sum_{i=1}^n (c + \epsilon)\epsilon/n + 2\epsilon \mu(K) = \epsilon[c + \epsilon + 2\mu(K)]
\end{aligned}$$

so $F(f) \leq \int_S f d\mu$. Replacing f by $-f$ gives the reverse inequality.

Corollary 5 *F is continuous if and only if μ is finite. In this case, $\|F\| \leq \mu(S) = \|\mu\|$.*

Proof: $\Leftarrow$: $|F(f)| \leq \int_S |f|\, d\mu \leq \|f\|_\infty \mu(S)$.
$\Rightarrow$: $\mu(S) = \sup\{F(f) : f \prec S\} \leq \|F\|$ since $\|f\|_\infty \leq 1$ when $f \prec S$.

We denote by $rca(S)$ the space of all finite, regular signed measures ν on $\mathcal{B}$ with the norm $\|\nu\| = |\nu|(S)$ [§6.4; recall a signed measure ν is regular if $|\nu|$ is regular]. Each $\nu \in rca(S)$ induces a linear functional F_ν on $C_c(S)$ via integration,

$$F_\nu(f) = \int_S f\, d\nu = \int_S f\, d\nu^+ - \int_S f\, d\nu^-$$

(see Exer. 3.2.31), and since

$$|F_\nu(f)| \leq \int_S |f|\, d|\nu| \leq \|f\|_\infty \|\nu\|,$$

F_ν is continuous with $\|F_\nu\| \leq \|\nu\|$. We say that ν is a *representing signed measure* for F_ν. Thus, we have a linear map $U : \nu \to F_\nu$ from $rca(S)$ into $C_c(S)'$. We show that U is an isometry.

Lemma 6 *Let $K_1, \ldots, K_n$ be pairwise disjoint compact subsets of S and $a_1, \ldots, a_n \in \mathbf{R}$. Then there exists $f \in C_c(S)$ such that*

(i) $f(t) = a_i$ *for* $t \in K_i$,

(ii) $\|f\|_\infty = \max\{|a_1|, \ldots, |a_n|\}$.

Proof: There exist pairwise disjoint $V_1, \ldots, V_n$ such that $K_i \subset V_i$ [for $n = 2$, this is easy; use induction]. By 3.5.5 there exist $f_i \in C_c(S)$ such that $K_i \prec f_i \prec V_i$. Set $f = \sum_{i=1}^n a_i f_i$.

Theorem 7 *Let $\nu \in rca(S)$. Then $\|F_\nu\| = \|\nu\|$.*

Proof: The inequality $\|F_\nu\| \leq \|\nu\|$ was observed above.
Let $\epsilon > 0$. There exists a partition $\{A_1, \ldots, A_n\}$ of S by elements of $\mathcal{B}$ such that

$$\sum_{i=1}^n |\nu(A_i)| > \|\nu\| - \epsilon$$

[2.2.1.7]. By regularity, choose compact sets $K_i \subset A_i$ such that

$$\|\nu\| - \epsilon < \sum_{i=1}^n |\nu(K_i)| \leq \sum_{i=1}^n |\nu|(K_i),$$

where we may assume $\nu(K_i) \neq 0$ for all i. By Lemma 6 choose $f \in C_c(S)$ such that $f = \operatorname{sign} \nu(K_i)$ on K_i and $\|f\|_\infty \leq 1$. Set $K = \bigcup_{i=1}^n K_i$. Then

$$\int_K f\, d\nu = \sum_{i=1}^n \int_{K_i} f\, d\nu = \sum_{i=1}^n |\nu(K_i)| > \|\nu\| - \epsilon$$

6.5. THE SPACE OF CONTINUOUS FUNCTIONS

and
$$\left|\int_{K^c} f d\nu\right| \leq |\nu|(K^c) < \epsilon.$$
Hence,
$$\|F_\nu\| \geq F_\nu(f) = \int_K f d\nu + \int_{K^c} f d\nu \geq \|\nu\| - 2\epsilon$$
so $\|F_\nu\| \geq \|\nu\|$.

We next show that U is a linear isometry onto $C_c(S)'$.

Theorem 8 $U : rca(S) \to C_c(S)'$ is onto.

Proof: Let $F \in C_c(S)'$. Then $F = F_1 - F_2$, where F_i are positive, continuous linear functionals on $C_c(S)$ [Remark 5.7.20]. Let μ_i be the regular Borel measure representing F_i [Theorem 4]. Since F_i is continuous, μ_i is finite so $\mu = \mu_1 - \mu_2 \in rca(S)$ is a representing signed measure for F.

Thus, as Banach spaces $rca(S)$ and $C_c(S)'$ are the same. We show that they are also equal as Banach lattices, i.e., that U is a lattice isomorphism.

Proposition 9 $\nu \in rca(S)$ is positive [i.e., a measure] if and only if F_ν is positive.

Proof: $\Rightarrow$: Clear. $\Leftarrow$: By the first part of the proof of Theorem 4 the representing measure for F_ν satisfies
$$\nu(V) = \sup\{F_\nu(f) : f \prec V\}$$
for V open. Hence, $\nu(V) \geq 0$ for every open V and $\nu \geq 0$ by regularity.

Thus, both U and U^{-1} are positive linear maps and U is a lattice isomorphism by Theorem 5.7.9. Summarizing, we have

Theorem 10 (Riesz Representation Theorem) U is an isometric lattice isomorphism from $rca(S)$ onto $C_c(S)'$ $[rca(S) = C_c(S)']$.

Concerning reflexivity, we have

Example 11 $C[0,1]$ is not reflexive since $C[0,1]$ is separable but $rca[0,1]$ is not separable [if $t \neq s$, $\|\delta_t - \delta_s\| = 2$ where δ_t is the point mass measure at t; Corollary 5.6.1.13].

For the case when S is an interval $I = [a, b]$ in $\mathbf{R}$ there is another description of the dual of $C(I)$ using the Riemann-Stieltjes integral. If g is a function of bounded variation on I, then $F_g(f) = \int_a^b f dg$ defines a linear functional on $C(I)$, and since
$$|F_g(f)| \leq \|f\|_\infty \operatorname{Var}(g : I),$$
F_g is continuous with $\|F_g\| \leq \operatorname{Var}(g : I)$. It can be shown that $\|F_g\| = \operatorname{Var}(g : I)$ so the map $\Psi : g \to F_g$ is a linear isometry from $BV[a,b]$ into $C(I)'$. This isometry is not one-one, but if $BV[a,b]$ is replaced by the space of normalized functions of bounded variation, $NBV[a,b]$, consisting of the right continuous functions which vanish at a,

then Ψ is a one-one, linear isometry from $NBV[a,b]$ onto $C(I)'$. For a description of this version of the Riesz Representation Theorem, see [RN] or [TL].

For a discussion of the history of the Riesz Representation Theorem, see [Gr].

Exercise 1. Let $x \in S$. Define $F : C_c(S) \to \mathbf{R}$ by $F(f) = f(x)$. Show F is a continuous, positive linear functional on $C_c(S)$. What is its representing measure?

Exercise 2. Let $g : [a,b] \to \mathbf{R}$ be continuous. Define $G : C[a,b] \to \mathbf{R}$ by $G(f) = \int_a^b fg\,dm$. Show $G \in C[a,b]'$. What is G's representing measure?

Exercise 3. Let $\mathbf{N}$ have the discrete topology. Show $C_c(\mathbf{N}) = c_{00}$. Define F on $C_c(\mathbf{N})$ by $F(f) = \sum_{j=1}^{\infty} jf(j)$. Show F is positive, linear and find its representing measure. Is F continuous?

Exercise 4. Let $\mathbf{R}_d$ be the real line with the discrete topology. Let $S = \mathbf{R}_d \times \mathbf{R}$ with the product topology.

(a) Show that a subset of S is open if and only if its intersection with every vertical line is open in $\mathbf{R}$.

(b) Show the topology on S is locally compact.

(c) Show $f \in C_c(S)$ if and only if $f(\cdot, y) \in C_c(\mathbf{R})$ for every y and $f(\cdot, y) = 0$ for all but finitely many y.

(d) Define F on $C_c(S)$ by: if $f \in C_c(S)$, let $x_1, \ldots, x_n$ be those values x for which there exists a y with $f(x,y) \neq 0$ [part (c)] and set

$$F(f) = \sum_{i=1}^{n} \int_{-\infty}^{\infty} f(x_i, y)dy = \sum_{x \in \mathbf{R}} \int_{-\infty}^{\infty} f(x,y)dy.$$

Show F is positive and linear.

(e) Define ν on $\mathcal{B}(S)$ by $\nu(E) = \sum_{x \in \mathbf{R}} m(\{y : (x,y) \in E\})$. Show ν is a measure which represents F.

(f) Show ν is not regular.

Exercise 5. Let $C_\infty(S)$ be the space of all continuous functions which vanish at ∞ in the sense that for every $\epsilon > 0$ there exists a compact set K such that $|f(t)| < \epsilon$ for $t \notin K$. Show $C_\infty(S)$ is a Banach space under the sup-norm and describe its dual space.

6.6 Hilbert Space

In this chapter we set down the basic properties of Hilbert spaces. Hilbert spaces are special cases of Banach spaces and have many important and special properties not shared by general Banach spaces.

Let X be a vector space over the field $\mathbf{F}$ of either real or complex numbers. If $z \in \mathbf{C}$, the complex conjugate of z is denoted by $\bar{z}$ and the modulus by $|z|$.

Definition 1 *An* inner product (scalar product, dot product) *on X is a function $\cdot : X \times X \to \mathbf{F}$, $(x,y) \to x \cdot y$, satisfying*

(i) $(x+y) \cdot z = x \cdot z + y \cdot z \ \forall x, y, z \in X$,

(ii) $\lambda(x \cdot y) = (\lambda x) \cdot y \ \forall x, y \in X, \lambda \in \mathbf{F}$,

(iii) $x \cdot y = \overline{y \cdot x} \ \forall x, y \in X$,

(iv) $x \cdot x \geq 0 \ \forall x \in X$,

(v) $x \cdot x = 0$ *if and only if* $x = 0$.

A vector space X with an inner product defined on it is called an *inner product space*.

It follows easily from the axioms that $0 \cdot x = x \cdot 0 = 0$, $x \cdot (\lambda y) = \bar{\lambda}(x \cdot y)$ and $x \cdot (y+z) = x \cdot y + x \cdot z$. We have the important inequality.

Theorem 2 (Cauchy-Schwarz Inequality) *If X is an inner product space, then*

$$|x \cdot y| \leq \sqrt{x \cdot x}\sqrt{y \cdot y} \ \forall x, y \in X. \tag{6.1}$$

Proof: If $y = 0$, the result is trivial so assume $y \neq 0$. In this case, (1) is equivalent to $\left|x \cdot y/\sqrt{y \cdot y}\right| \leq \sqrt{x \cdot x}$ so we may assume $y \cdot y = 1$. Then

$$\begin{aligned} 0 \leq (x - (x \cdot y)y) \cdot (x - (x \cdot y)y) &= x \cdot x + |x \cdot y|^2 - (x \cdot y)(y \cdot x) - (\overline{x \cdot y})(x \cdot y) \\ &= x \cdot x + |x \cdot y|^2 - (x \cdot y)(\overline{x \cdot y}) - |x \cdot y|^2 = x \cdot x - |x \cdot y|^2. \end{aligned} \tag{6.2}$$

Remark 3 Equality holds in (1) if x and y are linearly dependent. The converse also holds for if equality holds in (1) with $y \neq 0$, then (2) implies that $x - (x \cdot y)y = 0$. Note that axiom (v) was not used in the proof of (1).

Proposition 4 *If X is an inner product space, the map $x \to \sqrt{x \cdot x} = \|x\|$ defines a norm on X.*

Proof: Only the triangle inequality needs to be checked. For $x, y \in X$,

$$\|x+y\|^2 = (x+y) \cdot (x+y) \leq \|x\|^2 + \|y\|^2 + 2\|x\|\|y\| = (\|x\| + \|y\|)^2$$

by Theorem 2.

If X is an inner product space, we always assume that X is equipped with the norm induced by the inner product.

Proposition 5 *The inner product is a continuous function from $X \times X \to \mathbf{F}$.*

Proof: If $x_k \to x$ and $y_k \to y$, then

$$|x_k \cdot y_k - x \cdot y| \leq |x_k \cdot y_k - x_k \cdot y| + |x_k \cdot y - x \cdot y| \leq \|x_k\|\|y_k - y\| + \|x_k - x\|\|y\|$$

by the Cauchy-Schwarz Inequality.

We have the following important property of the norm in an inner product space.

Proposition 6 (Parallelogram Law) *If X is an inner product space, then*

$$\|x+y\|^2 + \|x-y\|^2 = 2\|x\|^2 + 2\|y\|^2 \quad \forall x, y \in X.$$

Proof:

$$\|x+y\|^2 + \|x-y\|^2 = (x+y) \cdot (x+y) + (x-y) \cdot (x-y)$$
$$= 2\|x\|^2 + 2\|y\|^2 + x \cdot y + y \cdot x - x \cdot y - y \cdot x.$$

Remark 7 The parallelogram law characterizes inner product spaces among the class of NLS. That is, if X is a NLS whose norm satisfies the parallelogram law, then the norm of X is induced by an inner product. If X is real, the inner product is defined by

$$4x \cdot y = \|x+y\|^2 - \|x-y\|^2,$$

while if X is complex, the inner product is defined by

$$4x \cdot y = \|x+y\|^2 - \|x-y\|^2 + i\|x+iy\|^2 - i\|x-iy\|^2.$$

We leave the (tedious!) verification to the reader.

See Exercise 7 for an example of a norm which is not induced by an inner product.

Definition 8 *An inner product space which is complete under the induced norm is called a* Hilbert space.

Example 9 $\mathbf{R}^n$ or $\mathbf{C}^n$ is a Hilbert space under the usual inner product

$$x \cdot y = \sum_{i=1}^{n} x_i \overline{y_i}, \quad x = (x_1, \ldots, x_n), \quad y = (y_1, \ldots, y_n).$$

The norm induced by the inner product is the Euclidean norm.

6.6. HILBERT SPACE

Example 10 In the case of complex scalars, ℓ^2 is the space of complex-valued sequences $\{x_i\}$ such that $\sum_{i=1}^{\infty} |x_i|^2 < \infty$. ℓ^2 under the inner product

$$x \cdot y = \sum_{i=1}^{\infty} x_i \overline{y_i}, \quad x = \{x_i\}, \quad y = \{y_i\},$$

is a (the original!) Hilbert space.

Example 11 Let (S, Σ, μ) be a measure space. The space of real-valued functions, $L^2(\mu)$, is a Hilbert space under the inner product $f \cdot g = \int_S fg\, d\mu$ since this inner product induces the usual L^2-norm. We also want to define the space $L^2(\mu)$ for complex-valued functions. For this we need to discuss integrating complex-valued functions.

Let $f: S \to \mathbf{C}$ with u and v the real and imaginary parts of f, respectively. We say that f is measurable (integrable) if and only if both u and v are measurable (integrable). If f is μ-integrable, we define the integral of f with respect to μ to be

$$\int_S f\, d\mu = \int_S u\, d\mu + i \int_S v\, d\mu.$$

If $L^1_\mathbf{C}(\mu)$ is the space of all $\mathbf{C}$-valued, μ-integrable functions, it is easily checked that $L^1(\mu)$ is a vector space and the integral is a linear functional on $L^1(\mu)$. Since $|u| \leq |f|$, $|v| \leq |f|$ and $|f| \leq |u| + |v|$, a measurable function f is integrable if and only if $|f|$ is integrable. Moreover, if f is integrable, then

$$\left| \int_S f\, d\mu \right| \leq \int_S |f|\, d\mu$$

[let $\int_S f\, d\mu = |\int_S f\, d\mu| e^{i\theta}$ so

$$\begin{aligned} |\int_S f\, d\mu| &= e^{-i\theta} \int_S f\, d\mu = \int_S e^{-i\theta} f\, d\mu = \int_S \mathcal{R}(e^{-i\theta} f)\, d\mu \\ &\quad + i \int_S \mathcal{I}(e^{-i\theta} f)\, d\mu = \int_S \mathcal{R}(e^{-i\theta} f)\, d\mu \\ &\leq \int_S \left| \mathcal{R}(e^{-i\theta} f) \right| d\mu \leq \int_S |f|\, d\mu, \end{aligned}$$

where $\mathcal{R}z$ ($\mathcal{I}z$) denotes the real (imaginary) part of $z \in \mathbf{C}$].

For $1 \leq p < \infty$, let $L^p_\mathbf{C}(\mu)$ be the space of all $\mathbf{C}$-valued measurable functions f such that $|f| \in L^p(\mu)$ and set $\|f\|_p = (\int_S |f|^p\, d\mu)^{1/p}$. Then $L^p_\mathbf{C}$ is a complex Banach space; we usually write simply $L^p(\mu)$ for this space and indicate whether real or complex scalars are being used.

For $p = 2$, $L^2(\mu)$ over the scalar field of complex numbers is a Hilbert space under the inner product $f \cdot g = \int_S f\overline{g}\, d\mu$.

Example 12 $C[a, b]$ with the inner product $f \cdot g = \int_a^b f(t)\overline{g}(t)\, dt$ is an inner product space which is not a Hilbert space.

We establish an important geometric property of Hilbert space.

Theorem 13 Let K be non-void, closed, convex subset of a Hilbert space H. If $x \in H$, then there is a unique $y \in K$ such that
$$\|x - y\| = \min\{\|x - z\| : z \in K\} = \operatorname{dist}(x, K).$$
Furthermore, y can be characterized by:
$$y \in K, \mathcal{R}(x - y) \cdot (z - y) \leq 0 \text{ for all } z \in K. \tag{6.3}$$

Proof: Set $d = \operatorname{dist}(x, K) \geq 0$. If $w, z \in K$, applying the parallelogram law to $(x - z)/2$ and $(x - w)/2$ gives
$$d^2 \leq \|(w + z)/2 - x\|^2 = \|w - x\|^2/2 + \|z - x\|^2/2 - \|(w - z)/2\|^2 \tag{6.4}$$
since $(w + z)/2 \in K$.

If $\|z - x\| = d$ and $\|w - x\| = d$, then (4) implies that $w = z$ so uniqueness holds. Pick $\{y_k\} \subseteq K$ such that $\|x - y_k\| \to d$. Set $w = y_k$, $z = y_j$ in (4) to obtain
$$\|(y_k - y_j)/2\|^2 \leq \|y_k - x\|^2/2 + \|y_j - x\|^2/2 - d^2 \to 0.$$
Thus, $\{y_k\}$ is a Cauchy sequence in H and converges to some $y \in K$ with $\|x - y\| = d$.
If $y \in K$ satisfies $\|y - x\| = d$, then for $z \in K$ and $0 < t < 1$,
$$\|x - y\| \leq \|x - tz - (1 - t)y\| = \|x - y - t(z - y)\|$$
so
$$\|x - y\|^2 \leq \|x - y\|^2 + t^2 \|z - y\|^2 - t(x - y) \cdot (z - y) - t(z - y) \cdot (x - y)$$
or
$$2\mathcal{R}(x - y) \cdot (z - y) \leq t \|z - y\|^2.$$
Letting $t \to 0$ gives (3).

On the other hand if y satisfies (3) and $z \in K$, computing $\|x - z\|^2 = \|(x - y) - (z - y)\|^2$ gives
$$\|x - y\|^2 - \|x - z\|^2 = 2\mathcal{R}(x - y) \cdot (z - y) - \|z - y\|^2 \leq 0$$
so $\|x - y\| = d$.

Let $P_K : H \to H$ be the "projection" map which sends x to y in Theorem 13. If $H = \mathbf{R}^2$, inequality (3) means that the vector from x to $P_K x$ makes an obtuse angle with the vector from z to $P_K x$. Moreover, this map is uniformly continuous on H since $\|P_K u - P_K v\| \leq \|u - v\|$. [From (3), $\mathcal{R}(u - P_K u) \cdot (P_K v - P_K u) \leq 0$ and $\mathcal{R}(v - P_K v) \cdot (P_K u - P_K v) \leq 0$ so adding gives $\mathcal{R}(u - v - (P_K u - P_K v)) \cdot (P_K v - P_K u) \leq 0$ so
$$\|P_K u - P_K v\|^2 \leq \mathcal{R}(u - v) \cdot (P_K v - P_K u) \leq \|u - v\| \|P_K v - P_K u\|$$
by the Cauchy-Schwarz Inequality.]

If X is an inner product space, then two elements $x, y \in X$ are *orthogonal*, written $x \perp y$, if $x \cdot y = 0$. A subset $E \subseteq X$ is said to be *orthogonal* if $x \perp y \ \forall x, y \in E$, $x \neq y$. If $E \subseteq X$ is orthogonal and $\|x\| = 1 \ \forall x \in E$, then E is said to be *orthonormal*.

6.6. HILBERT SPACE

Example 14 In ℓ^2 (over either $\mathbf{R}$ or $\mathbf{C}$), $\{e^k : k \in \mathbf{N}\}$ is orthonormal, where e^k is the sequence with a 1 in the kth coordinate and 0 in the other coordinates.

Example 15 In $L^2[-\pi, \pi]$ (over $\mathbf{C}$), $\{e^{int}/\sqrt{2\pi} : n = 0, \pm 1, \ldots\}$ is orthonormal.

Proposition 16 (Pythagorean Theorem) *If* $x \perp y$, *then* $\|x + y\|^2 = \|x - y\|^2 = \|x\|^2 + \|y\|^2$.

Proof: $(x + y) \cdot (x + y) = \|x + y\|^2 = \|x\|^2 + \|y\|^2 = (x - y) \cdot (x - y)$.

If X is an inner product space and $M \subseteq X$, we set $M^\perp = \{x \in X : x \perp y \,\forall y \in M\}$; $M^\perp$ is called the *orthogonal complement* of M. Note that $M^\perp$ is a closed linear subspace of X. We now use Theorem 13 to show that any closed linear subspace of a Hilbert space is complemented.

Theorem 17 *If M is a closed linear subspace of a Hilbert space H, then $H = M \oplus M^\perp$.*

Proof: For $x \in H$ let $y = P_M x \in M$ be as in Theorem 13. Then $x - y \in M^\perp$ since by (3) $\mathcal{R}(x - y) \cdot w \leq 0$ for every $w \in M$.

Hence, $x = (x - y) + y$ with $x - y \in M^\perp$ and $y \in M$. Since $M \cap M^\perp = \{0\}$, we have $H = M \oplus M^\perp$.

We can now establish the Riesz Representation Theorem for Hilbert space.

Proposition 18 *Let H be a Hilbert space and $y \in H$. If $f_y : H \to \mathbf{F}$ is defined by $\langle f_y, x \rangle = x \cdot y$, then $f_y \in H'$ and $\|f_y\| = \|y\|$.*

Proof: f_y is clearly linear and since $|\langle f_y, x \rangle| = |x \cdot y| \leq \|x\| \|y\|$, $f_y \in H'$ with $\|f_y\| \leq \|y\|$. Since $\langle f_y, y \rangle = y \cdot y = \|y\|^2$, $\|f_y\| = \|y\|$.

Thus, the map $y \to f_y$ is an isometry from H into its dual space H'. We show that this map is onto.

Theorem 19 (Riesz Representation Theorem) *If H is a Hilbert space and $f \in H'$, then $\exists$ a unique $y \in H$ such that $f = f_y$.*

Proof: If $f = 0$, put $y = 0$. Suppose $f \neq 0$. Set $M = \mathcal{N}(f)$, the kernel of f, so M is a proper closed subspace of H and $M^\perp \neq \{0\}$. Choose $z \in M^\perp$, $z \neq 0$. Then $\langle f, z \rangle \neq 0$. Set $y = \left(\overline{\langle f, z \rangle} / \|z\|^2\right) z$ so $y \in M^\perp$, $y \neq 0$, and $\langle f, y \rangle = |\langle f, z \rangle|^2 / \|z\|^2 = y \cdot y$. For $x \in H$, let

$$x_1 = x - \left(\langle f, x \rangle / \|y\|^2\right) y, \quad x_2 = \left(\langle f, x \rangle / \|y\|^2\right) y$$

so $x = x_1 + x_2$ and $\langle f, x_1 \rangle = 0$ so $x_1 \in M$ and $x_1 \cdot y = 0$. Hence $x \cdot y = x_2 \cdot y = \langle f, x \rangle = \langle f_y, x \rangle$, and $f = f_y$.

The map $\Phi : y \to f_y$ is an isometry from H onto H' which is additive but is only conjugate homogeneous in the sense that $\Phi(ty) = \bar{t}\Phi(y)$. From this it follows that Hilbert spaces are always reflexive.

As noted earlier, $L^2(\mu)$ is a Hilbert space so it follows from Theorem 19 that the dual of $L^2(\mu)$ can be identified with $L^2(\mu)$ [in the case of real scalars]. This proof of the Riesz Representation Theorem for $L^2(\mu)$ is independent of the proof given in Theorem 6.1.20 which depended on the use of the Radon-Nikodym Theorem. Indeed, an independent proof of the Radon-Nikodym Theorem can be based on Theorem 19; see [R2], Theorem 6.9.

If X is an inner product space and $E = \{x_a : a \in A\}$ is an orthonormal subset of X, then $\forall x \in X$ the scalars $\hat{x}(a) = x \cdot x_a$, $a \in A$, are called the *Fourier coefficients* of x with respect to E. We establish several important properties of the Fourier coefficients.

Proposition 20 *Let X be an inner product space and $\{x_1, \ldots, x_n\}$ an orthonormal set in X. Then for each $x \in X$,*

(i) $\sum_{i=1}^{n} |x \cdot x_i|^2 = \sum_{i=1}^{n} |\hat{x}(i)|^2 \leq \|x\|^2$,

(ii) $\left(x - \sum_{i=1}^{n}(x \cdot x_i)x_i\right) \perp x_i \; \forall i$.

Proof: (i):

$$
\begin{aligned}
0 \leq \left\| x - \sum_{i=1}^{n}(x \cdot x_i)x_i \right\|^2 &= \left(x - \sum_{i=1}^{n}(x \cdot x_i)x_i\right) \cdot \left(x - \sum_{i=1}^{n}(x \cdot x_i)x_i\right) \\
&= \|x\|^2 - \sum_i (x \cdot x_i)(\overline{x \cdot x_i}) - \sum_i (\overline{x \cdot x_i})(x \cdot x_i) \\
&+ \sum_i \sum_j (x \cdot x_i)(\overline{x \cdot x_j}) x_i \cdot x_j = \|x\|^2 - \sum_i |x \cdot x_i|^2.
\end{aligned}
$$

(ii): $\left(x - \sum_i (x \cdot x_i)x_i\right) \cdot x_j = x \cdot x_j - \sum_i (x \cdot x_i)(x_i \cdot x_j) = x \cdot x_j - x \cdot x_j = 0$.

We generalize the inequality in (i) to infinite orthonormal sets.

Proposition 21 *Let $E = \{x_a : a \in A\}$ be an orthonormal set in an inner product space X. For each $x \in X$ the set $E_x = \{a \in A : x \cdot x_a \neq 0\}$ is at most countable.*

Proof: For $n \in \mathbf{N}$, let

$$S_n = \{a \in A : |x \cdot x_a|^2 > \|x\|^2/n\}.$$

By Proposition 20 S_n contains at most $n - 1$ elements. Since $E_x = \bigcup_{n=1}^{\infty} S_n$, the result follows.

Theorem 22 (Bessel's Inequality) *Let $E = \{x_a : a \in A\}$ be an orthonormal subset of an inner product space X. For each $x \in X$,*

$$\sum_{a \in A} |x \cdot x_a|^2 = \sum_{a \in A} |\hat{x}(a)|^2 \leq \|x\|^2. \tag{6.5}$$

6.6. HILBERT SPACE

Proof: If A is finite, this is Proposition 20. If A is infinite, we must assign a meaning to the series in (5). Let $S = \{a \in A : x \cdot x_a \neq 0\}$. If $S = \emptyset$, we set $\sum_{a \in A} |x \cdot x_a|^2 = 0$, and if S is finite, we set

$$\sum_{a \in A} |x \cdot x_a|^2 = \sum_{a \in S} |x \cdot x_a|^2$$

and (5) follows from Proposition 20. If S is infinite, S is countable by Proposition 21 so the elements $\{x_a : a \in S\}$ can be arranged in a sequence, say $y_1, y_2, \ldots$. By Proposition 20,

$$\forall n \quad \sum_{i=1}^{n} |x \cdot y_i|^2 \leq \|x\|^2$$

so the series $\sum_{i=1}^{\infty} |x \cdot y_i|^2$ is absolutely convergent and its sum is independent of the ordering of the elements $\{x_a : a \in S\}$. Therefore, we may define

$$\sum_{a \in A} |x \cdot x_a|^2 = \sum_{i=1}^{\infty} |x \cdot y_i|^2$$

and

$$\sum_{a \in A} |x \cdot x_a|^2 \leq \|x\|^2$$

by Proposition 20.

We will now show that equality holds in Bessel's Inequality for certain orthonormal sets in a Hilbert space. An orthonormal subset E of a Hilbert space H is said to be *complete* (or a *complete orthonormal set*) if $E_1 \subseteq H$ orthonormal and $E_1 \supseteq E$ implies that $E = E_1$ (i.e., E is a maximal orthonormal set with respect to set inclusion). [See Exercise 1.] We give several criteria for an orthonormal set to be complete. First, we require a lemma.

Lemma 23 *Let* $\{x_1, \ldots, x_n\}$ *be an orthonormal set in an inner product space* X.

(i) *If* $x = \sum_{k=1}^{n} c_k x_k$, *then* $c_k = x \cdot x_k = \hat{x}(k)$ *and* $\|x\|^2 = \sum_{k=1}^{n} |c_k|^2$.

(ii) *For* $\{c_1, \ldots, c_n\} \subseteq \mathbf{F}$, $\left\| x - \sum_{k=1}^{n} c_k x_k \right\|$ *attains its minimum (as a function of* $(c_1, \ldots, c_n)$) *at* $c_k = x \cdot x_k = \hat{x}(k)$, $k = 1, \ldots, n$.

Proof: (i): That $c_k = x \cdot x_k$ is immediate;

$$\|x\|^2 = x \cdot x = \sum_k \sum_j c_k \bar{c}_j x_k \cdot x_j = \sum_{k=1}^{n} |c_k|^2.$$

(ii):

$$\begin{aligned}
0 \le \left\| x - \sum_{k=1}^{n} c_k x_k \right\|^2 &= \left(x - \sum_{k=1}^{n} c_k x_k\right) \cdot \left(x - \sum_{k=1}^{n} c_k x_k\right) \\
&= \|x\|^2 - \sum_{k=1}^{n} c_k(\overline{x \cdot x_k}) - \sum_{k=1}^{n} \bar{c}_k(x \cdot x_k) + \sum_{k=1}^{n} |c_k|^2 \\
&= \left(\|x\|^2 - \sum_{k=1}^{n} |x \cdot x_k|^2\right) + \sum_{k=1}^{n} (x \cdot x_k - c_k)(\overline{x \cdot x_k - c_k})
\end{aligned}$$

and the expression on the right is clearly minimal at $c_k = x \cdot x_k$.

Theorem 24 *Let $E = \{x_a : a \in A\}$ be an orthonormal set in a Hilbert space H. The following are equivalent:*

(i) *E is complete,*

(ii) *$x \perp x_a \ \forall a \in A$ implies $x = 0$,*

(iii) *span E is dense in H.*

(iv) *If $x \in H$, $\|x\|^2 = \sum_{a \in A} |x \cdot x_a|^2$ (equality in Bessel's Inequality),*

(v) *$x = \sum_{a \in A} (x \cdot x_a) x_a \ \forall x \in H$,*

(vi) *if $x, y \in H$, $x \cdot y = \sum_{a \in A} (x \cdot x_a)(\overline{y \cdot x_a})$ (Parseval's Equality).*

Proof: (i)$\Rightarrow$(ii): If (ii) is false, $\exists x \ne 0$ such that $x \perp x_a \ \forall a \in A$. Set $z = x/\|x\|$ so $\{z\} \cup E$ is an orthonormal set which properly contains E so (i) does not hold.

(ii)$\Rightarrow$(iii): Let M be the closure of span E. If $M \ne H$, $H = M \oplus M^\perp$ with $M^\perp \ne \{0\}$. If $x \ne 0$, $x \in M^\perp$, then $x \perp x_a \ \forall a \in A$ so (ii) fails.

(iii)$\Rightarrow$(iv): Let $\epsilon > 0$ and $x \in H$. $\exists x_{a_1}, \ldots, x_{a_n} \in E$ and $c_1, \ldots, c_n \in \mathbf{F}$ such that

$$\left\| x - \sum_{k=1}^{n} c_k x_{a_k} \right\| < \epsilon.$$

By Lemma 23 (ii),

$$\left\| x - \sum_{k=1}^{n} (x \cdot x_{a_k}) x_{a_k} \right\| < \epsilon. \tag{6.6}$$

By Lemma 23 (i) and (6),

$$(\|x\| - \epsilon)^2 \le \left\| \sum_{k=1}^{n} (x \cdot x_{a_k}) x_{a_k} \right\|^2 = \sum_{k=1}^{n} |x \cdot x_{a_k}|^2 \le \sum_{a \in A} |x \cdot x_a|^2.$$

Bessel's Inequality gives the reverse inequality.

6.6. HILBERT SPACE

(iv)$\Rightarrow$(v): As in Theorem 22 let $S = \{x_a : x_a \cdot x \neq 0\}$ and arrange the elements of S into a sequence $y_1, y_2, \ldots$. Then

$$\|x - \sum_{k=1}^n (x \cdot y_k) y_k\|^2 = \|x\|^2 - \sum_{k=1}^n |x \cdot y_k|^2$$
$$= \sum_{k=1}^\infty |x \cdot y_k|^2 - \sum_{k=1}^n |x \cdot y_k|^2 = \sum_{k=n+1}^\infty |x \cdot y_k|^2$$

by (iv). Hence,

$$x = \sum_{k=1}^\infty (x \cdot y_k) y_k = \sum_{a \in A} (x \cdot x_a) x_a.$$

(v)$\Rightarrow$(vi): By Proposition 5,

$$x \cdot y = \sum_{a \in A} \sum_{b \in A} (x \cdot x_a)(\overline{y \cdot x_b}) x_a \cdot x_b = \sum_{a \in A} (x \cdot x_a)(\overline{y \cdot x_a}).$$

(vi)$\Rightarrow$(i): If (i) fails, $\exists z \in H$ with $\|z\| = 1$ and $z \perp x_a$ $\forall a \in A$. Then $z \cdot z = 1$ while $\sum_{a \in A} |z \cdot x_a|^2 = 0$ so (vi) fails.

Theorem 25 *Let H be a Hilbert space with E and F complete orthonormal subsets. Then E and F have the same cardinality.*

Proof: Since orthonormal sets are linearly independent, we may assume that E and F are infinite.

For $e \in E$, let $F_e = \{f \in F : f \cdot e \neq 0\}$. By Theorem 24 (ii), $F = \bigcup_{e \in E} F_e$ and by Proposition 21 each F_e is at most countable. Hence, the cardinality of F is less than or equal to the cardinality of E. Symmetry gives the reverse inequality.

The cardinality of a (any) complete orthonormal set is called the *orthonormal dimension* of the Hilbert space.

Example 26 $\{e^k : k \in \mathbf{N}\}$ is a complete orthonormal subset of ℓ^2. More generally, let A be a non-empty set and let $e_a : A \to \mathbf{R}$ be the characteristic function of $\{a\}$ for $a \in A$. Then $\{e_a : a \in A\}$ is a complete orthonormal set in $\ell^2(A)$ [here $\ell^2(A)$ is $L^2(\mu)$ where μ is the counting measure on A as in §6.1].

Theorem 27 (Riesz-Fischer) *If H is a Hilbert space, then H is linearly isometric to $\ell^2(A)$ for some $A \neq 0$.*

Proof: Let $E = \{x_a : a \in A\}$ be a complete orthonormal subset of H. By Theorem 24 the map $U : H \to \ell^2(A)$ defined by $Ux = \{x \cdot x_a : a \in A\}$ is a linear isometry. Since U carries E onto the complete orthonormal set $\{e_a : a \in A\}$ in $\ell^2(A)$ [Example 26], U is onto $\ell^2(A)$.

Note that the map U also preserves inner products by Theorem 24 so H and $\ell^2(A)$ are isomorphic as inner product spaces.

Example 28 $\{e^{int}/\sqrt{2\pi} : n = 0, \pm 1, \ldots\} = E$ is a complete orthonormal subset of $L^2[-\pi, \pi]$.

This follows easily from condition (iii) of Theorem 24. By the Stone-Weierstrass Theorem, the span of E is dense in $C[-\pi,\pi]$ with respect to the sup-norm and $C[-\pi,\pi]$ is dense in $L^2[-\pi,\pi]$. Since convergence in the sup-norm implies converge in the L^2-norm, it follows that span E is dense in $L^2[-\pi,\pi]$.

If $f \in L^2[-\pi,\pi]$ and

$$c_k = c_k(f) = \frac{1}{2\pi} \int_{-\pi}^{\pi} f(t) e^{-ikt} dt,$$

the series $\sum_{k=-\infty}^{\infty} c_k e^{ikt}$ is the classical Fourier series of f [with respect to the complete orthonormal set $\{e^{ikt}/\sqrt{2\pi} : k \in \mathbf{Z}\}$] and the c_k are called the Fourier coefficients of f. If follows from Theorem 24 that this series always converges to f in the L^2-norm. Note that the formula for the Fourier coefficient c_k is meaningful when f is integrable over $[-\pi,\pi]$. One of the important problems in Fourier analysis is determining what integrable functions are such that their Fourier series converge to the function, at least a.e. duBois-Reymond gave an example of a continuous function whose Fourier series diverges at a single point. We will now show that the Uniform Boundedness Principle can be used to show the existence of such a function.

Let $f \in L^1[-\pi,\pi]$. The nth partial sum of the Fourier series for f is

$$s_n(f)(t) = \sum_{k=-n}^{n} c_k(f) e^{ikt} = \frac{1}{2\pi} \int_{-\pi}^{\pi} f(s) \sum_{k=-n}^{n} e^{ik(t-s)} ds.$$

The function $D_n(t) = \sum_{k=-n}^{n} e^{ikt}$ which appears in the integral above is called the *Dirichlet kernel*; we now compute a more useful form for D_n. We have $(e^{it}-1)D_n(t) = e^{i(n+1)t} - e^{-int}$ so $e^{-it/2}(e^{it}-1)D_n(t) = e^{i(n+1/2)t} - e^{-i(n+1/2)t}$ which implies

$$D_n(t) = \sin(n+1/2)t/\sin(t/2).$$

Thus,

$$s_n(f)(t) = \frac{1}{2\pi} \int_{-\pi}^{\pi} f(s) D_n(t-s) ds.$$

We define a linear functional F_n on $C[-\pi,\pi]$ by

$$F_n(f) = s_n(f)(0) = \frac{1}{2\pi} \int_{-\pi}^{\pi} f(s) D_n(s) ds,$$

i.e., $F_n(f)$ is the nth partial sum of the Fourier series for f evaluated at 0. Since

$$|F_n(f)| \leq \|f\|_\infty \frac{1}{2\pi} \int_{-\pi}^{\pi} |D_n(s)| ds,$$

F_n is continuous and $\|F_n\| \leq \frac{1}{2\pi} \int_{-\pi}^{\pi} |D_n(s)| ds$. We claim that $\|F_n\| = \frac{1}{2\pi} \int_{-\pi}^{\pi} |D_n(s)| ds$. This equality follows easily from the following lemma.

Lemma 29 *Let $g : [a,b] \to \mathbf{R}$ be continuous. Define $G : C[a,b] \to \mathbf{R}$ by $G(f) = \int_a^b f(t) g(t) dt$. Then G is linear and continuous with $\|G\| = \int_a^b |g(t)| dt$.*

6.6. HILBERT SPACE

Proof: Since $|G(f)| \leq \|f\|_\infty \int_a^b |g(t)|\,dt$, G is linear, continuous and $\|G\| \leq \int_a^b |g(t)|\,dt$. Fix $n \in \mathbf{N}$. Then

$$\begin{aligned}
\int_a^b |g| = \int_a^b |g|\frac{1+n|g|}{1+n|g|} &= \int_a^b \frac{|g|}{1+n|g|} + \int_a^b g\frac{ng}{1+n|g|} \\
&\leq \int_a^b \frac{1}{n} + G\left(\frac{ng}{1+n|g|}\right) \\
&\leq \frac{b-a}{n} + \|G\|\left\|\frac{ng}{1+n|g|}\right\|_\infty \leq \frac{b-a}{n} + \|G\|
\end{aligned}$$

so $\int_a^b |g(t)|\,dt \leq \|G\|$.

Next, we compute a lower estimate for $\int_{-\pi}^{\pi} |D_n(s)|\,ds$.
Since $\sin u \leq u$ for $0 \leq u \leq \pi$,

$$\begin{aligned}
\int_{-\pi}^{\pi} |D_n(t)|\,dt &\geq \int_0^{\pi} \left|\frac{\sin(2n+1)u}{u}\right|\,du \\
&= \sum_{k=0}^{2n} \int_{\frac{k\pi}{2n+1}}^{\frac{(k+1)\pi}{2n+1}} \left|\frac{\sin(2n+1)u}{u}\right|\,du \\
&\geq \sum_{k=0}^{2n} \frac{2n+1}{(k+1)\pi} \int_{\frac{k\pi}{2n+1}}^{\frac{(k+1)\pi}{2n+1}} |\sin(2n+1)u|\,du \\
&= \sum_{k=0}^{2n} \frac{1}{(k+1)\pi} \int_{k\pi}^{(k+1)\pi} |\sin v|\,dv = \frac{2}{\pi}\sum_{k=0}^{2n} \frac{1}{k+1}.
\end{aligned}$$

Thus, $\{\|F_n\| : n \in \mathbf{N}\}$ is unbounded. It follows from the UBP that there exists $f \in C[-\pi,\pi]$ such that $\{F_n(f) : n\}$ is unbounded, i.e., the Fourier series of f at 0 must diverge. Of course, the point 0 was chosen only for convenience, and the same result holds for any other point of $[-\pi,\pi]$ [the function f will depend upon the point chosen].

Banach and Steinhaus derived a method, called *condensation of singularities*, which can be used to construct a continuous function whose Fourier series diverges at any (arbitrary) countable subset of $[-\pi,\pi]$. We first give their result. The proof uses the Baire Category Theorem [see A2].

Theorem 30 *Let X be a Banach space, Y_k a NLS and $T_k \in L(X,Y_k)$ for each $k \in \mathbf{N}$. Then $B = \{x \in X : \varlimsup \|T_k x\| < \infty\}$ either coincides with X or is first category in X.*

Proof: Suppose B is second category in X. For each $x \in B$, $\limsup_k \|\frac{1}{k}T_n x\| = 0$.
Let $\epsilon > 0$. Then $B \subset \bigcup_{k=1}^{\infty} B_k$, where $B_k = \{x \in X : \sup_{n \geq 1} \|\frac{1}{k}T_n x\| \leq \epsilon\}$. Each B_k is closed so some B_k contains a sphere, i.e., there exist $x_0 \in X \cap B_k$, $r > 0$ such that $\|x - x_0\| \leq r$ implies $\sup_{n \geq 1}\|\frac{1}{k}T_n x\| \leq \epsilon$. Thus, for $\|z\| \leq r$, if $x = x_0 + z$, then

$$\left\|\frac{1}{k}T_n z\right\| \leq \left\|\frac{1}{k}T_n x\right\| + \left\|\frac{1}{k}T_n x_0\right\| \leq 2\epsilon$$

and $\sup_n \|T_n z\| \leq 2\epsilon k$. Hence, if $x \in X$, $\sup_n \|T_n x\| \leq 2\epsilon k \|x\|/r$ so $X = B$.

Corollary 31 (Condensation of Singularities) *For each $q \in \mathbf{N}$ let $\{T_{pq}\}_p$ be a sequence of operators in $L(X, Y_q)$, where X is a Banach space and Y_q is a NLS. Suppose for each p there exists $x_p \in X$ with $\varlimsup_q \|T_{pq} x_p\| = \infty$. Then $B = \{x : \varlimsup_q \|T_{pq} x\| = \infty \; \forall p \in \mathbf{N}\}$ is second category in X.*

Proof: For each p, $B_p = \{x \in X : \varlimsup_q \|T_{pq} x\| < \infty\}$ is first category in X by Theorem 30 and the hypothesis. Thus, $B = X \setminus \bigcup_{p=1}^{\infty} B_p$ is second category.

Let $\{t_j\}$ be a sequence of distinct points in $[-\pi, \pi]$. Let $s_{nj}(f)$ be the nth partial sum of the Fourier series of $f \in C[-\pi, \pi]$ evaluated at t_j. By the construction above, for each j there exists $f_j \in C[-\pi, \pi]$ such that $\varlimsup_n |s_{nj}(f_j)| = \infty$. By Corollary 31 there exists $f \in C[-\pi, \pi]$ such that $\varlimsup_n |s_{nj}(f)| = \infty$ for all j. That is, the Fourier series of f diverges at each t_j.

It can be shown that if the set $\{t_j\}$ is chosen to be dense in $[-\pi, \pi]$, then the set $P = \{t \in [-\pi, \pi] : \varlimsup_n |s_n(f)(t)| = \infty\}$ is second category in $[-\pi, \pi]$ and, hence, uncountable [see [Y] II.4].

The classical Fourier series of an integrable function can behave very badly with respect to pointwise convergence. For example, Kolmogorov gave an example of an integrable function whose Fourier series diverges a.e. in $[-\pi, \pi]$. It was an open problem for many years whether the Fourier series of an L^2 function must converge a.e. This was shown to be the case, only in 1966, by L. Carleson. Hunt later extended Carleson's result to L^p for $1 < p < \infty$. For a discussion of the results by Hunt, see [As].

Whereas the classical Fourier series of integrable functions can display poor convergence behavior, the Cesaro averages of the partial sums of the Fourier series behave much better. For $f \in L^1[-\pi, \pi]$ the Cesaro averages of the Fourier series of f are defined to be $\sigma_n(f) = \frac{1}{n+1} \sum_{k=0}^{n} s_k(f)$, where $s_n(f)$ is the nth partial sum of the Fourier series of f. From the discussion above,

$$\sigma_n(f)(t) = \frac{1}{2\pi} \int_{-\pi}^{\pi} \frac{1}{n+1} \sum_{k=0}^{n} D_k(t-s) f(s) \, ds.$$

The function $K_n(t) = \frac{1}{n+1} \sum_{k=0}^{n} D_k(t)$ is called the *Fejér kernel*. Thus, we have

$$\sigma_n(f)(t) = \frac{1}{2\pi} \int_{-\pi}^{\pi} K_n(t-s) f(s) \, ds.$$

If we assume that all functions in $L^1[-\pi, \pi]$ are extended periodically (with period 2π) to $\mathbf{R}$ and if we replace integration over $\mathbf{R}$ by integration over $[-\pi, \pi]$, we may interpret the formula above as a convolution integral,

$$\sigma_n(f)(t) = \frac{1}{2\pi} K_n * f(t)$$

6.6. HILBERT SPACE

as in §3.11. We will now show that the sequence $\{\frac{1}{2\pi}K_n\}$ satisfies the properties of an approximate identity as defined in 3.11.2. For this it is convenient to derive another formula for the Fejér kernel. Substituting $(e^{it}-1)D_n(t) = e^{i(n+1)t} - e^{-int}$ into the definition of K_n gives

$$(n+1)K_n(t)(e^{it}-1)(e^{-it}-1) = (e^{-it}-1)\sum_{k=0}^{n}\left(e^{i(n+1)t}-e^{-int}\right)$$
$$= 2 - e^{i(n+1)t} - e^{-i(n+1)t}$$

so that

$$K_n(t) = \frac{1}{n+1}\frac{1-\cos(n+1)t}{1-\cos t} \qquad (6.7)$$

From (7), we obtain

Lemma 32 (i) $K_n \geq 0$,

(ii) $\int_{-\pi}^{\pi} K_n = 2\pi$,

(iii) For $0 < \delta \leq |t| \leq \pi$, $K_n(t) \leq \frac{2}{(n+1)(1-\cos\delta)}$.

From (i), (ii) and (iii), it follows that $\{\frac{1}{2\pi}K_n\}$ satisfies the conditions of an approximate identity given in 3.11.2. [Note that the Dirichlet kernel fails property (i).] From the analogue of Theorem 6.1.24, it follows that if $f \in L^p[-\pi,\pi]$, then $\{\sigma_n(f)\}$ converges in L^p-norm to f. Also, from the analogue of Exercise 3.11.10, it follows that if f is a continuous function on **R** of period 2π, the Cesaro averages $\{\sigma_n(f)\}$ converge to f uniformly on $[-\pi,\pi]$; this result is known as Fejér's Theorem. It s also the case that if $f \in L^1[-\pi,\pi]$, then $\{\sigma_n(f)\}$ converges a.e. to f; see [HS] V.18.29.

For a discussion of the historical development of Fourier series by Zygmund, see [As]. Zygmund's book ([Zy]) contains a detailed discussion of Fourier series.

Exercise 1. Show that any orthonormal subset of a Hilbert space is contained in a complete orthonormal set.

Exercise 2. Show a Hilbert space is separable if and only if its orthonormal dimension is countable.

Exercise 3. If D is a dense subset of an inner product space and $x \perp D$, show $x = 0$.

Exercise 4. If E is a linear subspace of a Hilbert space H, Y is a B-space and $T: E \to Y$ is a continuous linear operator, show T has a continuous linear extension $\hat{T}: H \to Y$.

Exercise 5 (Gram-Schmidt). Let $x_1, \ldots, x_n$ be linearly independent. Set $y_1 = x_1$,
$$y_k = x_k - \sum_{j=1}^{k-1}(x_k \cdot y_j) y_j / \|y_j\|^2$$
for $k > 1$ and $z_k = y_k / \|y_k\|$. Show $\{z_k\}$ is orthonormal and span$\{x_k\}$ = span$\{z_k\}$.

Exercise 6 (Riemann-Lebesgue Lemma). Show that if $f \in L^1[-\pi, \pi]$, then
$$c_k = \frac{1}{\sqrt{2\pi}} \int_{-\pi}^{\pi} f(t) e^{-ikt} dt \to 0$$
as $|k| \to \infty$. [Hint: First prove this for characteristic functions of intervals.]

Exercise 7. Show that the sup-norm on $C[0,1]$ does not satisfy the parallelogram law.

Exercise 8. Let M be a closed subspace of a Hilbert space H.

(i) Show $M^{\perp\perp} = M$.

(ii) Show H/M is a Hilbert space.

(iii) If $x \in H$, $x = m + m'$ where $m \in M$, $m' \in M^\perp$ as in Theorem 17, show $Px = m$ defines a continuous linear operator on H such that $P^2 = P$.

Exercise 9. Show a linear isometry onto an inner product space preserves inner products.

Exercise 10. Let M, N be closed subspaces of H with $M \perp N$. Show $M + N$ is closed.

Exercise 11. Show a linear subspace M of a Hilbert space is dense if and only if $M^\perp = \{0\}$. Give an example of a proper closed subspace M of an inner product space such that $M^\perp = \{0\}$.

Exercise 12. let $E = \{e_a : a \in A\}$ be an orthonormal family in a Hilbert space H. Show the closed linear subspace generated by E is $\left\{\sum_{a \in A} t_a e_a : \sum_{a \in A} |t_a|^2 < \infty\right\}$.

Exercise 13. Show Theorem 19 is false for inner product spaces.

Exercise 14 (Hellinger-Toeplitz) Let $T : H \to H$ be linear and satisfy $Tx \cdot y = x \cdot Ty$ for all x, y. Show T is continuous. Hint: Use UBP or CGT.

6.6.1 The Fourier Transform

In this section we define and study some of the basic properties of the Fourier transform. Although the Fourier transform is defined for functions on $\mathbf{R}^n$, we restrict our attention to functions on $\mathbf{R}$. In dealing with the Fourier transform, there is always a factor of 2π involved. We will take care of this by using the normalized measure $\mu = m/\sqrt{2\pi}$. We denote by $\underline{L}^p$ the space $L^p(\mu)$, where we are considering complex-valued functions. All statements about measurability refer to the Lebesgue measure, and any integral over $\mathbf{R}$ is denoted by $\int f d\mu = \int_{\mathbf{R}} f d\mu$. In this section, we agree that the convolution of two measurable functions f and g is defined by

$$f * g(x) = \int f(x-y) g(y) d\mu(y)$$

[see §3.11].

If $f \in L^1$, we define the *Fourier transform* of f, $\hat{f}$, by $\hat{f}(t) = \int e^{-itx} f(x) d\mu(x)$. Note that since e^{-itx} is bounded and is continuous as a function of x, $\hat{f}$ is defined for all $t \in \mathbf{R}$. We list some of the basic properties of the Fourier transform.

Theorem 1 *Let $f, g \in L^1$*

(i) *If $h(x) = e^{iax} f(x)$, then $\hat{h}(t) = \hat{f}(t-a)$.*

(ii) *If $h(x) = f(x-a)$, then $\hat{h}(t) = e^{-iat} \hat{f}(t)$.*

(iii) *If $h(x) = \overline{f}(-x)$, then $\hat{h}(t) = (\hat{f}(t))^{-}$.*

(iv) *If $h(x) = f(ax)$, $a > 0$, then $\hat{h}(t) = \hat{f}(t/a)/a$.*

(v) $(f * g)^{\wedge} = \hat{f} \hat{g}.$

(vi) *If $h(x) = -ixf(x)$ and $h \in L^1$, then $\hat{f}$ is differentiable and $\left(\hat{f}\right)'(t) = \hat{h}(t)$.*

(vii) *If $f' \in L^1$ and $\lim\limits_{|x| \to \infty} f(x) = 0$, then $(f')^{\wedge}(t) = it \hat{f}(t)$.*

(viii) *(Riemann-Lebesgue Lemma) $\hat{f}$ is continuous, bounded $\left[\left\| \hat{f} \right\|_{\infty} \leq \|f\|_1 \right]$ and* $\lim\limits_{|t| \to \infty} \hat{f}(t) = 0.$

(ix) $\int f \hat{g} = \int \hat{f} g.$

Proof: (i)-(iv) are easily checked. (v) follows from Fubini's Theorem [recall 3.11.1] since

$$\begin{aligned}(f * g)^{\wedge}(t) &= \int e^{-itx} \int f(x-y) g(y) d\mu(y) d\mu(x) \\ &= \int g(y) e^{-ity} \int e^{-it(x-y)} f(x-y) d\mu(x) d\mu(y) \\ &= \int g(y) e^{-ity} d\mu(y) \int e^{-itx} f(x) d\mu(x) = \hat{f}(t) \hat{g}(t).\end{aligned}$$

250   CHAPTER 6. FUNCTION SPACES

(vi) follows from 3.4.3. (vii) follows by integration by parts.

(viii): $\hat{f}$ is bounded with $\left|\hat{f}(t)\right| \leq \|f\|_1$ for every t. Since $e^{\pi i} = -1$,

$$\hat{f}(t) = -\int f(x)e^{-it(x+\pi/t)}d\mu(x) = -\int f(x - \pi/t)e^{-itx}d\mu(x)$$

so

$$\left|2\hat{f}(t)\right| = \left|\int \{f(x) - f(x - \pi/t)\}e^{-itx}d\mu(x)\right| \leq \left\|f - f_{-\pi/t}\right\|_1 \to 0$$

as $t \to \infty$ by 3.11.10. Finally, for $t, s \in \mathbf{R}$, we have

$$\left|\hat{f}(t+s) - \hat{f}(t)\right| \leq \int |f(x)|\left|e^{-isx} - 1\right|d\mu(x) = \int |f(x)||2\sin(sx/2)|\,dx,$$

and since $\lim\limits_{s\to 0}\sin(sx/2) = 0$ for each x, the DCT implies that the last term is small for s small independent of t. Hence, $\hat{f}$ is uniformly continuous.

(ix) follows from Fubini's Theorem.

Conditions (vii) shows that the Fourier transform converts differentiation into multiplication by it, and this property makes the Fourier transform useful in the study of differential equations. Conditions (vi) and (vii) also point out another important property of the Fourier transform: growth properties of f are reflected in smoothness properties of $\hat{f}$ and vice-versa.

From (viii), the Fourier transform is a continuous linear transform from L^1 into the space, $C_0(\mathbf{R})$, of bounded, continuous functions on $\mathbf{R}$ which vanish at ∞ when $C_0(\mathbf{R})$ is equipped with the sup-norm. The Fourier transform is not onto $C_0(\mathbf{R})$ [Exercise 7]; there is not an intrinsic characterization of the functions which are Fourier transforms of L^1 functions.

Let $\mathcal{S} = \{f : \mathbf{R} \to \mathbf{R} : f \text{ is infinitely differentiable and for each } j, k, \sup\{|x^j|\left|f^{(k)}(x)\right| : x \in \mathbf{R}\} < \infty\}$. $\mathcal{S}$ is called the space of *rapidly decreasing functions* or the *Schwartz space* after Laurent Schwartz. By Leibniz' rule for the differentiation of products, it is easy to see that an infinitely differentiable function f belongs to $\mathcal{S}$ if and only if $(x^j f(x))^{(k)}$ is bounded for each j, k. The function $f(x) = e^{-x^2}$ furnishes an example of a function in $\mathcal{S}$ [see Example 3 below]; indeed, $C_c^\infty(\mathbf{R}) \subset \mathcal{S}$ so $\mathcal{S}$ is dense in L^1 (3.11.12).

Corollary 2 $\mathcal{S}^\wedge \subset \mathcal{S}$.

Proof: If $f \in \mathcal{S}$, for each j, k the function $x \to x^j f^{(k)}(x)$ is integrable and vanishes at ∞ so $\hat{f}$ is infinitely differentiable by (vi), and by (vii),

$$\left((-ix)^j f^{(k)}(x)\right)^\wedge (t) = \left((it)^k \hat{f}(t)\right)^{(j)}$$

so $\left((it)^k \hat{f}(t)\right)^{(j)}$ is bounded by (viii). Hence, by the observation above, $\hat{f} \in \mathcal{S}$.

It will follow from the Fourier Inversion Theorem below that $\mathcal{S}^\wedge = \mathcal{S}$.

Example 3 If $f(x) = e^{-x^2/2}$, then $f = \hat{f}$.

6.6. HILBERT SPACE

From (vi),
$$\left(\hat{f}\right)'(t) = \int -ixe^{-ixt}e^{-x^2/2}d\mu(x)$$

and integration by parts gives $\left(\hat{f}\right)'(t) = -t\hat{f}(t)$. Hence, the derivative of the quotient $\hat{f}(t)/f(t)$ is 0 so $\hat{f}(t)/f(t) = c$, a constant. Putting $t=0$ gives $c=1$ from 3.4.4 [because of the normalization factor $\frac{1}{\sqrt{2\pi}}$ in the measure].

We now consider inverting the Fourier transform. For this we define the *inverse Fourier transform* of a function $f \in L^1$ to be $f^\vee(t) = \int e^{itx} f(x) d\mu(x)$. Since $f^\vee(x) = \hat{f}(-x)$, the inverse Fourier transform shares most of the properties of the Fourier transform given in Theorem 1. We do not bother to record these properties.

Let $g(x) = e^{-x^2/2}$ and $g_a(x) = e^{-(xa)^2/2}$. By Example 3 and Theorem 1 (iv), $\hat{g}_a(t) = (1/a)g_{1/a}(t)$.

Theorem 4 (Fourier Inversion Theorem) *Let $f, \hat{f} \in L^1$. Then $\left(\hat{f}\right)^\vee = f$ a.e.*

Proof: By Theorem 1 (ix) and (i),
$$\begin{aligned} \int \hat{f}(t)e^{ixt}g_a(t)d\mu(t) &= \int f(t)\left(e^{ixt}g_a(t)\right)^\wedge d\mu(t) \\ &= \int f(t)\hat{g}_a(t-x)d\mu(t) \\ &= \int f(t)\tfrac{1}{a}g_{1/a}(t-x)d\mu(t) = f * \tfrac{1}{a}g_{1/a}(x). \end{aligned} \quad (6.1)$$

Since $\lim_{a\to 0} g_a(x) = 1$ and the integrand on the left hand side of (1) is bounded by $|\hat{f}|$, the DCT implies that the left hand side of (1) goes to $\left(\hat{f}\right)^\vee(x)$ as $a \to 0$. If $k = \frac{1}{a}$, $k \in \mathbb{N}$, $\frac{1}{a}g_{1/a}$ is the approximate identity in Example 3.11.5 so the right hand side of (1) converges in L^1-norm to f [3.11.11]. Since any sequence which converges in L^1-norm has a subsequence which converges pointwise a.e. [3.7.1, 3.6.3], $\left(\hat{f}\right)^\vee = f$ a.e.

Remark 5 Thus, if $f, \hat{f} \in L^1$, then f can be made into a continuous function which vanishes at ∞ by modification on a null set and $\left(\hat{f}\right)^\vee = f$ everywhere [Theorem 1 (viii)].

From the computation in Corollary 2, we have

Corollary 6 $\mathcal{S}^\wedge = \mathcal{S}$.

Concerning uniqueness for the Fourier transform, we have

Corollary 7 *If $f \in L^1$ and $\hat{f} = 0$, then $f = 0$ a.e.*

We now consider extending the Fourier transform to L^2; the resulting extension is sometimes called the *Plancherel transform*.

Lemma 8 *Let $X = \{f \in L^1 : \hat{f} \in L^1\}$. If $f \in X$, then $\hat{f} \in X \subset L^2$ and the Fourier transform restricted to X preserves inner products on X.*

Proof: By Theorem 4, $X^\wedge = X$ and if $f \in X$, f is bounded so f and $\hat{f}$ belong to L^2 [Exercise 6.2.9]. If $f, g \in X$, set $h = (\hat{g})^-$. Then

$$\hat{h}(t) = \left(\int e^{itx}\hat{g}(x)d\mu(x)\right)^- = (\hat{g})^{\vee -}(t) = \overline{g}(t)$$

almost everywhere. Hence,

$$f \cdot g = \int f\overline{g}d\mu = \int f\hat{h}d\mu = \int \hat{f}h d\mu = \int \hat{f}(\hat{g})^- d\mu = \hat{f} \cdot \hat{g}$$

so the Fourier transform on X preserves inner products.

It follows from Lemma 8 that if $f \in X$, then $\|f\|_2 = \|\hat{f}\|_2$, so the Fourier transform is a linear isometry of X onto X. Since $S \subset X$, X is dense in L^2 so the Fourier transform can be extended to a linear isometry from L^2 onto L^2 which preserves inner products. We need to show that this extension agrees with the Fourier transform on $L^1 \cap L^2$.

Let $f \in L^1 \cap L^2$ and let $h_k(t) = kg_k(t)$ be the approximate identity used in the Fourier Inversion Theorem above. Then $f * h_k \to f$ in L^1 [3.11.11] so $(f * h_k)^\wedge \to \hat{f}$ uniformly on **R** [Exercise 1]. By Theorem 6.1.24, $f * h_k \to f$ in L^2-norm so $\hat{f}$ is equal to the L^2-extension of the Fourier transform of f.

We continue to denote the extension of the Fourier transform of a function $f \in L^2$ by $\hat{f}$.

Let $f \in L^2$ and set $f_k = C_{[-k,k]}f$. Then $f_k \to f$ in L^2 so $\hat{f}_k \to \hat{f}$ in L^2. Since $f_k \in L^1 \cap L^2$, its Fourier transform is given by

$$\int_{-k}^{k} e^{-itx}f(x)d\mu(x) = \hat{f}_k(t)$$

and the Fourier transform of f is the L^2-limit of the sequence $\{\hat{f}_k\}$. [A similar formula holds for the inverse Fourier transform.] Note that $\hat{f}$ is only determined up to a.e. since it is an L^2-limit whereas the Fourier transform of an L^1-function is unambiguously defined everywhere.

Exercise 1. If $f_k \to f$ in L^1, show $\hat{f}_k \to \hat{f}$ uniformly on **R**.

Exercise 2. If $f \in L^2$ and $\hat{f} \in L^1$, show

$$f(x) = \int e^{ixt}\hat{f}(t)d\mu(t)$$

for almost all $x \in \mathbf{R}$.

Exercise 3. If $f, g \in L^2$, show $\left(\hat{f}\hat{g}\right)^\vee = f * g$. [Note $\hat{f}\hat{g} \in L^1$ by Hölder's Inequality and $f * g$ makes sense by Exercise 6.2.6.]

6.6. HILBERT SPACE

Exercise 4. Let $\varphi_a = C_{[-a,a]}$. Show $\widehat{\varphi}_a(t) = 2\frac{\sin at}{t}$, $\|\varphi_a\|_2^2 = 2a$, and $\int \left(\frac{\sin ay}{y}\right)^2 d\mu(y) = \sqrt{\frac{\pi}{2}}a$.

Exercise 5. If $f(x) = e^{-|x|}$, find $\check{f}$. Show $(f(x/k))^{\vee}$ defines an approximate identity.

Exercise 6. If $f \in L^1$ and $f * f = f[0]$ a.e., show $f = 0$ a.e.

Exercise 7. Let $f \in L^1$ be odd. Use the fact that $\left\{\int_\alpha^\beta \frac{\sin x}{x} dx : |\alpha| < |\beta|\right\}$ is bounded and Fubini's Theorem to show that $\left\{\int_1^b \widehat{f}(t)/t\, dt : b > 1\right\}$ is bounded. Use this to show that $g(t) = 1/\ln|t|$ for $|t| \geq e$ and $g(t) = 0$ for $|t| < e$ is not the Fourier transform of an L^1 function.

A.1. FUNCTIONS OF BOUNDED VARIATIONS

A1: Functions of Bounded Variations

For the reader who may not be familiar with the basic properties of functions of bounded variation, we record them in this appendix.

Let $f : [a,b] \to \mathbf{R}$. If $\pi = \{a = x_0 < x_1 < \cdots < x_n = b\}$ is a partition of $[a,b]$, the *variation of f over π* is

$$var(f : \pi) = \sum_{i=0}^{n-1} |f(x_{i+1}) - f(x_i)|,$$

and the *variation of f over $[a,b]$* is

$$Var(f : [a,b]) = \sup var(f : \pi),$$

where the supremum is taken over all possible partitions, π, of $[a,b]$. If

$$Var(f : [a,b]) < \infty,$$

f is said to have *bounded variation*; the class of all such functions is denoted by $BV[a,b]$. The variation measures the amount the function oscillates in $[a,b]$.

As the example below illustrates, even a continuous function can fail to belong to $BV[a,b]$.

Example 1 Let $f(t) = t\sin(1/t)$ for $0 < t \leq 1$ and $f(0) = 0$. Set $x_n = 1/(n+1/2)\pi$. Then $f(x_n) = 1/(n+1/2)\pi$ if n is even, and $f(x_n) = -1/(n+1/2)\pi$ if n is odd. If π_n is the partition $\{0 < x_n < x_{n-1} < \cdots < x_1 < 1\}$, then

$$\sum_{i=1}^{n-1} |f(x_i) - f(x_{i-1})| \geq \frac{2}{\pi} \sum_{i=1}^{n-1} 1/(i+1)$$

so $Var(f : [0,1]) = \infty$.

Proposition 2 *If $f \in BV[a,b]$, then f is bounded on $[a,b]$.*

Proof: Let $x \in (a,b)$. Then

$$|f(x) - f(a)| + |f(b) - f(x)| \leq Var(f : [a,b])$$

so

$$2|f(x)| \leq |f(a)| + |f(b)| + Var(f : [a,b])$$

and f is bounded.

We consider properties of $Var(f : I)$ as a function of the interval I. First, as a consequence of the triangle inequality, we have

Lemma 3 *Let $f : [a,b] \to \mathbf{R}$. If π and π' are partitions of $[a,b]$ with $\pi \subset \pi'$, then $var(f : \pi) \leq var(f : \pi')$.*

Proposition 4 *Let $f : [a,b] \to \mathbf{R}$ and $a < c < b$. Then*
$$Var(f : [a,b]) = Var(f : [a,c]) + Var(f : [c,b]).$$

Proof: Let π be a partition of $[a,b]$ and π' the partition obtained by adding the point c to π. Let π_1 and π_2 be the partitions of $[a,c]$ and $[c,b]$, respectively, induced by π'. Then by Lemma 3,
$$var(f : \pi) \leq var(f : \pi') = var(f : \pi_1) + var(f : \pi_2) \leq Var(f : [a,c]) + Var(f : [c,b])$$
so
$$Var(f : [a,b]) \leq Var(f : [a,c]) + Var(f : [c,b]).$$

If π_1 and π_2 are partitions of $[a,c]$ and $[c,b]$, respectively, then $\pi = \pi_1 \cup \pi_2$ is a partition of $[a,b]$ so
$$var(f : \pi) = var(f : \pi_1) + var(f : \pi_2) \leq Var(f : [a,b]).$$

Hence,
$$Var(f : [a,c]) + Var(f : [c,b]) \leq Var(f : [a,b]).$$

Next, we consider how the variation, $Var(f : I)$, depends on the function f.

Proposition 5 *Let $f, g \in BV[a,b]$. Then*

(i) $f + g \in BV[a,b]$ with $Var(f + g : [a,b]) \leq Var(f : [a,b]) + Var(g : [a,b])$,

(ii) *for $t \in \mathbf{R}$, $tf \in BV[a,b]$ with $Var(tf : [a,b]) = |t| Var(f : [a,b])$.*

Proof: (i) follows from the triangle inequality and (ii) is clear.

Thus, $BV[a,b]$ is a vector space under the usual pointwise addition and scalar multiplication of functions.

Let $f \in BV[a,b]$. We define the *total variation* of f by $V_f(t) = Var(f : [a,t])$ if $a < t \leq b$ and $V_f(a) = 0$.

Proposition 6 *V_f and $V_f - f$ are increasing on $[a,b]$.*

Proof: V_f is increasing by Proposition 4.
Let $a \leq x < y \leq b$ and $g = V_f - f$. Then
$$g(y) = V_f(x) + Var(f : [x,y]) - f(y)$$
implies
$$g(y) - g(x) = V_f(x) - f(y) + Var(f : [x,y]) - V_f(x) + f(x) = Var(f : [x,y]) - (f(y) - f(x)) \geq 0.$$

A.1. FUNCTIONS OF BOUNDED VARIATIONS

Proposition 7 *If $f \in BV[a,b]$ is (right, left) continuous at x, then V_f is (right, left) continuous at x.*

Proof: Let $\epsilon > 0$. Suppose f is right continuous at $x < b$. There is a partition π of $[x, b]$ such that $var(f : \pi) > Var(f : [x, b]) - \epsilon$. Since f is right continuous at x, we may add a point x_1 to π to obtain a partition

$$\pi' = \{x < x_1 < \cdots < x_n\}$$

of $[x, b]$ such that $|f(x) - f(x_1)| < \epsilon$. Then

$$\epsilon + var(f : \pi') = \epsilon + |f(x_0) + f(x_1)| + \sum_{i=1}^{n-1} |f(x_{i+1}) - f(x_i)| < 2\epsilon + Var(f : [x_1, b])$$

so

$$Var(f : [x, b]) < var(f : \pi) + \epsilon \leq var(f : \pi') + \epsilon < 2\epsilon + Var(f : [x_1, b]).$$

Thus, $0 \leq V_f(x_1) - V_f(x) < 2\epsilon$, and since $V_f \uparrow$, $\lim_{y \to x^+} V_f(y) = V_f(x)$, i.e., V_f is right continuous.

The statement about left continuity is similar.

Since $f = V_f - (V_f - f)$, Propositions 6 and 7 along with Exercise 1 and Proposition 5 give the following characterization of functions of bounded variation.

Theorem 8 *Let $f : [a, b] \to \mathbf{R}$. Then $f \in BV[a, b]$ if and only if $f = g - h$, where $g, h \uparrow$. If f is (right, left) continuous, then g and h can be chosen to be (right, left) continuous.*

Exercise 1. If $f \uparrow$ on $[a, b]$, show $Var(f : [a, b]) = f(b) - f(a)$.

Exercise 2. Give a necessary and sufficient condition for f to satisfy

$$Var(f : [a, b]) = 0.$$

Exercise 3. Let $f, g \in BV[a, b]$. Show fg and $|f|$ belong to $BV[a, b]$, and, hence, $f \wedge g$, $f \vee g \in BV[a, b]$ [formula 1.2 of §1.1]. That is, $BV[a, b]$ is a vector lattice and also an algebra; in the terminology of Example 5.7.7 $BV[a, b]$ is a function space.

Exercise 4. Let $\varphi = (x, y) : [a, b] \to \mathbf{R}^2$ be continuous. If

$$\pi = \{a = t_0 < t_1 < \ldots < t_n = b\}$$

is a partition of $[a, b]$, set

$$L(\varphi, \pi) = \sum_{i=0}^{n-1} \left(|x(t_{i+1}) - x(t_i)|^2 + |y(t_{i+1}) - y(t_i)|^2\right)^{1/2}.$$

Then $L(\varphi) = \sup L(\varphi, \pi)$, where the supremum is taken over all partitions π, is the *arclength* of the path φ. Show $L(\varphi) < \infty$ if and only if both $x, y \in BV[a, b]$.

A2: The Baire Category Theorem

The Baire Category Theorem has been used several times in the text. For the convenience of the reader unfamiliar with this result, we give a statement and proof in this appendix. References to further applications of the Baire Category Theorem are given at the end of the appendix.

Let (S, d) be a metric space. A subset $E \subset S$ is *nowhere dense* if the interior of $\overline{E}$ is empty. For example, the Cantor set is nowhere dense in $[0,1]$ [Exercise 2]. A subset $E \subset S$ is *first category* in S if E is a countable union of nowhere dense sets. For example, $\mathbf{Q}$ is first category in $\mathbf{R}$. A subset $E \subset S$ is *second category* in S if E is not first category in S. Baires' Theorem asserts that every complete metric space is second category in itself.

Proposition 1 *Let $\{F_k\}$ be a sequence of closed sets contained in the complete metric space (S, d) such that $F_k \supset F_{k+1}$ and $d_k = \text{diameter} F_k \to 0$. Then $\bigcap_{k=1}^{\infty} F_k$ is a singleton.*

Proof: It suffices to show $\bigcap_{k=1}^{\infty} F_k \neq \emptyset$. Pick $x_k \in F_k$. Then $d(x_k, x_j) \leq d_k$ for $j \geq k$ so $\{x_j\}$ is Cauchy and, therefore, convergent to some $x \in S$. Clearly $x \in \bigcap_{k=1}^{\infty} F_k$.

Theorem 2 (Baire Category Theorem) *A complete metric space is second category in itself.*

Proof: Let A_k be nowhere dense in S for every $k \in \mathbf{N}$ and set $E = \bigcup_{k=1}^{\infty} A_k$. We show $S \backslash E \neq \emptyset$. Let $x_0 \in S$. Since A_1 is nowhere dense, there is a closed ball B_1 of radius less than $1/2$ inside the closed ball $B_0 = \{x : d(x, x_0) \leq 1\}$ such that $B_1 \cap A_1 = \emptyset$ (Exercise 1). Since A_2 is nowhere dense, there is a closed ball B_2 of radius less than $1/2^2$ inside B_1 such that $B_2 \cap A_2 = \emptyset$. Continuing this construction gives a decreasing sequence of closed balls $\{B_k\}$ of radius less than $1/2^k$ such that $B_k \cap A_k = \emptyset$ for all k. By Proposition 1, $\bigcap_{k=1}^{\infty} B_k = \{x\}$. Clearly $x \in S \backslash E$.

We give a corollary of the Baire Category Theorem which is particularly useful in applications.

Corollary 3 *Let (S, d) be a complete metric space. If $S = \bigcup_{k=1}^{\infty} A_k$, then some $\overline{A}_k$ must have a non-void interior.*

Despite its esoteric appearance the Baire Category Theorem has a surprising number of applications to various areas of analysis. For example, Banach used the theorem to show the existence of a continuous, nowhere differentiable function. For this and other interesting examples, see [DeS] and [Boa].

A.2. THE BAIRE CATEGORY THEOREM

Exercise 1. Show that E is nowhere dense if and only if every sphere S contains a sphere S' such that $S' \cap E = \emptyset$.

Exercise 2. Show the Cantor set is nowhere dense in $[0,1]$.

Exercise 3. Let $f_k : \mathbf{R} \to \mathbf{R}$ be continuous, nonnegative and such that $\sum_{k=1}^{\infty} f_k(t)$ converges for every $t \in \mathbf{R}$. Show that there is an interval in $\mathbf{R}$ where the convergence is uniform.

Exercise 4. Show that $\mathbf{R}^2$ is not a countable union of lines.

Exercise 5. Show that if S is a complete metric space without isolated points, then S is uncountable.

A3: The Arzela-Ascoli Theorem

Let S be a compact Hausdorff space and assume $C(S)$ has the sup-norm. We give a characterization of the compact subsets of $C(S)$ due to Arzela and Ascoli. Of course, any compact subset of $C(S)$ is closed and bounded, but by Theorem 5.1.18 the converse cannot hold in general. We first give an additional necessary condition that a compact subset of $C(S)$ must satisfy.

Definition 1 *A subset $K \subset C(S)$ is equicontinuous at $s \in S$ if for every $\epsilon > 0$ there exists a neighborhood V of s such that $|f(s) - f(t)| < \epsilon$ for every $t \in V$ and $f \in K$. K is equicontinuous (on S) if K is equicontinuous at every point of S.*

Proposition 2 *If $K \subset C(S)$ is compact, then K is equicontinuous.*

Proof: Let $\epsilon > 0$ and $s \in S$. There exist $f_1, \ldots, f_k \in K$ such that $K \subset \bigcup_{i=1}^{k} S(f_i, \epsilon)$ where $S(f, \epsilon) = \{g : \|g - f\| < \epsilon\}$. Pick an open neighborhood V of s such that $|f_i(s) - f_i(t)| < \epsilon$ for $t \in V$ and $i = 1, \ldots, k$. Suppose $t \in V$ and $f \in K$. Choose i such that $f \in S(f_i, \epsilon)$. Then

$$|f(t) - f(s)| \le |f(t) - f_i(t)| + |f_i(t) - f_i(s)| + |f_i(s) - f(s)| < 3\epsilon$$

so K is equicontinuous.

Thus, any compact subset of $C(S)$ is closed, bounded and equicontinuous. We consider the converse of this statement.

Lemma 3 *Let $\{f_k\}$ be a sequence of real-valued functions defined on a countable set $E = \{x_k : k \in \mathbf{N}\}$ which is pointwise bounded on E. Then there is a subsequence $\{g_k\}$ of $\{f_k\}$ such that $\{g_k\}$ converges pointwise on E.*

Proof (Diagonalization Procedure): The sequence $\{f_k(x_1)\}$ is bounded in $\mathbf{R}$ and, therefore, has a convergent subsequence $\{f_{1,k}(x_1)\}_{k=1}^{\infty}$ [the reason for this slightly unorthodox notation will become apparent]. The sequence $\{f_{1,k}(x_2)\}_{k=1}^{\infty}$ is bounded and has a convergent subsequence $\{f_{2,k}(x_2)\}_{k=1}^{\infty}$. Note that $\{f_{2,k}(x_1)\}_{k=1}^{\infty}$ also converges. Proceeding by induction produces a sequence $\{S_k\}$ as follows:

$$\begin{array}{cccccc} S_1 & : & f_{1,1} & f_{1,2} & f_{1,3} & \cdots \\ S_2 & : & f_{2,1} & f_{2,2} & f_{2,3} & \cdots \\ \vdots & & & & & \end{array}$$

Note that (i) S_{k+1} is a subsequence of S_k and (ii) $\{f_{j,k}(x_i)\}_{k=1}^{\infty}$ converges for $1 \le i \le j$. Now let $\{g_k\}$ be the diagonal sequence $\{f_{k,k}\}$.

Theorem 4 (Arzela-Ascoli) *Let $K \subset C(S)$. Then K is compact if and only if K is closed, bounded and equicontinuous.*

A.3. THE ARZELA-ASCOLI THEOREM

Proof: $\Rightarrow$: Proposition 2.

$\Leftarrow$: It suffices to show that any sequence $\{f_k\} \subset K$ has a subsequence which converges uniformly on S. By equicontinuity, for each k there exists a finite set $F_k \subset S$ and open neighborhoods $\{V_t : t \in F_k\}$ such that $S = \cup\{V_t : t \in F_k\}$ and $|f(s) - f(t)| < 1/k$ when $s \in V_t$ and $f \in K$.

Set $E = \bigcup_{k=1}^{\infty} F_k$. By Lemma 3 there is a subsequence $\{g_k\}$ of $\{f_k\}$ such that $\{g_k\}$ converges pointwise on E. We claim that $\{g_k\}$ is a Cauchy sequence in $C(S)$. Let $\epsilon > 0$ and choose k such that $1/k < \epsilon$. There exists N such that $|g_i(t) - g_j(t)| < 1/k$ for all $t \in F_k$ and $i, j \geq N$. If $s \in S$, there exists $t \in F_k$ such that $s \in V_t$ so if $i, j \geq N$,

$$|g_i(s) - g_j(s)| \leq |g_i(s) - g_i(t)| + |g_i(t) - g_j(t)| + |g_j(t) - g_j(s)| < 3/k < 3\epsilon.$$

Hence, $\{g_k\}$ is a Cauchy sequence in $C(S)$ and the proof is complete.

If $K \subset C(S)$ is bounded and equicontinuous, then $\overline{K}$ is equicontinuous [Exercise 1] so $\overline{K}$ is compact by Theorem 4. Hence, every sequence in K has a subsequence which converges uniformly on S.

The Arzela-Ascoli Theorem has many applications in differential equations, integral equations and calculus of variations. See, for example, [CL].

Exercise 1. If $K \subset C(S)$ is equicontinuous, show $\overline{K}$ is equicontinuous.

Exercise 2. Give a specific example of a subset of $C[0,1]$ which is closed, bounded but not compact.

Exercise 3. Let $K \subset C[a,b]$ be such that each $f \in K$ has a continuous derivative and $K' = \{f' : f \in K\}$ is bounded in $C[a,b]$. Show K is equicontinuous.

Exercise 4. Let $K \subset C[a,b]$ be bounded. Let $F(t) = \int_a^t f(s)ds$. Show $\{F : f \in K\}$ is equicontinuous.

Exercise 5. Let $K \subset C(S)$ be equicontinuous and pointwise bounded on S. Show K is uniformly bounded on S.

Exercise 6. Let $\{f_k\} \subset C(S)$ be equicontinuous. If $\{f_k\}$ converges pointwise on a dense subset of S, show $\{f_k\}$ converges pointwise on S. Is compactness important?

A4: The Stone-Weierstrass Theorem

In this appendix we prove Stone's far reaching generalization of the Weierstrass Approximation Theorem. Let S be a compact Hausdorff space and $C(S)$ the space of all real-valued continuous functions on S equipped with the sup-norm [complex-valued continuous functions are considered at the end of this section]. Stone's Theorem gives algebraic conditions, modeled on the polynomials on the line, which insure that a subset of $C(S)$ is dense in $C(S)$.

A subset $\mathcal{A} \subseteq C(S)$ is called an *algebra* if

(i) $f, g \in \mathcal{A}$ imply $f + g$ and fg belong to $\mathcal{A}$.

(ii) $f \in \mathcal{A}$ implies $tf \in \mathcal{A}$ $\forall t \in \mathbf{R}$.

Example 1 $C(S)$ is an algebra. The set of polynomials, $\mathcal{P}$, is an algebra in $C[a,b]$. The set of even polynomials, $\mathcal{E}$, is an algebra in $C[a,b]$; the set of odd polynomials, $\mathcal{O}$, is not an algebra in $C[a,b]$. The polynomials with 0 constant term is an algebra in $C[a,b]$. If $K \subseteq \mathbf{R}^n$ is compact, the set of polynomials in n real variables forms an algebra in $C(K)$.

A subset $\mathcal{B} \subseteq C(S)$ *separates the points of* S if and only if for $t, s \in S$, $t \neq s$, $\exists f \in \mathcal{B}$ such that $f(t) \neq f(s)$.

Example 2 $\mathcal{P}$ separates the points of $[a,b]$; $\mathcal{E}$ does not separate the points of $[-1,1]$; $\mathcal{O}$ separates the points of $[a,b]$.

Lemma 3 *Let $\mathcal{L}$ be a vector subspace of $C(S)$ that separates the points of S and is such that $1 \in \mathcal{L}$. Then given $t, s \in S$, $t \neq s$, and $a, b \in \mathbf{R}$ $\exists f \in \mathcal{L}$ such that $f(s) = a$ and $f(t) = b$.*

Proof: There is $g \in \mathcal{L}$ such that $g(s) \neq g(t)$. Put $c = g(s) - g(t)$. Then the function
$$f = \frac{(a-b)g + (bg(s) - ag(t)) \cdot 1}{c} \in \mathcal{L}$$
and satisfies $f(s) = a$, $f(t) = b$.

Definition 4 *A subset $\mathcal{L}$ of $C(S)$ is called a function space if $\mathcal{L}$ is a vector space and if $f, g \in \mathcal{L}$ implies $f \vee g \in \mathcal{L}$ and $f \wedge g \in \mathcal{L}$.*

Note that if $\mathcal{L}$ is a function space, then whenever $f \in \mathcal{L}$, $f^+ = f \vee 0$, $f^- = (-f) \vee 0$ and $|f|$ also belong to $\mathcal{L}$.

Example 5 Let $\mathcal{PL}$ be the collection of all piecewise linear, continuous functions on $[0,1]$ (i.e., $f \in \mathcal{PL}$ if and only if $f \in C[0,1]$ and $\exists$ a partition $x_0 < x_1 < \ldots < x_n$ of $[0,1]$ such that f is linear on each subinterval $[x_{i-1}, x_i]$.) Then $\mathcal{PL}$ is a function space but is not an algebra.

A.4. THE STONE-WEIERSTRASS THEOREM

Lemma 6 *Let $\mathcal{L}$ be a function space in $C(S)$ that contains the constant function 1 and separates the points of S. Then given $g \in C(S)$ and $t_0 \in S$ and $\epsilon > 0$, $\exists f \in \mathcal{L}$ such that $f(t_0) = g(t_0)$ and $f(t) > g(t) - \epsilon$ $\forall t \in S$.*

Proof: By Lemma 3, $\forall t \in S$ $\exists f_t \in \mathcal{L}$ such that $f_t(t_0) = g(t_0)$ and $f_t(t) = g(t)$. Since f_t and g are continuous, there is a neighborhood V_t of t such that $f_t(s) > g(s) - \epsilon$ $\forall s \in V_t$.

Then $\{V_t : t \in S\}$ is an open cover of S and, therefore, there are $t_1, \ldots, t_n \in S$ such that $\bigcup_{i=1}^{n} V_{t_i} = S$. Let $f = f_{t_1} \vee \cdots \vee f_{t_n}$. Then $f \in \mathcal{L}$ and $f(t_0) = g(t_0)$. Also, if $s \in S$, then $\exists k$ such that $s \in V_{t_k}$. Then $f(s) \geq f_{t_k}(s) > g(s) - \epsilon$; so f is the desired function.

We can now give the lattice version of the Stone-Weierstrass theorem.

Theorem 7 *Let $\mathcal{L}$ be a function space in $C(S)$ that contains the constant function 1 and separates the points of S. Then $\mathcal{L}$ is dense in $C(S)$.*

Proof: Let $g \in C(S)$ and $\epsilon > 0$. By Lemma 6, $\forall t \in S$ $\exists f_t \in \mathcal{L}$ such that $f_t(t) = g(t)$ and $f_t(s) > g(s) - \epsilon$ $\forall s \in S$. By the continuity of f_t and g, $\exists$ a neighborhood U_t of t such that $f_t(s) < g(s) + \epsilon$ $\forall s \in U_t$. Since S is compact, $\exists t_1, \ldots, t_n \in S$ such that $\bigcup_{i=1}^{n} U_{t_i} = S$. Put $f = f_{t_1} \wedge \cdots \wedge f_{t_n}$. Then $f \in \mathcal{L}$.

Since $f_{t_i} > g - \epsilon$, $f > g - \epsilon$. If $s \in S$, then $s \in U_{t_k}$ for some k, so that $f(s) \leq f_{t_k}(s) < g(s) + \epsilon$. Thus

$$g(s) - \epsilon \leq f(s) \leq g(s) + \epsilon \quad \forall s \in S \text{ or } \|f - g\|_\infty \leq \epsilon.$$

In order to state the algebraic version of the Stone-Weierstrass theorem, a lemma is needed.

Lemma 8 *There is a sequence of polynomials $\{p_n\}$ such that $p_n(t) \to \sqrt{t}$ uniformly for $t \in [0, 1]$.*

Proof: Set $p_1 = 0$ and $p_{n+1}(t) = p_n(t) + \frac{1}{2}(t - p_n(t)^2)$ for $n \geq 1$. Clearly, each p_n is a polynomial.

We first claim that: $0 \leq p_n(t) \leq \sqrt{t}$, $0 \leq t \leq 1$. This certainly holds for $n = 1$; assume that it holds for $n \leq k$. $p_{k+1}(t) \geq 0$ and

$$\sqrt{t} - p_{k+1}(t) = \sqrt{t} - p_k(t) - \frac{1}{2}\left[t - p_k(t)^2\right] = \left[\sqrt{t} - p_k(t)\right]\left\{1 - \frac{1}{2}\left[\sqrt{t} + p_k(t)\right]\right\} \geq 0.$$

Thus, the claim is established by induction.

Since $p_n(t)^2 \leq t$ $\forall t \in [0, 1]$, it follows that $\{p_n(t)\} \uparrow$ $\forall t \in [0, 1]$. Put $p(t) = \lim p_n(t)$. Since $p(t) \geq 0$ and $p(t)^2 = t$, we have $p(t) = \sqrt{t}$, that is, $p_n(t) \to \sqrt{t}$ $\forall t \in [0, 1]$. The convergence is uniform on $[0, 1]$ by Dini's theorem ([DeS] 11.18).

Of course, the conclusion of Lemma 8 follows directly from the Weierstrass approximation theorem. We gave an independent proof in order to show that the Weierstrass approximation theorem truly is a corollary of the Stone-Weierstrass theorem.

Theorem 9 (Stone-Weierstrass) *Let $\mathcal{A}$ be an algebra in $C(S)$ such that $\mathcal{A}$ contains the constant function 1 and separates the points of S. Then $\mathcal{A}$ is dense in $C(S)$.*

Proof: By Theorem 7, it suffices to show that $\overline{\mathcal{A}}$ is a function space.

We first claim that $f \in \overline{\mathcal{A}}$ implies $|f| \in \overline{\mathcal{A}}$. Let $\{p_n\}$ be the polynomials in Lemma 8, let $f \in \overline{\mathcal{A}}$ be such that $f \neq 0$, and put $a = \|f\|_\infty > 0$. Then $g_n = p_n \circ (f^2/a^2) \in \overline{\mathcal{A}}$ $\forall n$ (Exercise 1) and, since $p_n(t) \to \sqrt{t}$ uniformly for $0 \le t \le 1$, $g_n \to \sqrt{f^2/a^2} = |f|/a$ in $\|\ \|_\infty$. Hence, $|f| = a(|f|/a) \in \overline{\mathcal{A}}$.

But $f \vee g = \frac{1}{2}(f + g + |f - g|)$, and $f \wedge g = \frac{1}{2}(f + g - |f - g|)$; thus, $\overline{\mathcal{A}}$ is indeed a function space.

Corollary 10 *(Weierstrass Approximation Theorem) The polynomials are dense in $C[a, b]$.*

A more general statement is given by Corollary 11.

Corollary 11 *Let $K \subseteq \mathbf{R}^n$ be compact. The polynomials in n-variables are dense in $C(K)$.*

Finally, we should check the necessity of the various hypotheses in Theorem 9. The algebra $\mathcal{E}$ in $C[-1, 1]$ does not separate the points of $[-1, 1]$ and is not dense in $C[-1, 1]$, so this condition cannot be dropped. Because the algebra of polynomials in $C[0, 1]$ that vanish at 0 is not dense in $C[0, 1]$, the condition that the algebra contains the constant function 1 cannot be dropped.

As stated above the Stone-Weierstrass Theorem is not valid for complex-valued continuous functions. For an example, let $T = \{e^{it} : 0 \le t \le 2\pi\}$ be the unit circle in the complex plane and consider the function $f(z) = \overline{z}$ and any complex polynomial $p(z) = \sum_{k=0}^{n} c_k z^k$. Write $f[p]$ for $f(e^{it})[p(e^{it})]$. Then

$$\int_0^{2\pi} \overline{f}(e^{it}) p(e^{it}) dt = \sum_{k=0}^n c_k \int_0^{2\pi} e^{i(k+1)t} dt = 0$$

so

$$2\pi = \int_0^{2\pi} f(e^{it}) \overline{f}(e^{it}) dt \le \left| \int_0^{2\pi} (f - p)\overline{f} \right| + \left| \int_0^{2\pi} p\overline{f} \right| \le 2\pi \|f - p\|_\infty.$$

Hence, f is distance at least 1 from any polynomial so the polynomials cannot be dense in the continuous complex-valued valued functions on T. [Those readers familiar with complex variables know that the uniform limit of any sequence of polynomials on $\{z : |z| \le 1\}$ must be analytic.]

For the complex form of the Stone-Weierstrass Theorem, we need to add an additional condition, namely, that whenever a function f belongs to the algebra $\mathcal{A}$ then so does its conjugate $\overline{f}$. [Note this property is missing in the example above.]

We denote by $C_\mathbf{C}(S)$ the space of all continuous, complex-valued functions on S. We assume that $C_\mathbf{C}(S)$ is equipped with the sup-norm.

A.4. THE STONE-WEIERSTRASS THEOREM

Theorem 12 *(Complex Form of Stone-Weierstrass).* Let $\mathcal{A}$ be an algebra in $C_{\mathbf{C}}(S)$ such that $\mathcal{A}$ contains the constant function 1, separates the points of S and is such that when $f \in \mathcal{A}$, $\overline{f} \in \mathcal{A}$. Then $\mathcal{A}$ is dense in $C_{\mathbf{C}}(S)$.

Proof: Let $\mathcal{A}_{\mathbf{R}}$ be the set of all real and imaginary parts of functions which belong to $\mathcal{A}$. Since $\mathcal{R}f = (f + \overline{f})/2$ and $\mathcal{I}f = (f - \overline{f})/2i$, $\mathcal{A}_{\mathbf{R}}$ is an algebra in $C(S)$ which contains 1 and separates the points of S. Hence, $\mathcal{A}_{\mathbf{R}}$ is dense in $C(S)$. Since $\mathcal{A} = \{f + ig : f, g \in \mathcal{A}_{\mathbf{R}}\}$, $\mathcal{A}$ is dense in $C_{\mathbf{C}}(S)$.

For further remarks on the Stone-Weierstrass theorem, see the article by M. Stone, in *Studies in Modern Analysis*, Volume 1 in *Mathematical Association of America Studies in Mathematics*, edited by R.C. Buck ([Bu]).

Exercise 1. If $\mathcal{A}$ is an algebra in $C(S)$, show that $\overline{\mathcal{A}}$ is an algebra.

Exercise 2. Is $\mathcal{P}$ a function space in $C[a, b]$?

Exercise 3. If $\mathcal{L}$ is a function space in $C(S)$, show that $\overline{\mathcal{L}}$ is a function space.

Exercise 4. Show that $\mathcal{PL}$ (see Example 5) is dense in $C[0, 1]$.

Exercise 5. Let $\mathcal{A}$ be the vector space in $C[0, 1]$ generated by the functions 1, $\sin^1 t, \sin^2 t, \ldots$. [$f \in \mathcal{A}$ if and only if $f(t) = \sum\limits_{k=0}^{n} a_k \sin^k t$ for some $a_i \in \mathbf{R}$.] Show that $\mathcal{A}$ is dense in $C[0, 1]$.

Exercise 6. Show that the algebra generated by the functions $\{1, t^2\}$ is dense in $C[0, 1]$ but is not dense in $C[-1, 1]$.

Exercise 7. Let S, T be compact Hausdorff spaces. If $f \in C(S)$, $g \in C(T)$, write $f \otimes g$ for the function $(s, t) \to f(s)g(t)$. Show that the functions of the form $\sum f_k \otimes g_k$ (finite sum) are dense in $C(S \times T)$.

Exercise 8. Give an example of a situation where Theorem 7 is applicable but Theorem 9 is not.

Exercise 9. If $f_k \in \mathcal{BV}[a, b]$ and $f_k \to f$ uniformly on $[a, b]$, is it necessarily true that $f \in \mathcal{BV}[a, b]$?

Exercise 10. If $g \in C[0, 1]$ and $\epsilon > 0$, show that $\exists \alpha_0, \alpha_1, \ldots, \alpha_k \in \mathbf{R}$ such that

$$\left| g(t) - \sum_{j=0}^{k} \alpha_j e^{jt} \right| < \epsilon, \qquad \forall t \in [0, 1].$$

Exercise 11. Show that the polynomials with rational coefficients are dense in $C[a,b]$.

Exercise 12. Let $\mathcal{A}$ be the set of all functions of the form $\sum\limits_{k=0}^{n} c_k e^{kt}$, $n \in \mathbf{N}$, $c_k \in \mathbf{R}$. Show that $\mathcal{A}$ is dense in $C[a,b]$.

Exercise 13. Let $\mathcal{A}$ be the set of all functions of the form $\sum\limits_{k=0}^{n} c_k \cos kt$, $n \in \mathbf{N}$, $c_k \in \mathbf{R}$. Show that $\mathcal{A}$ is dense in $C[0,\pi]$. Is $\mathcal{A}$ dense in $C[-\pi,\pi]$?

References

[AS] P. Antosik and C. Swartz, *Matrix Methods in Analysis*, Springer-Verlag, Heidelberg, 1985.

[Ap] T. Apostol, *Mathematical Analysis*, Addison-Wesley, Reading, 1975.

[As] J.M. Ash, *Studies in Harmonic Analysis*, Math. Assoc. Amer., 1976.

[B1] S. Banach, *Oeuvres*, PWN, Warsaw, 1967.

[B2] S. Banach, *Oeuvres II*, PWN, Warsaw, 1979.

[Ba] R. Bartle, *The Elements of Real Analysis*, Wiley, N.Y., 1976.

[Ba2] R. Bartle, An Extension of Egorov's Theorem, *Amer. Math. Monthly*, 87 (1980), 628-633.

[Bi] G. Birkoff, *Lattice Theory*, Amer. Math. Soc., N.Y., 1940.

[Boa] R.P. Boas, *A Primer of Real Functions*, Math. Assoc. Amer., Providence, 1960.

[Bo] S. Bochner, Additive set functions on groups, *Ann. Math.*, 40 (1939), 769-799.

[Bb] N. Bourbaki, *Integration*, Hermann, Paris, 1952.

[Br] J. Brooks, The Lebesgue Decomposition Theorem for Measures, *Amer. Math. Monthly*, 78 (1971), 660-662.

[Bu] R.C. Buck, Studies in Modern Analysis I, *Math. Assoc. Amer.*, 1962.

[Ca] C. Caratheodory, *Measure and Integration*, Chelsea, New York, 1963.

[Cat] F. Cater, Most Monotone Functions are not Singular, *Amer. Math. Monthly* 89 (1982), 466-469.

[Ch] S.B Chae, *Lebesgue Integration*, Marcel Dekker, N.Y., 1980.

[Cob] S. Cobzas, Hahn decompositions of finitely additive measures, *Arch. Math.*, 27 (1976), 620-621.

[CL] E. Coddington and N. Levinson, *Theory of Ordinary Differential Equations*, McGraw-Hill, N.Y., 1955.

[Co] D. Cohn, *Measure Theory*, Birkhäuser, Boston, 1980.

[DG] R. Darst and C. Goffman, A Borel Set which Contains no Rectangles, *Amer. Math. Monthly*, 77 (1970), 728-729.

[DM] L. Debnath and P. Mikusinski, *Introduction to Hilbert Spaces with Applications*, Academic Press, N.Y., 1990.

[DeS] J. DePree and C. Swartz, *Introduction to Real Analysis*, Wiley, N.Y., 1987.

[DU] J. Diestel and J. Uhl, *Vector Measures*, Amer. Math. Soc., Providence, 1977.

[Di] J. Dieudonne, *Foundations of Modern Analysis*, Academic Press, N.Y., 1960.

[Do] R. Doss, The Hahn Decomposition Theorem, *Proc. Amer. Math. Soc.*, 80 (1980), 377.

[Dr] L. Drewnowski, Equivalence of Brooks-Jewett, Vitali-Hahn-Saks and Nikodym Theorems, *Bull. Acad. Polon. Sci.*, 20 (1972), 725-731.

[D] L.E. Dubins, An Elementary Proof of Bochner's Finitely Additive Radon-Nikodym Theorem, *Amer. Math. Monthly*, 76 (1969), 520-523.

[DS] N. Dunford and J. Schwartz, *Linear Operators I*, Wiley, N.Y. 1958.

[Fe] M. Feldman, A Proof of Lusin's Theorem, *Amer. Math. Monthly*, 88 (1981), 191-192.

[Fre] G. Freilich, Increasing Continuous Singular Functions, *Amer. Math. Monthly* 1973 (80), 918-919.

[Fr] R. French, The Banach-Tarski Theorem, *Math. Intell.*, 10 (1988), 21-28.

[Ga] G. Gaudry, Sets of Positive Product Measure in which every Rectangle is Null, *Amer. Math. Monthly*, 81 (1974), 889-890.

[Gi] D.P. Giesy, A Finite-Valued Finitely Additive Unbounded Measure, *Amer. Math. Monthly*, 77 (1970), 508-510.

[Gr] J.D. Gray, The Shaping of the Riesz Representation Theorem, *Arch. History Exact Sci.*, 31 (1984), 127-187.

[Hah] H. Hahn, *Set Functions*, University of New Mexico Press, Albuquerque, 1948.

[Hal] P. Halmos, *Measure Theory*, Van Nostrand, Princeton, 1950.

[Ha] T. Hawkins, *Lebesgue's Theory of Integration*, Univ. of Wisconsin Press, Madison, 1970.

[He] R. Henstock, Definitions of Riemann Type of Variational Integral, *Proc. London Math. Soc.*, 11 (1962), 402-418.

REFERENCES

[HS] E. Hewitt and K. Stromberg, *Real and Abstract Analysis*, Springer-Verlag, N.Y., 1965.

[HY] E. Hewitt and K. Yosida, Finitely Additive Measures, *Trans. Amer. Math. Soc.*, 72 (1952), 46-66.

[Ho] H. Hochstadt, Eduard Helly, Father of the Hahn-Banach Theorem, *Math. Intell.* 2 (1979), 123-125.

[J] R.C. James, Reflexivity and the sup of linear functionals, *Israel J. Math.*, 13 (1972), 289-300.

[KO] S. Kakutani and J.C. Oxtoby, Construction of a non-separable invariant extension of the Lebesgue measure space, *Annals of Math.*, 52 (1950), 580-590.

[Kn] K. Knopp, *Infinite Sequences and Series*, Dover, N.Y., 1956.

[KF] Kolmogorov and Fomin, *Introductory Real Analysis*, Dover, N.Y., 1970.

[Ku] J. Kurzweil, Generalized Ordinary Differential Equations and Continuous Dependence on a Parameter, *Czech. Math. J.*, 82 (1957), 418-449.

[L1] H. Lebesgue, Intégrale, longueur, aire, *Ann. Mat.* (3), 7, 1902, 231-359.

[L2] H. Lebesuge, *Measure and the Integral*, Holden-Day, San Francisco, 1966.

[Lew] J.W. Lewin, A Truly Elementary Approach to the Bounded Convergence Theorem, *Amer. Math. Monthly* 93 (1986), 395-397.

[Lo] L. Loomis, *An Introduction to Abstract Harmonic Analysis*, Van Nostrand, Princeton, 1953.

[Mac] H.M. MacNeille, A Unified Theory of Integration, *Proc. Natl. Acad. Sci. U.S.A.*, 27 (1941), 71-76.

[Mc] R. McLeod, *The Generalized Riemann Integral*, Math. Assoc. Amer., Providence, 1980.

[M1] J. Mikusinski, Sur une definition de l'integrale de Lebesgue, *Bull. Acad. Polon. Sci.*, 12 (1964), 203-204.

[M2] J. Mikusinski, *The Bochner Integral*, Academic Press, N.Y., 1978.

[Mu] M. Munroe, *Introduction to Measure and Integration*, Addison-Wesley, Reading, 1953.

[Na] I. Natanson, *Theory of Functions of a Real Variable*, Ungar, N.Y., 1955.

[N1] O. Nikodym, Sur les suites de fonctions parfaitement additives d'ensemble abstraits, *C.R.A.S.* 192 (1931), 727.

[N2] O. Nikodym, Sur les familles borneés de fonctions parfaitement additives d'ensemble abstrait, *Monatsch. für Math. und Phys.*, 40 (1933), 427-432.

[P] A. Peressini, *Ordered Topological Vector Spaces*, Harper and Row, N.Y., 1967.

[Pe] I. Pesin, *Classical and Modern Integration Theories*, Academic Press, N.Y., 1970.

[Ra] J. Randolph, *Basic Real and Abstract Analysis*, Academic Press, N.Y., 1968.

[RR] K.P.S. Bhaskara Rao and M. Bhaskara Rao, *Theory of Charges*, Academic Press, N.Y., 1983.

[RN] F. Riesz and B. Nagy, *Functional Analysis*, Ungar, N.Y., 1955.

[Ro] W.W. Rogosinski, *Volume and Integral*, Oliver and Boyd, London, 1952.

[R] J. Romero, When is $L^p(\mu)$ contained in $L^q(\mu)$?, *Amer. Math. Monthly*, 90 (1983), 203-206.

[Roy] H. Royden, *Real Analysis*, MacMillen, N.Y., 1988.

[R1] W. Rudin, *Principles of Mathematical Analysis*, McGraw-Hill, N.Y., 1976.

[R2] W. Rudin, *Real and Complex Analysis*, McGraw-Hill, N.Y., 1966.

[Sc] J. Schwartz, A note on the space L_p^*, *Proc. Amer. Math. Soc.*, 2 (1951), 270-275.

[Sir] W. Sierpinski, Sur les rapports entre l'existence des integrales, *Fund. Math.*, 1 (1920), 142-147.

[Si] G. Simmons, *Introduction to Topology and Modern Analysis*, McGraw-Hill, N.Y., 1963.

[St] K. Stromberg, The Banach-Tarski Paradox, *Amer. Math. Monthly*, 86 (1979), 151-161.

[St2] K. Stromberg, *An Introduction to Classical Real Analysis*, Wadsworth, Belmont, 1981.

[Sw1] C. Swartz, The Evolution of the Uniform Boundedness Principle, *Math. Chron.*, 19 (1990), 1-18.

[Sw2] C. Swartz, *An Introduction to Functional Analysis*, Dekker, N.Y., 1992.

[Sw3] C. Swartz, The Nikodym and Vitali-Hahn-Saks Theorems for Algebras, *Amer. Math. Monthly*, 82 (1975), 833-834.

[Ta] A. Taylor, *General Theory of Functions and Integration*, Ginn, Waltham, 1965.

REFERENCES

[TL] A. Taylor and D. Lay, *Introduction to Functional Analysis*, Wiley, N.Y., 1980.

[TT] A. Tulcea and C. Tulcea, *Topics in the Theory of Lifting*, Springer-Verlag, N.Y., 1969.

[Vu] B. Vulich, *Introduction to the Theory of Partially Ordered Spaces*, Wolters-Noordhoff, Groningen, 1967.

[Wa] P. Walker, On Lebesgue Integrable Derivatives, *Amer. Math. Monthly*, 84 (1977), 287-288.

[Wi] J.H. Williamson, *Lebesgue Integration*, Holt, Rinehart, and Winston, New York, 1962.

[Z] A. Zaanen, *Linear Analysis*, North Holland, Amsterdam, 1956.

[Za] T. Zamfirescu, Most Monotone Functions are Singular, *Amer. Math. Monthly* 88 (1981), 47-49.

[Zy] A. Zygmund, *Trigonometric Series*, Cambridge University Press, 1968.

Index

$
$[x]$, 181
$\|\ \|$, 103
$\hat{f}$, 152
ℓ^∞, 168
$\frac{d\nu}{d\mu}$, 133
$\Gamma(x)$, 100
$\int_S f d\mu$, 82
$\lim E_k$, 2
$\lim t_k$, 2
$|\mu|$, 30
$|f|$, 3
$|x|$, 198
$\mu \perp \nu$, 34
$\mu \vee \nu$, 138
$\mu \wedge \nu$, 135
μ_f, 24
μ-a.e., 74
μ^*, 36
μ^+, 28
μ^-, 28
Ω_f, 124
$\overline{\int}_a^b f$, 96
$\overline{\lim} E_k$, 2
$\overline{\lim} t_k$, 2
$\overline{D} f(x)$, 154
$\|\ \|_p$, 207
$\|\ \|$, 165
$\|\ \|_\infty$, 218
$\mathbf{R}^*$, 4
$\mathcal{C}_\delta$, 2
$\mathcal{C}_\sigma$, 2
$\ell^p(S)$, 209
$\underline{\ell^p}$, 209
$\underline{\ell^1}$, 167
$\inf E$, 2

$\underline{\int}_a^b f$, 96
$\underline{\lim} E_k$, 2
$\underline{\lim} t_k$, 2
$\mu \perp \nu$, 139
$\mu \times \nu$, 116
$\nu \ll \mu$, 130
$\sigma(\mathcal{S} \times \mathcal{T})$, 116
$\sup E$, 2
$\mathcal{F}_\sigma$, 49
$\mathcal{G}_\delta$, 49
$\mathcal{P}(S)$, 1
$\overline{ba(\mathcal{A})}$, 223
$\overline{ba}$, 224
$BV[a,b]$, 255
$\overline{C_c^\infty(\mathbf{R}^n)}$, 127
$\overline{C_C(S)}$, 104
$\overline{D\nu(x)}$, 149
$\overline{Df(x)}$, 154
$f \prec V$, 228
$f * g$, 125
$\underline{f^+}$, 3
$\underline{f^-}$, 3
$K \prec f$, 228
$\overline{L(X)}$, 174
$\overline{L^\infty(\mu)}$, 218
$\overline{L^\infty(E)}$, 218
$\overline{L^0(I)}$, 110
$\overline{L^1(\mu)}$, 103
$\overline{L^1(I)}$, 103
$\overline{L^p(\mu)}$, 207
$\overline{L^p(E)}$, 207
$\overline{L^p}$, 249
$spt(f)$, 104
$t_k \downarrow$, 2
$t_k \uparrow$, 2
X', 175

INDEX

$\hat{f}$, 249
$\hat{T}$, 182
$\hat{x}(a)$, 240
$\mathbf{C}^n$, 1
$\mathbf{N}$, 1
$\mathbf{Q}$, 1
$\mathbf{R}$, 1
$\mathbf{R}^n$, 1
$\mathbf{R}_+$, 1
$\mathbf{Z}$, 1
$\mathcal{B}(S)$, 39
$\mathcal{F}(S)$, 197
$\mathcal{I}z$, 237
$\mathcal{L}\{f\}(s)$, 128
$\mathcal{M}(\mu^*)$, 36
$\mathcal{PL}$, 263
$\mathcal{R}\int_a^b f$, 96
$\mathcal{R}z$, 237
$\mathcal{S}$, 250
$\mathcal{S}(\Sigma)$, 193
$\mathcal{S}(\mathcal{A})$, 193
$A_r f(x)$, 146
$B(p,q)$, 123
$B(S)$, 167
$ba(\Sigma : \mu)$, 223
$ba(\mathcal{A})$, 135
$BV[a,b]$, 199
$BV_0[a,b]$, 204
c, 168
$C(S)$, 167
$C^k[0,1]$, 199
c_0, 169
$C_0(\mathbf{R})$, 250
c_{00}, 169
C_A, 1
$C_C(S)$, 265
$ca(\Sigma)$, 226
$d^+ f(x)$, 154
$d^- f(x)$, 154
$d_+ f(x)$, 154
$d_- f(x)$, 154
D_n, 244
E^t, 119
E_s, 119

$f \vee g$, 3
$f \wedge g$, 3
$f(\cdot, t)$, 117
$f(s, \cdot)$, 117
$f^\vee(x)$, 251
f_*, 151
f_h, 127
$J = J_X$, 191
$L(f, \pi)$, 96
$L(X, Y)$, 174
$L^0(\mu)$, 109
$L^1(\mu)$, 167
$L_C^1(\mu)$, 237
$L_{loc}^1(\mathbf{R}^n)$, 145
m, 24
$M^\perp$, 239
m_n, 25
$Mf(x)$, 146
$NBV[a,b]$, 233
$rca(S)$, 232
s, 168
$S(x, r)$, 145
T', 192
$U(f, \pi)$, 96
$Var(f : [a, b])$, 255
$var(f : \pi)$, 255
$x \cdot y$, 235
$x \perp y$, 238
$x \vee y$, 197
$x \wedge y$, 197
X/M, 181
x^+, 198
x^-, 198
X'', 190
$X^\sim$, 200

A

α-Cantor set, 53
μ-almost everywhere, 74
μ-almost uniformly, 74
absolute value, 198
absolutely continuous, 130, 159
absolutely convergent, 166
adjoint, 192
Alexanderoff, 63

INDEX

algebra, 16, 263
anti-symmetry, 197
Antosik-Mikusinski, 65
approximate identity, 126
arclength, 258
Arzela, 97
Arzela-Ascoli, 261
average value, 146

B
BCT, 92
Bounded Convergence Theorem, 92
B-space, 166
Baire Category Theorem, 259
Banach integral, 194
Banach lattice, 202
Banach limit, 195
Banach space, 166
Banach-Steinhaus, 179
Banach-Tarski Paradox, 11
Bartle, 117
below, 2
Bessel's Inequality, 240
Beta Function, 123
bidual, 190
Bochner, 137
Bohnenblust-Sobczyk, 187
Borel function, 73
Borel measure, 51, 61
Borel sets, 39
bound, 2
bounded, 2, 171
bounded above, 2
bounded variation, 255

C
complete measure, 46
σ-compact, 61
canonical map, 191
Cantor Set, 52
Cauchy in μ-measure, 107
Cauchy-Riemann integral, 97
Cauchy-Schwarz Inequality, 235
CGT, 183
change of variable, 161

characteristic functions, 1
Closed Graph Theorem, 183
closed intervals, 3
complement, 1
complete, 166
complete orthonormal set, 241
Condensation of Singularities, 246
converge, 166
converges, 2, 23
converges pointwise, 3
converges to f in μ-measure, 106
converges uniformly, 3
convolution product, 125
countable additivity, 10
countably additive, 20
countably subadditive, 21, 36
Counting Measure, 23

D
Daniell, 89
DCT, 87
decomposition property, 200
decreasing, 2-3
Diagonal Theorem, 66
Diagonalization, 261
Dini derivates, 154
Dirac Measure, 23
Dirichlet kernel, 244
distribution function, 56
Dominated Convergence Theorem, 87
dot product, 235
double sequence, 5
Drewnowski's Lemma, 35
dual, 175
dual norm, 175

E
μ-essential sup, 218
μ-essentially bounded functions, 218
Egoroff, 75
equicontinuous, 180, 261
equicontinuous from above at $\emptyset$, 211
equivalent, 169
Euler, 101

INDEX

F
σ-finite, 44
F. Riesz, 107–108, 200
Fatou, 87
Fejér kernel, 246
Fejér's Theorem, 247
finitely additive, 20
first category, 259
Fourier coefficients, 240
Fourier Inversion Theorem, 251
Fourier transform, 249
Frechet metric, 168
FTC, 98, 145, 149, 157
Fubini, 117, 153
function, 20
function space, 199, 263
Fundamental Theorem of Calculus, 98, 145, 157

G
Gamma Function, 100
Gelfand integral, 221
greatest lower bound, 2

H
Hölder Inequality, 221
Hölder's Inequality, 207
Hahn Decomposition, 33
Hahn-Banach Theorem, 186
Hardy-Littlewood maximal function, 146
Hellinger-Toeplitz, 248
Hewitt-Yosida Decomposition, 48
Hewitt-Yosida decomposition, 59
Hilbert space, 236

I
μ-integrable, 81
μ-integral, 84
μ-integrable, 85
ideal, 202
improper Riemann integral, 97
increasing, 2–3
infimum, 2, 197
inner product, 235
inner product space, 235
inner regular, 51, 61
integral, 81–82, 96
integral operator, 217
intersection, 1
intervals, 2
inverse Fourier transform, 251
isometry, 175
iterated series, 6

J
Jordan Decomposition, 29–30

K
kernel, 217

L
Laplace transform, 128
lattice homomorphism, 199
lattice isomorphism, 199
lattice semi-norm, 202
least upper bound, 2
Lebesgue, 152
Lebesgue Decomposition, 139, 161
Lebesgue integrable, 85
Lebesgue integral, 85
Lebesgue measure, 24–25, 43, 49
Lebesgue point, 147
Lebesgue set, 147
Lebesgue-Stieltjes, 24
Lebesgue-Stieltjes measure, 43, 56
Lebesgue-Stieltjes signed measure, 160
left shift, 176
Leibniz, 100
limit inferior, 2
limit superior, 2
linear functional, 175
linear topology, 165
linearly isometric, 175
Lipschitz condition, 159
locally integrable, 145
lower, 2, 96
lower variation, 28
Lusin, 78

M

μ-negative, 33
μ-null, 33
μ-positive, 33
$\sum$-measurable, 71
$\mu \times \nu$-measurable , 116
Maximal Theorem, 146
MCT, 83
measurable, 71
measurable rectangle, 116
measure, 20
measure space, 20
metric linear space, 165
metric outer measure, 39
Mikusinski, 113
Minkowski Inequality, 208
Minkowski's Inequality for Integrals, 214
monotone, 2–3, 20, 36, 204
monotone class, 17
Monotone Class Lemma, 18
Monotone Convergence Theorem, 83
mutually singular, 34

N
negative part, 28, 198
Nikodym Boundedness Theorem, 68
Nikodym Convergence Theorem, 67
NLS, 166
non-atomic, 222
norm, 165
normed space, 166
nowhere dense, 259

O
OMT, 184
open intervals, 3
Open Mapping Theorem, 184
operator norm, 174
order bounded, 200
order completeness, 2
order dual, 200
order ideal, 202
ordered vector space, 197
ordinate set, 124
orthogonal, 238
orthogonal complement, 239

orthonormal, 238
orthonormal dimension, 243
outer measure, 36
outer regular, 51, 61

P
pairwise disjoint, 1
Parallelogram Law, 236
Parseval's Equality, 242
partial order, 197
partition of unity, 228
permutation, 1
Pettis, 193
Pettis integral, 221
Plancherel transform, 251
Point Mass, 23
positive, 197, 199–200
positive part, 28, 198
power set, 1
premeasure, 20
product, 116
purely finitely additive, 48
Pythagorean Theorem, 239

R
Radon-Nikodym, 131, 138
Radon-Nikodym derivative, 133
rapidly decreasing functions, 250
rearrangement convergent, 5
rectangle, 116
reflexive, 191, 197
regular, 51, 61, 63
representing measure, 228
representing signed measure, 232
Reverse Hölder Inequality, 212
Riemann integrable, 96
Riemann-Lebesgue Lemma, 249
Riesz, 170
Riesz Decomposition, 200
Riesz Representation Theorem, 175, 213, 219, 224, 229, 233, 239
Riesz space, 197
Riesz-Fischer, 208, 243
right shift, 176
ring, 16

S

σ-set, 15
σ-algebra, 17
$\sum$-simple, 78
s-section, 119
t-section, 119
δ-sequence, 126
scalar product, 235
Schwartz space, 250
second category, 259
second dual, 190
semi-algebra, 15
semi-norm, 166
semi-normed (normed) vector lattice, 202
semi-ring, 15
separates the points, 263
shrinks regularly, 148
sides, 116
Sierpinski, 121
signed measure, 20
simple, 78
singular, 34, 139, 153
standard representation, 78
Stone-Weierstrass, 265
sublinear functional, 186
subseries convergent, 7
sum, 96
sup-norm, 167–168
supremum, 2, 197

T

Tonelli, 119
topological vector space, 165
total variation, 30, 256
transitive, 197
translation invariant, 51
transpose, 192

U

UBP, 178
unconditionally convergent, 5
Uniform Bounded Principle, 178
uniformly μ-continuous, 134, 142
uniformly μ-continuous, 211
uniformly countably additive, 66

union, 1
upper, 2, 96
upper variation, 28

V

variation, 30, 255
vector lattice, 197
vector sublattice, 202
vector topology, 165
Vitali, 211
Vitali-Hahn-Saks Theorem, 142

W

Weierstrass Approximation Theorem, 265